Stefania Rizzo

Experimental characterization and modelling of textile reinforced masonry structures with the Equivalent frame method

Karlsruher Reihe

Massivbau
Baustofftechnologie
Materialprüfung

Heft 90

Institut für Massivbau und Baustofftechnologie
Materialprüfungs- und Forschungsanstalt, MPA Karlsruhe

Prof. Dr.-Ing. Frank Dehn
Prof. Dr.-Ing. Lothar Stempniewski

Experimental characterization and modelling of textile reinforced masonry structures with the Equivalent frame method

by
Stefania Rizzo

KIT Scientific Publishing

Karlsruher Institut für Technologie
Institut für Massivbau und Baustofftechnologie

Experimental characterization and modelling of textile reinforced masonry structures with the Equivalent frame method

Zur Erlangung des akademischen Grades einer Doktor-Ingenieurin von der KIT-Fakultät für Bauingenieur-, Geo- und Umweltwissenschaften des Karlsruher Instituts für Technologie (KIT) genehmigte Dissertation

von Stefania Rizzo, M.Sc.

Tag der mündlichen Prüfung: 10. Dezember 2021
Referent: Prof. Dr.-Ing. Lothar Stempniewski
Korreferent: Prof. Sergio Lagomarsino, PhD

Impressum

KIT Scientific Publishing

Karlsruher Institut für Technologie (KIT)
KIT Scientific Publishing
Straße am Forum 2
D-76131 Karlsruhe

KIT Scientific Publishing is a registered trademark of Karlsruhe Institute of Technology.
Reprint using the book cover is not allowed.

www.ksp.kit.edu

Print on Demand 2022 – Gedruckt auf FSC-zertifiziertem Papier

ISSN 1869-912X
ISBN 978-3-7315-1214-1
DOI 10.5445/KSP/1000148218

Abstract

Masonry is a construction material commonly adopted to build residential structures worldwide. It is characterized by several advantages such as high thermal insulation, fire resistance and cost-effectiveness. However, past seismic events like the earthquakes in Kashmir (Pakistan, 2005), Chile (2010), L'Aquila (Italy, 2009), Amatrice (Italy, 2016) have shown that existing masonry buildings can suffer extensive damage under seismic actions until collapse. This poor structural response is generally caused by several factors, such as lack of good connections between the structural elements (e.g. walls, floors) and low masonry tensile, shear strength. Therefore, the seismic strengthening of existing structures is fundamental to sustain the potential ground motion due to the earthquake.

In this regard, composite materials are widely used for repairing and reinforcing masonry buildings. A common technique consists in the application of high-performance textile strips or fabrics to the masonry substrate with organic matrices (fibre-reinforced polymers (FRP)) or open textile meshes applied with inorganic matrices (fabric-reinforced cementitious matrix (FRCM) or textile-reinforced mortar (TRM)). The most adopted continuous fibres to create the composite materials are alkali-resistant (AR) glass, carbon, aramid, basalt, or polyparaphenylene benzobisoxazole (PBO).

It is important to stress that the FRPs have limitations, like heat sensitivity and problematic application on wet substrates. They also need surface preparation, and the impermeability of the polymeric matrix may induce bond problems with the masonry substrate. Finally, polymers are considered hazardous materials that require special handling processes before, during, and after their use. Therefore, cement- or lime mortar-based composites have been developed. The organic matrix of the FRP has been replaced with an inorganic one to overcome the mentioned application limits. In this regard, the FRCM/TRM have many advantages compared to the FRPs, e.g. physical/chemical compatibility with the masonry substrate, vapour permeability, high resistance to elevated temperatures and ultraviolet radiation, possible application on irregular or wet substrates with minimal or no surface preparation.

The FRCM/TRM textile component usually is an open mesh of continuous fibre yarns arranged in two or more directions. They can be dry, coated, or preimpregnated. The textile grids are completely embedded in the matrix, protecting the fibre yarns and ensuring the stress transfer between masonry substrate and textile component. The matrices are generally made of fine-grained mortar with Portland cement and dry organic polymers. The latter are usually added to improve the setting time, workability, and bond behaviour, but they reduce the fire resistance and vapour permeability at the same time. Therefore, their quantity is generally lower than 5% by weight.

The reinforcing system presented in this work is named EQ-GRID. It has been developed at the Karlsruhe Institute of Technology in Germany to strengthen and retrofit masonry structures. The textile component is a multi-axial hybrid grid made of alkali-resistant glass and polypropylene fibres. The matrix is a natural hydraulic lime mortar (NHL) explicitly developed for this system. It can penetrate the mesh openings and encapsulate the yarns very well. This property is crucial for the bond at the textile-matrix and matrix-support interface.

Since the EQ-GRID system aims to improve masonry structures' strength and inelastic deformation capacity, a broad experimental campaign has been performed at the Karlsruhe Institute of Technology in Germany. The results are presented and discussed in this work. The test program included tensile tests on bare textile samples and composite specimens for each main direction of the grid (vertical, horizontal and diagonal) and compression and bending tests on matrix specimens. In this way, the mechanical properties of the EQ-GRID system have been determined. Furthermore, double- and single-lap shear-bond tests have been performed to investigate the bonding behaviour of the system applied to standard masonry support. The results have shown no premature debonding from the substrate, although the system was applied without any mechanical anchorage. Finally, cyclic lateral shear tests have been performed on masonry panels in original conditions and strengthened with EQ-GRID to compare the performance of the reinforced structural element with the original one. The results have been depicted in terms of shear-drift curves, from which equivalent multilinear curves have been determined.

Moreover, the in-plane shear strength and failure domains of all the tested masonry panels has been analytically determined. The increase of in-plane load-bearing capacity due to the EQ-GRID system is considered through the proposed modification of the masonry initial shear strength and brick tensile strength as well as after the Italian technical standard CNR-DT 215/2018 "Guide for the Design and Construction of Externally Bonded Fibre Reinforced Inorganic Matrix Systems for Strengthening Existing Structures".

Furthermore, the cyclic lateral shear tests have been modelled through the Equivalent frame method implemented in the TREMURI program. This modelling technique consists of idealizing the walls in a frame. Each load-bearing masonry wall is subdivided into a set of deformable masonry panels where the nonlinear behaviour and deformation are concentrated. Then, the structural elements are connected by rigid nodes that are parts of the wall not usually subjected to damage.

All the tested masonry panels have been modelled through the nonlinear beam element with lumped inelasticity and piecewise behaviour implemented in the program. During the modelling, the strength increase due to the system has been calculated after CNR-DT 215/2018. The formulations proposed by this standard are implemented in the program and are compatible with the adopted nonlinear beam element.

Finally, two shaking table tests performed on a masonry building model at the Institute of Earthquake Engineering and Engineering Seismology (IZIIS) in Skopje have also been modelled with the Equivalent frame method. The aim was to reproduce the observed behaviour of the masonry building in its original condition and after the seismic retrofitting with EQ-GRID. Therefore, modal analyses, nonlinear monotonic, cyclic static and time-history analyses have been performed. The numerical results have been compared to the experimental ones in terms of structural behaviour, achieved damage level, measured displacements and accelerations. The results have confirmed the Equivalent frame method as a modelling technique suitable for unreinforced masonry buildings and textile reinforced masonry structures.

Kurzfassung

Mauerwerk wird häufig als Material für den Bau von Wohngebäuden auf der ganzen Welt verwendet. Es zeichnet sich durch mehrere Vorteile wie hohe Wärmedämmung, Feuerbeständigkeit und Kosteneffizienz aus. Allerdings haben vergangene seismische Ereignisse wie die Erdbeben in Kaschmir (Pakistan, 2005), Chile (2010), L'Aquila (Italien, 2009), Amatrice (Italien, 2016) gezeigt, dass bestehende Gebäude aus Mauerwerk unter seismischen Einwirkungen erhebliche Schäden bis hin zum Einsturz erleiden können. Dieses schlechte strukturelle Verhalten wird im Allgemeinen durch mehrere Faktoren verursacht, wie z. B. das Fehlen guter Verbindungen zwischen den tragenden Bauteilen (z. B. Wände, Decken) und eine geringe Zug- sowie Scherfestigkeit des Mauerwerks. Daher ist die seismische Verstärkung bestehender Tragwerke von grundlegender Bedeutung, um den potenziellen Bodenbewegungen infolge eines Erdbebens standzuhalten.

In dieser Hinsicht werden Verbundwerkstoffe häufig für die statische Sanierung und Verstärkung von Mauerwerksbauten verwendet. Eine gängige Technik besteht in der Aufbringung von hochleistungsfähigen Textilstreifen oder Geweben auf das Mauerwerkssubstrat mit organischen Matrizen (faserverstärkte Polymere (FRP)) oder offenen Textilgeweben, die mit anorganischen Matrizen (gewebeverstärkte zementäre Matrix (FRCM) oder textilverstärkter Mörtel (TRM)) aufgebracht werden. Die am häufigsten verwendeten Fasermaterialien zur Herstellung der Verbundwerkstoffe sind alkalibeständiges (AR) Glas, Kohlenstoff, Aramid, Basalt oder Polyparaphenylenbenzobisoxazol (PBO).

Allerdings haben die FRPs einige Einschränkungen, wie z.B. Hitzeempfindlichkeit, problematische Anwendung auf nassen Substraten. Außerdem benötigen sie eine Oberflächenvorbereitung und die Undurchlässigkeit der Polymermatrix kann zu Haftungsproblemen mit dem Mauersubstrat führen. Schließlich gelten Polymere als Gefahrstoffe, die vor, während und nach ihrer Verwendung besondere Handhabungsprozesse erfordern. Daher wurden Verbundwerkstoffe auf Zement- oder Kalkmörtelbasis entwickelt. Um die genannten Anwendungsgrenzen zu überwinden, wurde die organische Matrix des FRP durch eine anorganische ersetzt. In dieser Hinsicht haben die FRCM/TRM viele Vorteile gegenüber den FRPs, z. B. physikalische/chemische Kompatibilität mit dem Mauerwerkssubstrat, Dampfdurchlässigkeit, hohe Beständigkeit gegenüber erhöhten Temperaturen und Ultraviolettstrahlung, mögliche Anwendung auf unregelmäßigen oder nassen Substraten mit minimaler oder keiner Oberflächenvorbereitung.

Die FRCM/TRM-Textilkomponente ist in der Regel ein offenes Gewebe aus Fasergarnen, die in zwei oder mehr Richtungen angeordnet sind. Sie können trocken, beschichtet oder vorimprägniert sein. Die Textilgitter werden vollständig in die Matrix eingebettet, wodurch die Fasergarne geschützt werden können und die Spannungsübertragung zwischen Substrat und Textilkomponente gewährleistet ist. Die Matrizen bestehen in der Regel aus feinkörnigem Mörtel mit Portlandzement und trockenen organischen Polymeren. Letztere werden in der Regel zugesetzt, um die Abbindezeit, die Verarbeitbarkeit und das Verbundverhalten zu verbessern, aber sie reduzieren gleichzeitig den Feuerwiderstand und die Dampfdurchlässigkeit. Daher ist im Allgemeinen ihre Menge geringer als 5% im Gewicht.

Das in dieser Arbeit vorgestellte Verstärkungssystem heißt EQ-GRID. Es wurde am Karlsruher Institut für Technologie in Deutschland zur seismischen Verstärkung und Retrofitting von Mauerwerksbauten entwickelt. Die Textilkomponente ist ein multiaxiales Hybridgitter aus alkalibeständigen Glas- und Polypropylenfasern. Die Matrix ist ein speziell für dieses System entwickelter natürlicher hydraulischer Kalkmörtel (NHL). Sie kann die Maschenöffnungen des Gitters durchdringen und die Garne sehr gut einkapseln. Diese Eigenschaft ist entscheidend für den Verbund an der Schnittstelle Textil-Matrix und Matrix-Substrat.

Da das EQ-GRID-System darauf abzielt, die Festigkeit und das unelastische Verformungsvermögen von Mauerwerkskonstruktionen zu verbessern, wurde am Karlsruher Institut für Technologie in Deutschland eine umfangreiche Versuchskampagne durchgeführt. Die Ergebnisse werden in dieser Arbeit vorgestellt und diskutiert. Das Versuchsprogramm umfasste Zugversuche an Textilproben und Verbundwerkstoffproben für jede Hauptrichtung des Gitters (vertikal, horizontal und diagonal) sowie Druck- und Biegeversuche an Matrixproben. Daher wurden die mechanischen Eigenschaften des EQ-GRID Systems bestimmt. Darüber hinaus wurden doppel- und einlagige Scherverbundversuche durchgeführt, um das Verbundverhalten des Systems auf Standardmauerwerksziegel zu untersuchen. Die Ergebnisse haben keine vorzeitige Ablösung vom Untergrund gezeigt, obwohl das System ohne mechanische Verankerung aufgebracht wurde. Schließlich wurden zyklische Schubversuche an Mauerwerkswände im Originalzustand und mit EQ-GRID verstärkt durchgeführt, um die Leistungsfähigkeit des verstärkten Bauteils mit dem Original zu vergleichen. Die Ergebnisse wurden in Form von Schub-Drift-Kurven dargestellt, aus denen äquivalente multilineare Kurven ermittelt wurden.

Außerdem wurden der Schubwiderstand und die Tragfähigkeitsdiagramme aller getesteten Mauerwerkswände analytisch bestimmt. Die durch das EQ-GRID System Erhöhung der Tragfähigkeit wird durch die vorgeschlagene Modifikation der Anfangsscherfestigkeit des Mauerwerks und der Steinzugfestigkeit sowie nach der italienischen technischen Norm CNR-DT 215/2018 "Guide for the Design and Construction of Externally Bonded

Fibre Reinforced Inorganic Matrix Systems for Strengthening Existing Structures" berechnet.

Darüber hinaus wurden die zyklischen Schubversuche durch die im Programm TREMURI implementierte "Equivalent Frame" Methode modelliert. Diese Modellierungstechnik besteht darin, die Wände in einem Rahmen zu idealisieren. Jede tragende Mauerwerkswand wird in verformbaren Mauerwerkselementen unterteilt, in denen das nichtlineare Verhalten und die Verformung konzentriert sind. Dann werden die Mauerwerkselemente durch starre Knoten verbunden, die Teile der Wand sind, die normalerweise nicht beschädigt werden.

Alle untersuchten Mauerwerkswände wurden durch das im Programm implementierte nichtlineare Balkenelement mit konzentrierter Unelastizität und multilinearem Verhalten modelliert. Die durch das System erhörte Tragfähigkeit wurde nach CNR-DT 215/2018 berechnet. Die von dieser Norm vorgeschlagenen Formeln sind im Programm implementiert und sind mit dem angewendeten nichtlinearen Balkenelement kompatibel.

Schließlich wurden zwei Rütteltischversuche, die an einem Gebäudemodell aus Mauerwerk am Institut für Erdbebeningenieurwesen und Ingenieurseismologie (IZIIS) in Skopje durchgeführt wurden, ebenfalls mit der "Equivalent Frame" Methode modelliert. Ziel war es, das beobachtete Verhalten des Gebäudemodells im Originalzustand und nach der seismischen Ertüchtigung mit EQ-GRID numerisch zu reproduzieren. Dazu wurden Modalanalysen, nichtlineare monotone, zyklische statische und dynamische Analysen durchgeführt. Die numerischen Ergebnisse wurden mit den Experimentellen hinsichtlich des Strukturverhaltens, des erreichten Schadensniveaus und der gemessenen Verschiebungen und Beschleunigungen verglichen. Die Ergebnisse konnten bestätigen, dass die "Equivalent Frame" Methode eine geeignete Modellierungstechnik für unbewehrte sowie textilbewehrte Mauerwerksbauten ist.

List of contents

Notation and list of abbreviations

Capital Latin letters

A	cross-sectional area of the masonry panel
A_{AR}	equivalent cross-sectional area of the EQ-GRID AR glass yarns
$A_{env.3}$	area below the experimental in-plane shear force-horizontal displacement curve until DL3
$A_{env.4}$	area below the experimental in-plane shear force-horizontal displacement curve between DL3 and DL4
$A_{env.5}$	area below the experimental in-plane shear force-horizontal displacement curve between DL4 and DL5
A_{eq}	equivalent cross-sectional area of EQ-GRID homogenized to the glass fibre
A_f	equivalent cross-sectional area of the FRCM mesh reinforcement
$A_{floor,1}$	area of the 1st floor of the masonry building model tested on the shaking table in Skopje
$A_{floor,2}$	area of the 2nd floor of the masonry building model tested on the shaking table in Skopje
$A_{mult.3}$	area below the equivalent multilinear curve until DL3
$A_{mult.4}$	area below the equivalent multilinear curve between DL3 and DL4
$A_{mult.5}$	area below the equivalent multilinear curve between DL4 and DL5
A_{PP}	equivalent cross-sectional area of the EQ-GRID polypropylene yarns
A_t	area of the three-nodes orthotropic membrane element implemented in TREMURI
$[B_i]$	shape matrix defined in node *i* of the three-nodes orthotropic membrane element implemented in TREMURI
$[B_j]$	shape matrix defined in node *j* of the three-nodes orthotropic membrane element implemented in TREMURI
$[\hat{D}]$	deformability matrix of the three-nodes orthotropic membrane element implemented in TREMURI

$[D]$	deformability matrix of the three-nodes orthotropic membrane element in the rotated configuration to take into account the actual orientation of the diaphragm
$[D_e]$	rigid end matrix of the masonry element implemented in TREMURI
E	Young's modulus perpendicular to the bed joints of the masonry element implemented in TREMURI
E_{AR}	elastic tensile modulus of AR glass fibre yarns of the EQ-GRID mesh (avg.)
E_{craked}	tensile modulus of elasticity of the cracked FRCM (avg.)
E_f	tensile elastic modulus of the FRCM mesh (avg.)
$E_{f,d}$	secant tensile elastic modulus of the EQ-GRID mesh in the diagonal direction (avg.)
$E_{f,h}$	secant tensile elastic modulus of the EQ-GRID mesh in the horizontal direction (avg.)
$E_{f,v}$	secant tensile elastic modulus of the EQ-GRID mesh in the vertical direction (avg.)
E_{KS}	Young's modulus perpendicular to the bed joints of the tested calcium-silicate brick masonry panels (avg.)
E_M	compressive elastic modulus of the EQ-GRID mortar matrix (avg.)
E_{PP}	elastic tensile modulus of polypropylene fibre yarns of the EQ-GRID mesh (avg.)
E_{sec}	secant Young's modulus of the tested masory panels
$E_{uncracked}$	tensile modulus of elasticity of uncracked FRCM (avg.)
E_{WZI}	Young's modulus perpendicular to the bed joints of the tested hollow clay brick masonry panels (avg.)
E_1	Young's modulus of the horizontal diaphragms in the floor spanning orientation in TREMURI
E_2	Young's modulus of the horizontal diaphragms perpendicular to the floor spanning orientation in TREMURI
$E_{//}$	Young's modulus parallel to the bed joints of the masonry element implemented in TREMURI
F	axial tensile force

F_{AR}	axial tensile force in the AR glass fibre yarns of EQ-GRID
$F_{cr,m}$	first cracking tensile load of the EQ-GRID composite specimens (avg.)
$F^{1}{}_{h}$	reactive force in the x-direction acting on the 2D fictitious node of the wall 1 in TREMURI
$F^{2}{}_{h}$	reactive force in the x-direction acting on the 2D fictitious node of the wall 2 in TREMURI
F_m	resultant compressive load during the in-plane bending failure of an FRCM strengthened panel
F_m^*	resultant compressive load when both masonry and FRCM attain their ultimate tensile strain during the in-plane bending failure of an FRCM strengthened panel
$F_{m,u,0\ldots4}$	resultant compressive load at the in-plane bending failure modes 0, 1, 2, 3, 4 of a masonry panel strengthened with EQ-GRID
F_{PP}	axial tensile force in the polypropylene fibre yarns of EQ-GRID
F_t	resultant tensile load during the in-plane bending failure of an FRCM strengthened panel
F_t^*	resultant tensile load when both masonry and FRCM attain their ultimate tensile strain during the in-plane bending failure of an FRCM strengthened panel
$F_{t,u,0\ldots4}$	resultant tensile load at the in-plane bending failure modes 0, 1, 2, 3, 4 of a masonry panel strengthened with EQ-GRID
F_u	ultimate axial tensile force
$F^{1}{}_{v}$	reactive force in the z-direction acting on the 2D fictitious node of the wall 1 in TREMURI
$F^{2}{}_{v}$	reactive force in the z-direction acting on the 2D fictitious node of the wall 2 in TREMURI
F_x	reactive force in the global X-direction transmitted by the masonry element to the 2D fictitious node in TREMURI
F_y	reactive force in the global Y-direction transmitted by the masonry element to the 2D fictitious node in TREMURI
F_z	reactive force in the global Z-direction transmitted by the masonry element to the 2D fictitious node in TREMURI
G	elastic shear modulus

G_{KS}	elastic shear modulus of the tested calcium-silicate brick masonry panels (avg.)
G_{WZI}	elastic shear modulus of the tested hollow clay brick masonry panels (avg.)
G_w	masonry panel's self-weight
G_{12}	elastic shear modulus of the horizontal diaphragms in TREMURI
H	limit dimension of the design length l_f (CNR-DT 215/2018)
H_p	minimum between the tensile resistance of the stretched interposed element inside the spandrel and $0.4 \cdot f_{mh,u} \cdot h \cdot t$
$[K_e]$	elastic stiffness matrix of the masonry element implemented in TREMURI
$[K^e]$	elastic stiffness matrix of the three-nodes orthotropic membrane element implemented in TREMURI
$[K_{ij}^e]$	component ij of the elastic stiffness matrix $[K^e]$
I	moment of inertia of the masonry panel
L	length of the masonry wall
M	in-plane bending moment
M_f	contribution of the FRCM system to nominal flexural strength of the panel according to ACI 459.4R-13
M_i	bending moment acting in node i of the masonry element implemented in TREMURI
M_I^X	mass component in the global X-direction of the 3D rigid node I of the Equivalent frame in TREMURI
M_I^Y	mass component in the global Y-direction of the 3D rigid node I of the Equivalent frame in TREMURI
M_j	bending moment acting in node j of the masonry element implemented in TREMURI
M_n	nominal flexural strength of the FRCM reinforced masonry panel according to ACI 459.4R-13
M_x	reactive moment about the global X-direction transmitted by the masonry element to the 2D fictitious node in TREMURI
M_y	reactive moment about the global Y-direction transmitted by the masonry element to the 2D fictitious node in TREMURI

M_{Rd}	ultimate in-plane bending moment of a URM panel
$M_{Rd,s}$	ultimate in-plane bending moment of a masonry panel strengthened with FRCM according to CNR DT-215/2018
$M_{Rd,s,0...4}$	ultimate in-plane bending moment of a masonry panel strengthened with EQ-GRID at the flexural failure modes 0, 1, 2, 3, 4
M^1	reactive moment acting on the 2D fictitious node of wall 1 in TREMURI
M^2	reactive moment acting on the 2D fictitious node of wall 2 in TREMURI
N	normal force
N_i	normal force acting in node *i* of the masonry element implemented in TREMURI
N_j	normal force acting in node *j* of the masonry element implemented in TREMURI
$N_{Rd,s}$	ultimate normal force of a masonry panel strengthened with FRCM under flexural failure
$N_{Rd,s,0...4}$	ultimate normal force of a masonry panel strengthened with EQ-GRID at the flexural failure modes 0, 1, 2, 3, 4
N_v	normal force applied by the vertical actuator during the cyclic lateral shear tests
N_{v1}	vertical force developed in one threaded bar positioned on the left of the specimen during the cyclic lateral shear tests
N_{v2}	vertical force developed in one threaded bars positioned on the rigth of the specimen during the cyclic lateral shear tests
$[R]$	rotation matrix
T	period
T_x	Tex, yarn count expressed in [g/km]
T_1	first period of the masonry building model tested on the shaking table in Skopje measured before testing
T_1^*	first period of the masonry building model tested on the shaking table in Skopje measured after testing
T_4	fourth period of the masonry building model tested on the shaking table in Skopje

V	in-plane shear force
V_{Base}	shear force at the base of the masonry building model in TREMURI
V_{DL3}	experimental residual lateral strength of the tested masonry panels at the achievement of DL3
V_{DL4}	experimental residual lateral strength of the tested masonry panels at the achievement of DL4
V_{DL5}	experimental residual lateral strength of the tested masonry panels at the achievement of DL5
V_{Ed}	experimental values of the in-plane shear force recorded during the cyclic lateral shear tests
V_f	contribution of FRCM to nominal shear strength according to ACI 459.4R-13
V_m	contribution of the unreinforced masonry to the nominal shear strength according to ACI 459.4R-13
V_{max}	experimental maximum in-plane shear force achieved by the tested "WZI" and "KS" masonry panels
V_i	in-plane shear force acting in node *i* of the masonry element implemented in TREMURI
V_j	in-plane shear force acting in node *j* of the masonry element implemented in TREMURI
$V_{j,max}$	maximum value of the in-plane shear force acting in node *j* of the masonry element implemented in TREMURI
V_n	nominal shear strength after ACI 459.4R-13
$V^F{}_n$	in-plane ultimate shear force of an FRCM strengthened panel under bending failure (ACI 459.4R-13)
V_t	in-plane shear strength of the unreinforced masonry panel (diagonal cracking)
$V_{t,f}$	increase of the in-plane shear strength of a masonry panel due to the FRCM after CNR-DT 215/2018
$V_{t,lim}$	upper limit of the in-plane shear strength of the unreinforced masonry panel (diagonal cracking)
$V_{t,lim,mod}$	modified upper limit of the in-plane shear strength of a masonry panel strengthened with EQ-GRID (diagonal cracking)

$V_{t,mod}$	modified in-plane shear strength of a masonry panel strengthened with EQ-GRID (diagonal cracking)
V_{Rd}	in-plane ultimate shear strength of a URM panel
$V^F{}_{Rd}$	in-plane ultimate shear force of a URM panel under bending failure
$V_{Rd,mod}$	modified in-plane ultimate shear strength of a masory panel strengthened with EQ-GRID
$V_{Rd,s}$	in-plane ultimate shear strength of the FRCM reinforced masonry panel according to CNR-DT 215/2018
$V^F{}_{Rd,s}$	in-plane ultimate shear force of an FRCM strengthened panel under bending failure (CNR-DT 215/2018)
$V_{res,i}$	residual lateral strength in the multilinear constitutive law of the ML-BEAM element at the achievement of the ith DL
$V_{res,3}$	residual lateral strength in the multilinear constitutive law of the ML-BEAM element at the achievement of DL3
$V_{res,4}$	residual lateral strength in the multilinear constitutive law of the ML-BEAM element at the achievement of DL4
$V_{t,c}$	in-plane masonry crushing capacity of a textile-reinforced panel after CNR-DT 215/2018
V_u	in-plane ultimate shear force of the equivalent bilinear curve
$V_{u,3}$	in-plane ultimate shear force of the equivalent multilinear curve (DL2-DL3)
$V_{u,4}$	in-plane ultimate shear force of the equivalent multilinear curve (DL3-DL4)
$V_{u,5}$	in-plane ultimate shear force of the equivalent multilinear curve (DL4-DL5)
V_{URM}	in-plane ultimate shear strength of a URM masonry panel under diagonal cracking (CNR-DT 215/2018)
$V_{walls,x}$	volume of the masonry walls positioned in the x-direction of the building model tested on the shaking table in Skopje
$V_{walls,y}$	volume of the masonry walls positioned in the y-direction of the building model tested on the shaking table in Skopje
V_1	in-plane shear force equal to $0.7V_{max}$

Lower-case Latin letters

a	distance between the threaded bars and the centre of the masonry panels in the cyclic lateral shear tests
a_{input}	maximum absolute value of the earthquake acceleration applied at the base of the masonry building model tested on the shaking table in Skopje
a_1	first factor that defines the admissible range in the V – N domain of a masonry panel, in which the hybrid failure mode is possible
a_2	second factor that defines the admissible range in the V – N domain of a masonry panel, in which the hybrid failure mode is possible
b	shear stress distribution factor at the centre of a masonry panel
b_f	width of the textile grid
c	mortar joints' cohesion
c_1	coefficient that aims to degrade the stiffness k_U^+ of the the ML-BEAM element with respect to the secant one k_{sec}
c_2	coefficient that can further degrade the stiffness k_U^+ through the progressive strength decay $\beta_{E,i}$ reached by the ML-BEAM element
c_3	coefficient that fixes the point B^+ of the horizontal branch in the unloading curve of the ML-BEAM element
c_4	coefficient that defines the extension of point B^+ - C^+ branch in the unloading curve of the ML-BEAM element
d	drift value
d_f	distance between the compressed edge of the masonry panel and the fibre of the reinforcement that attains the highest tensile strain
d_i	drift value of a masonry panel at the attainment of the ith DL
d_2	drift value of the tested masonry panels at the achievement of the DL2
$d_{2,S}$	drift value of the masonry elements in TREMURI at the achievement of DL2 (shear failure)
$d_{2,F}$	drift value of the masonry elements in TREMURI at the achievement of DL2 (flexural failure)
d_3	drift value of the tested masonry panels at the achievement of the DL3
$d_{3,S}$	drift value of the masonry elements in TREMURI at the achievement of DL3 (shear failure)

$d_{3,F}$	drift value of the masonry elements in TREMURI at the achievement of DL3 (flexural failure)
d_4	drift value of the tested masonry panels at the achievement of the DL4
$d_{4,S}$	drift value of the masonry elements in TREMURI at the achievement of DL4 (shear failure)
$d_{4,F}$	drift value of the masonry elements in TREMURI at the achievement of DL4 (flexural failure)
d_5	drift value of the tested masonry panels at the achievement of the DL5
$d_{5,S}$	drift value of the masonry elements in TREMURI at the achievement of DL5 (shear failure)
$d_{5,F}$	drift value of the masonry elements in TREMURI at the achievement of DL5 (flexural failure)
e	load eccentricity
e_{bottom}	eccentricity of the normal force acting at the bottom toe of a masonry pier
$\{e\}^T$	transpose of the strain vector of the three-nodes orthotropic membrane element implemented in TREMURI
f	frequency
f_b	brick compressive strength (avg.)
$f_{b,KS}$	calcium-silicate brick compressive strength
$f_{bond,double,max}$	maximum shear-bond stress obtained from each double-lap shear-bond test
$f_{bond,single,m}$	maximum shear-bond stress obtained from the single-lap shear-bond test (avg.)
$f_{b,WZI}$	hollow clay brick compressive strength
f_{bt}	brick tensile strength (avg.)
$f_{bt,cal}$	brick tensile strength according to DIN EN 1996-1-1/NA (char.)
$f_{bt,KS}$	calcium-silicate brick tensile strength
$f_{bt,KS,s1}$	calcium-silicate brick tensile strength modified through the EQ-GRID system's tensile strength in the case of one side application (avg.)
$f_{bt,KS,s2}$	calcium-silicate brick tensile strength modified through the EQ-GRID system's tensile strength in the case of two sides application (avg.)

$f_{bt,s}$ brick tensile strength modified through the EQ-GRID system's tensile strength (avg.)

$f_{bt,WZI}$ hollow clay brick tensile strength

$f_{bt,WZI,s1}$ hollow clay brick tensile strength modified through the EQ-GRID system's tensile strength in the case of one side application (avg.)

$f_{bt,WZI,s2}$ hollow clay brick tensile strength modified through the EQ-GRID system's tensile strength in the case of two sides application (avg.)

$f_{clamping,cr}$ first cracking tensile stress of the EQ-GRID composite specimens tested with clamping grips (avg.)

$f_{clamping,u}$ ultimate tensile strength of the EQ-GRID composite specimens tested with clamping grips (avg.)

$f_{clevis,cr}$ first cracking tensile stress of the EQ-GRID composite specimens tested with clevis-type grips (avg.)

$f_{clevis,u}$ ultimate tensile strength of the EQ-GRID composite specimens tested with clevis-type grips (avg.)

$f_{c,M}$ EQ-GRID matrix compressive strength after 28 days (avg.)

f_{cr} cracking tensile stress in the equivalent bilinear diagram of an FRCM system after ACI 459.4R-13

$f_{cr,M}$ first cracking tensile stress of the EQ-GRID matrix (avg.)

f_d design tensile strength of the FRCM

f_e design tensile strength of the FRCM reinforcement after ACI 459.4R-13 (flexural failure)

f_{fv} design tensile strength of the FRCM reinforcement after ACI 459.4R-13 (shear failure)

f_M mortar compressive strength (avg.)

$f_{M,t}$ mortar flexural tensile strength (avg.)

$f_{mh,u}$ masonry ultimate compressive strength in the direction parallel to the bed joints

$f_{m,k,KS}$ ultimate compressive strength in the direction normal to the bed joints of the tested calcium-silicate brick masonry panels (char.)

$f_{m,k,WZI}$ ultimate compressive strength in the direction normal to the bed joints of the tested hollow clay brick masonry panels (char.)

$f_{m,t}$	masonry referential (diagonal) tensile strength
$f_{mt,u}$	ultimate equivalent tensile strength of a URM spandrel
$f_{mt,u1}$	equivalent tensile strength of a URM spandrel associated with the block's tensile strength
$f_{mt,u2}$	equivalent tensile strength of a URM spandrel associated with the mortar joints' failure
$f_{m,u}$	masonry ultimate compressive strength in the direction normal to the bed joints
$f_{m,u,KS}$	ultimate compressive strength in the direction normal to the bed joints of the tested calcium-silicate brick masonry panels (avg.)
$f_{m,u,WZI}$	ultimate compressive strength in the direction normal to the bed joints of the tested hollow clay brick masonry panels (avg.)
f_p	EQ-GRID matrix pull-off strength
f_t	conventional tensile strength of the EQ-GRID system in the case of flexural failure of a masonry pier
$f_{t,M}$	EQ-GRID matrix tensile bending strength stress after 28 days (avg.)
$f_{t,u}$	conventional tensile strength limit of the EQ-GRID system in the case of flexural failure of a masonry pier
f_u	ultimate tensile strength of an FRCM obtained from the direct tensile tests with clevis-type grips (ACI 459.4R-13)
f_v	masonry shear strength
f_{v0}	masonry initial shear strength
$f_{v0,red}$	reduced masonry initial shear strength
f_{vlt1}	masonry shear strength (mortar joints' failure) according to DIN EN 1996-1-1/NA (char.)
f_{vlt2}	masonry shear strength (brick's tensile failure) according to DIN EN 1996-1-1/NA (char.)
f_{vlt}	ultimate masonry shear strength according to DIN EN 1996-1-1/NA (char.)
f_{vm0}	masonry initial shear strength (avg.)
$f_{vm0,KS}$	initial shear strength of the calcium-silicate brick masonry (avg.)

$f_{vm0,KS,s1}$	initial shear strength of the calcium silicate brick masonry modified through the tensile strength of the EQ-GRID system in the case of one side application (avg.)
$f_{vm0,KS,s2}$	initial shear strength of the calcium silicate brick masonry modified through the tensile strength of the EQ-GRID system in the case of two sides application (avg.)
$f_{vm0,red,KS}$	reduced initial shear strength of the calcium-silicate brick masonry (avg.)
$f_{vm0,red,WZI}$	reduced initial shear strength of the hollow clay brick masonry (avg.)
$f_{vm0,s}$	masonry initial shear strength modified through the tensile strength of the EQ-GRID system (avg.)
$f_{vm0,WZI}$	initial shear strength of the hollow clay brick masonry (avg.)
$f_{vm0,WZI,s1}$	initial shear strength of the hollow clay brick masonry modified through the tensile strength of the EQ-GRID system in the case of one side application (avg.)
$f_{vm0,WZI,s2}$	initial shear strength of the hollow clay brick masonry modified through the tensile strength of the EQ-GRID system in the case of two sides application (avg.)
g	mortar joint thickness
h	height of the masonry panel
h'	shear length defined as the distance between the masonry panel's end section and that with zero moment
$h_{floor,1}$	thickness of the 1st floor of the masonry building model tested on the shaking table in Skopje
$h_{floor,2}$	thickness of the 2nd floor of the masonry building model tested on the shaking table in Skopje
h_s	total thickness of the EQ-GRID composite specimens
h_1	heigth of the 1st floor of the masonry building model tested on the shaking table in Skopje
h_2	height of the 2nd floor of the masonry building model tested on the shaking table in Skopje
k	coefficient that accounts for the boundary condition of a masonry panel
k_{el}	elastic stiffness of the ML-BEAM element implemented in TREMURI

k_L	stiffness of the ML-BEAM element between the unloading and loading phases (from C^+ to A^-)
k_n	factor recommended by Annex D of Eurocode 0 to calculate the characteristic values of mechanical properties
k_{sec}	elastic secant stiffness
k_U^+	ML-BEAM element's stiffness in the unloading phase (positive quadrant from A^+ to C^+)
k_U-	ML-BEAM element's stiffness in the unloading phase (negative quadrant from A^- to C^-)
k_O	ratio between the shear force at the end of the initial elastic phase and the shear strength V_{Rd} ($V_{Rd,s}$) of the ML-BEAM element implemented in TREMURI
i	node of the masonry element implemented in TREMURI
j	node of the masonry element implemented in TREMURI
l	length of the masonry panel
l_c	uncracked section length of the masonry panel
l_f	design length of the applied FRCM effective in shear after CNR-DT 215/2018
l_I	distance between the 2D node with mass m and the 3D node I of the Equivalent frame in TREMURI
m	mass of the 2D node belonging to the Equivalent frame in TREMURI
$m_{floor,1}$	mass of the 1st floor of the masonry building model tested on the shaking table in Skopje
$m_{floor,2}$	mass of the 2nd floor of the masonry building model tested on the shaking table in Skopje
m_{tot}	total mass of the masonry building model tested on the shaking table in Skopje
$m_{walls,x}$	mass of the masonry walls positioned in the x-direction of the building model tested on the shaking table in Skopje
$m_{walls,y}$	mass of the masonry walls positioned in the y-direction of the building model tested on the shaking table in Skopje
m_x	modal mass in the x-direction

m_y	modal mass in the y-direction
m_o	seismic mass at the base of the masonry building model tested on the shaking table in Skopje
m_1	seismic mass at the 1st floor of the masonry building model tested on the shaking table in Skopje
m_2	seismic mass at the 2nd floor of the masonry building model tested on the shaking table in Skopje
m_{12}	ratio between E_1 and E_2
n	homogenization coefficient defined as the ratio between the elastic modulus of the polypropylene E_{PP} and that of the AR glass E_{AR}
n_d	homogenization coefficient for the diagonal direction of the EQ-GRID system
n_f	number of layers of mesh reinforcement
n_h	homogenization coefficient for the horizontal direction of the EQ-GRID system
$n_{specimen}$	number of tested specimens
n_v	homogenization coefficient for the vertical direction of the EQ-GRID system
n_y	number of yarns per unit width expressed in [n°/cm]
$\{q\}$	vector of the nodal actions
s	equivalent thickness assumed for the three-nodes orthotropic membrane element implemented in TREMURI
$\{s\}^T$	transpose of the stress vector of the three-nodes orthotropic membrane element implemented in TREMURI
t	thickness of a masonry panel
t_{eq}	equivalent thickness of EQ-GRID homogenized to the glass fibre
$t_{eq,d}$	equivalent thickness in the diagonal direction of EQ-GRID homogenized to the glass fibre
$t_{eq,h}$	equivalent thickness in the horizontal direction of EQ-GRID homogenized to the glass fibre
$t_{eq,v}$	equivalent thickness in the vertical direction of EQ-GRID homogenized to the glass fibre

t_f	equivalent thickness of the FRCM mesh reinforcement
$t_{f,b}$	equivalent thickness of one textile layer with fibres arranged parallel to panel's axis, which is effective during the in-plane bending failure
$t_{f,s}$	equivalent thickness of one textile layer with fibres arranged parallel to the shear force (CNR-DT 215/2018
t_M	thickness of the vertical and horizontal mortar joints
v_{max}	maximum vertical displacement recorded during the double-lap shear-bond tests
u	horizontal displacement of the fictitious 2D node belonging to the generic wall in the Equivalent frame method implemented in TREMURI
$u_{el,1}$	experimental horizontal displacement at V_1, measured at the top of the tested masonry panels
$u_{el,2}$	elastic limit of the horizontal displacement in the equivalent bilinear or multilinear curve
u_i	horizontal displacement of node i of the masonry element implemented in TREMURI
u_j	horizontal displacement of node j of the masonry element implemented in TREMURI
$u_{j,e}$	elastic horizontal displacement of node j of the masonry element implemented in TREMURI
$u_{j,e,F}$	elastic horizontal displacement of node j of the masonry element implemented in TREMURI considering the flexural stiffness
$u_{j,e,S}$	elastic horizontal displacement of node j of the masonry element implemented in TREMURI considering the shear stiffness
u_u	ultimate horizontal displacement in the equivalent bilinear curve
u_x	displacement in the x-direction
u_y	displacement in the y-direction
u_z	displacement in the z-direction
u_1	horizontal displacement of the barycenter of the rigid node 1 in the equivalent frame schematization in TREMURI
u_2	horizontal displacement of the barycenter of the rigid node 2 in the equivalent frame schematization in TREMURI

u_3	horizontal displacement at the achievement of DL3 in the equivalent multilinear curve
u_4	horizontal displacement at the achievement of DL4 in the equivalent multilinear curve
u_5	horizontal displacement at the achievement of DL5 in the equivalent multilinear curve
w	vertical displacement of the fictitious 2D node belonging to the generic wall in the Equivalent frame method implemented in TREMURI
w_i	vertical displacement of node i of the masonry element implemented in TREMURI
w_j	vertical displacement of node j of the masonry element implemented in TREMURI
w_s	width of the EQ-GRID composite samples
w_1	vertical displacement of the barycenter of the rigid node 1 in the equivalent frame schematization in TREMURI
w_2	vertical displacement of the barycenter of the rigid node 2 in the equivalent frame schematization in TREMURI
x_j	x coordinate of node j
x_k	x coordinate of node k
y_j	y coordinate of node j
y_k	y coordinate of node k
y_n	neutral axis
y^*_n	neutral axis when both masonry and FRCM attain their ultimate tensile strain during the in-plane bending failure of an FRCM strengthened panel
$y_{n,0...4}$	neutral axis of a masonry pier's cross-section strengthened with EQ-GRID at the flexural failure modes 0, 1, 2, 3, 4

Capital Greek letters

$\Delta\sigma$	additional vertical stress normal to the bed joint
Δ_x	width of the masonry block

Δ_{xi}	eccentricity in the x-direction of the masonry element's extremity i from the barycenter of the rigid node in the equivalent frame schematization in TREMURI
Δ_{xj}	eccentricity in the x-direction of the masonry element's extremity j from the barycenter of the rigid node in the equivalent frame schematization in TREMURI
Δ_{y}	height of the masonry block
Δ_{yi}	eccentricity in the y-direction of the masonry element's extremity i from the barycenter of the rigid node in the equivalent frame schematization in TREMURI
Δ_{yj}	eccentricity in the y-direction of the masonry element's extremity j from the barycenter of the rigid node in the equivalent frame schematization in TREMURI
ϕ	interlocking parameter
ϕ_{KS}	interlocking parameter of the tested calcium-silicate brick masonry panels
ϕ_{m}	strength reduction factor for flexure according to ACI 459.4R-13
ϕ_{v}	strength reduction factor for shear according to ACI 459.4R-13
ϕ_{WZI}	interlocking parameter of the tested hollow clay brick masonry panels
ψ	coefficient equal to $\frac{1.2EI^2}{Gh^2}$

Lower-case Greek letters

α	coefficient for the masonry compressive stress-block formulation
α'	angle formed by the barycentric axis of the masonry elements with the horizontal direction in the Equivalent frame schematization in TREMURI
α^*	angle formed by the main load-bearing direction of the floor with the global X-axis in the Equivalent frame modelling with TREMURI
α_t	coefficient that considers the reduced tensile strength of the fibres under shear actions (CNR-DT 215/2018)
α_1	Rayleigh coefficient
β	coefficient for the masonry compressive stress-block formulation

$\beta_{E,i}$	residual shear strength of a panel with respect to V_{Rd} for the URM one or $V_{Rd,s}$ for the FRCM strengthened one
β_1	Rayleigh coefficient
γ_c	material density of the reinforced concrete
γ_m	partial safety factor
$\gamma_{masonry}$	material density of the masonry
γ_{Rd}	model safety factor (CNR-DT 215/2018)
γ_{12}	shear strain of the three-nodes orthotropic membrane element implemented in TREMURI
ε_{AR}	tensile strain of the EQ-GRID AR glass fibre yarns
$\varepsilon_{clamping,cr}$	first cracking tensile strain of the EQ-GRID composite specimens tested with clamping grips (avg.)
$\varepsilon_{clamping,u}$	ultimate tensile strain of the EQ-GRID composite specimens tested with clamping grips (avg.)
$\varepsilon_{clevis,cr}$	first cracking tensile strain of the EQ-GRID composite specimens tested with clevis-type grips (avg.)
$\varepsilon_{clevis,u}$	ultimate tensile strain of the EQ-GRID composite specimens tested with clevis-type grips (avg.)
ε_{cr}	cracking tensile strain in the equivalent bilinear diagram of an FRCM system after ACI 459.4R-13
ε_e	design tensile strain of the FRCM reinforcement after ACI 459.4R-13 (flexural failure)
$\varepsilon_{el,c}$	elastic compressive strain of a URM spandrel
$\varepsilon_{el,t}$	elastic equivalent tensile strain of a URM spandrel
ε_{fd}	design strain of the yarns arranged parallel to the shear force after CNR-DT 215/2018
ε_{fv}	design tensile strain of the FRCM reinforcement after ACI 459.4R-13 (shear failure)
$\varepsilon_{lim,conv}$	limit conventional tensile strain of the FRCM according to CNR-DT 215/2018
$\varepsilon_{lim,conv,h}$	limit conventional tensile strain in the horizontal direction of the EQ-GRID system

$\varepsilon_{lim,conv,v}$	limit conventional tensile strain in the vertical direction of the EQ-GRID system
ε_m	masonry compressive strain
ε_{m0}	elastic limit of the masonry compressive strain
$\varepsilon_{m,u}$	masonry ultimate compressive strain
ε_{PP}	tensile strain of the EQ-GRID polypropylene fibre yarns
ε_u	ultimate tensile strain of an FRCM obtained from the direct tensile tests with clevis-type grips (ACI 459.4R-13)
$\varepsilon_{u,c}$	ultimate compressive strain of a URM spandrel
$\varepsilon_{u,k}$	characteristic failure tensile strength of the EQ-GRID's textile component
$\varepsilon_{u,m}$	average failure tensile strain of the EQ-GRID's textile component
$\varepsilon_{u,t}$	ultimate equivalent tensile strain of a URM spandrel
ε_v	tensile strain in the vertical direction an FRCM system
ε_1	strain in the floor spanning orientation of the three-nodes orthotropic membrane element implemented in TREMURI (in 1.5.3)
ε_2	strain perpendicular to the floor spanning orientation of the three-nodes orthotropic membrane element implemented in TREMURI (in 1.5.3)
ε_1	maximum tensile strain in the strain distribution diagram of a masonry panel's cross-section strengthened with EQ-GRID (in 4.2.3)
ε_2	maximum compressive strain in the strain distribution diagram of a masonry panel's cross-section strengthened with EQ-GRID (in 4.2.3)
ε_I	first limit value of the strain range assumed for the calculation of the tensile elastic modulus of the EQ-GRID's textile component
ε_{II}	second limit value of the strain range assumed for the calculation of the tensile elastic modulus of the EQ-GRID's textile component
$\varepsilon_{@0.6fu}$	tensile strain at $0.6f_u$ in the stress-strain diagram of an FRCM after ACI 459.4R-13
$\varepsilon_{@0.9fu}$	tensile strain at $0.9f_u$ in the stress-strain diagram of an FRCM after ACI 459.4R-13
η	ratio between f_{tu} and $f_{mh,u}$

$\eta_{clamping}$	exploitation ratio of the average tensile strength of the EQ-GRID textile component (direct tensile tests on composite specimens performed with clamping grips)
η_{clevis}	exploitation ratio of the average tensile strength of the EQ-GRID textile component (direct tensile tests on composite specimens performed with clevis-type grips)
θ	angle of the generic wall formed with the global X-axis in TREMURI
θ_1	angle of the wall 1 formed with the global X-axis in TREMURI
θ_2	angle of the wall 2 formed with the global X-axis in TREMURI
λ	masonry panel's slenderness
$[\lambda]$	vector of the load multipliers λ_i in the nonlinear static analysis
λ_1	load multiplier applied at the first floor of the masonry building model in the nonlinear static analysis
λ_2	load multiplier applied at the second floor of the masonry building model in the nonlinear static analysis
μ	mortar joints' friction coefficient
μ_c	ductility in compression of a URM spandrel
μ_{DL3}	ductility reached by the tested masonry panels at the achievement of DL3
μ_{DL4}	ductility reached by the tested masonry panels at the achievement of DL4
μ_{DL5}	ductility reached by the tested masonry panels at the achievement of DL5
μ_{KS}	mortar joints' friction coefficient of the calcium-silicate brick masonry
μ_{red}	reduced mortar joints' friction coefficient
$\mu_{red,KS}$	reduced mortar joints' friction coefficient of the calcium-silicate brick masonry
$\mu_{red,WZI}$	reduced mortar joints' friction coefficient of the calcium-silicate brick masonry
μ_t	ductility in tension of a URM spandrel
μ_{WZI}	mortar joints' friction coefficient of the calcium-silicate brick masonry
μ^+	ductility reached by the ML-BEAM element in the positive quadrant
μ^-	ductility reached by the ML-BEAM element in the negative quadrant

ν	Poisson ratio
ξ	elastic damping ratio
ξ^*	lower elastic damping ratio in consideration of an expected displacement ductility μ
ρ_{fib}	fibre density [g/cm²]
σ	tensile stress
σ_{AR}	tensile stress of the EQ-GRID AR glass fibre yarns
σ_b	reduced vertical compressive stress in one half of the masonry block
$\sigma_{lim,conv}$	limit conventional tensile strength of the FRCM according to CNR-DT 215/2018
σ_n	horizontal compressive stress acting in a masonry spandrel
σ_{PP}	tensile stress of the EQ-GRID polypropylene fibre yarns
σ_u	failure tensile strength of the EQ-GRID's textile component
$\sigma_{u,k}$	characteristic failure tensile strength of the EQ-GRID's textile component
$\sigma_{u,m}$	average failure tensile strength of the EQ-GRID's textile component
σ_v	vertical compressive stress
σ_x	stress parallel to the bed joint
σ_y	stress normal to the bed joint
σ_0	the greater compressive stress between the horizontal and vertical one acting in a masonry spandrel
σ_1	stress in the floor spanning orientation of the three-nodes orthotropic membrane element implemented in TREMURI
σ_2	stress perpendicular to the floor spanning orientation of the three-nodes orthotropic membrane element implemented in TREMURI
σ_I	tensile stress at ε_I
σ_{II}	tensile stress at ε_{II}
σ_η	principal compressive stress acting in the horizontal middle section of a masonry panel due to vertical and shear actions (Mohr's circle)
σ_ξ	principal tensile stress acting in the horizontal middle section of a masonry panel due to vertical and shear actions (Mohr's circle)

τ	mean shear stress acting in the horizontal middle section of a masonry panel due to shear actions
τ_{max}	maximum shear stress acting in the horizontal middle section of a masonry panel due to shear actions
τ_{12}	shear stress of the three-nodes orthotropic membrane element implemented in TREMURI
φ	rotation of the fictitious 2D node belonging to the generic wall in the Equivalent frame method implemented in TREMURI
φ_i	rotation about the z-axis of node *i* of the masonry element implemented in TREMURI
φ_j	rotation about the z-axis of node *j* of the masonry element implemented in TREMURI
φ_v	shear strength reduction after ACI 459.4R-13
φ_x	rotation about the x-axis
φ_y	rotation about the y-axis
φ_1	rotation about the z-axis of the barycenter of the rigid node 1 in the equivalent frame schematization in TREMURI
φ_2	rotation about the z-axis of the barycenter of the rigid node 2 in the equivalent frame schematization in TREMURI
χ	shear coefficient for rectangular cross-sections
ω_i	frequency of the ith mode
ω_j	frequency of the jth mode
ω_1	first frequency of the masonry building model measured before testing
ω_4	fourth frequency of the masonry building model
ω_1^*	first frequency of the masonry building model measured after testing

Abbreviations

AR	alkali-resistant
BM	unreinforced masonry building model tested on the shaking table in Skopje

BM-SR	masonry building model retrofitted with EQ-GRID and tested on the shaking table in Skopje
Co.V.	coefficient of variation
CF	confidence factor
D	diagonal
DL	damage level
d.o.f.	degree of freedom
EF	equivalent frame
FRCM	fabric-reinforced cementitious matrix
FRP	fibre-reinforced polymer
H	horizontal
KS	calcium-silicate masonry brick
ML-BEAM	nonlinear beam element with lumped inelasticity idealization and multilinear constitutive law implemented in TREMURI
PTFE	polytetrafluoroethylene
R.c.	reinforced concrete
TRC	textile-reinforced concrete
TRM	textile-reinforced mortar
SSWP	strong spandrels-weak piers
STD	standard deviation
URM	unreinforced masonry
V	vertical
WSSP	weak spandrels-strong piers
WZI	hollow clay masonry brick

Preface

This thesis was written during my work as a research assistant at the Institute for Reinforced Concrete and Building Materials at the Karlsruhe Institute of Technology (KIT).

I would like to give my special thanks to Prof. Dr.-Ing. Lothar Stempniewski for suggesting this work, supporting and taking over the PhD supervision. My special thanks also go to Prof. Sergio Lagomarsino, PhD and Prof. Serena Cattari, PhD, whose support, suggestions, and discussions played a very important role in the success of this work. Moreover, I would like to thank Prof. Dr.-Ing. Werner Fuchs, Prof. Dr. Friedemann Wenzel, Prof. Dr. Frank Schilling and Prof. Dr.-Ing. Alexander Stark for their interest to participate in the commission.

Then, I would like to thank all my colleagues at the Institute for Reinforced Concrete and Building Materials for the pleasant working atmosphere and the exchange of ideas, which was not always only technical. I have also found friends with whom I spent many good times.

Furthermore, I would like to thank the Dr. Günther Kast GmbH & Co. KG and RÖFIX AG companies for their good professional cooperation during my research activity.

Finally, I would like to express my sincere thanks to my family for sustaining me. The constant encouragement and support of my husband Giuseppe Mercorio were decisive in finishing this work.

Karlsruhe, December 2021

Stefania Rizzo

Chapter 1

Structural behaviour and modelling of unreinforced masonry buildings under seismic actions

1.1 Behaviour of masonry structures under seismic actions

A significant portion of the residential structures in Europe and large areas of Asia and South America consists of masonry buildings. Today, it is common to construct residential buildings composed of load-bearing masonry brick walls. However, many masonry structures are built in the past and often in high seismic hazard areas. Despite the different technological aspects, such as materials and geometric proportions, which depend on the construction place, historical period, and intended original use, defining a common structural behaviour is possible. Therefore, the fundamental step to perform the modelling of a masonry structure as close as possible to reality and effective for seismic verification is to identify the main characteristics of the masonry buildings' seismic response. From observing the damage, it is possible to understand how the behaviour under seismic actions strongly depends on different typological-constructive aspects.

The damage mechanisms observed in buildings can be essentially divided into two categories depending on the response of the walls (Figure 1.1):

- Mode I;
- Mode II [1].

Mode I concerns all the failure kinematics connected to a masonry wall's behaviour perpendicular to its plane (flexural and overturning behaviour). On the contrary, Mode II regards the wall's response in its plane with damage typically due to shear and bending (rocking) failure.

The activation of these collapse modes strongly depends on the building's global behaviour, which is a function of its typological and technological characteristics.

Figure 1.1: Damage modes of masonry walls [2]

Therefore, identifying the main structural elements resistant to seismic actions is important to investigate their influence on global behaviour. Several load-bearing structural elements are common in masonry construction: walls, floors, vaults and roofs. It is important to stress that almost all the existing masonry buildings are designed to resist in an optimal way to the vertical actions. However, a seismic event generates horizontal forces that cause bending stresses in the walls, arches and vaults, which work mainly in compression. If no suitable elements are present to withstand these actions, serious damages may occur in the whole building or only in a part of it (local collapses).

Moreover, the global seismic behaviour of a building is strongly influenced by each structural elements' mechanical properties and the degree of connection between them. An earthquake generates complex dynamic actions due to the ground's motion and the structure's response during the seismic event. Under the same acceleration measured at the basis, very different stress levels can occur in the structural elements based on the building's characteristics (stiffness, damping, strength). It is important to specify that these properties can also change during the seismic event due to the cracking and partialization phenomena of the resistant sections.

Moreover, low values of the fundamental period and damping generally lead to a strong amplification of the ground's accelerations. Since the masonry buildings are usually squat and have high translational stiffness, they can show very low values of the fundamental period, causing an amplification of the input accelerations and the inertia forces that stress the structure. However, the building inevitably cracks during a seismic event, increasing its deformability. In this case, the period shifts towards higher values attenuating the initial amplification. Moreover, the development of the crack pattern generally contributes to raising the damping levels. Despite the masonry elements' limited ductility, the structure can progressively adapt to the stresses induced from the seismic input accompanying the ground's motion and limiting its catastrophic effects. Therefore, if the masonry building's design is carried out by considering this adaptability, the structure's reliability in resisting seismic actions can be improved.

Furthermore, the concept of box-like behaviour is the basis of a correct structural design of masonry buildings. The vertical resistant elements, i.e. the walls, must be effectively connected by the horizontal components, i.e. the floors, to guarantee a box-like behaviour. The good performance shown by the masonry buildings properly constructed confirms that this design concept allows reaching an excellent global resistance. The structure can resist the actions coming from any direction.

Moreover, observing the damage suffered by masonry constructions during seismic events, it is possible to conclude that an earthquake does not randomly damage a building. It rather selects the weakest structural elements, causing their cracking or collapse. For example, a bad connection between the parts that constitute the whole structure can cause a low global stiffness that prevents withstanding the seismic actions correctly. Furthermore, flexible floors not properly connected to the walls and inadequate connections between the vertical walls make it impossible or insufficient to distribute the forces acting on the building. Thus, local actions are generated on the walls, making those arranged orthogonally to the earthquake's direction more vulnerable and, if they are not properly connected to the adjacent ones, they can risk overturning. In this regard, Figure 1.2 shows the variation of the structural behaviour induced by different degrees of connection between the structural parts and floors' stiffness.

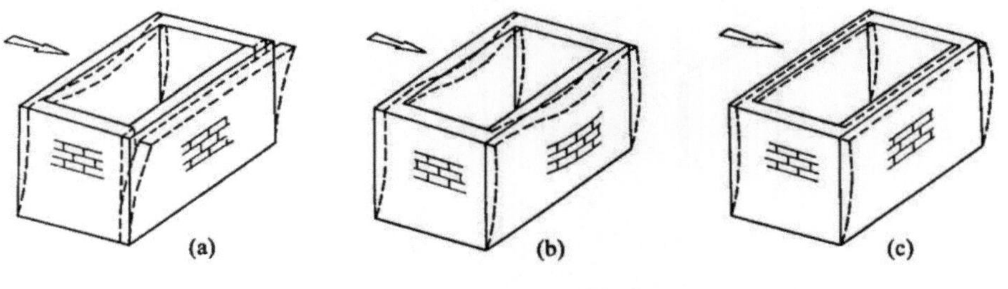

Figure 1.2: Influence of the constraint degree between the walls on the seismic response: (a) unconstrained; (b) constrained with flexible horizontal diaphragm; (c) constrained with stiff horizontal diaphragm [3]

It is important to highlight that the whole structure, the elements' size and shape can be decisive in activating the global and local failure mechanisms. The presence of altimetric and planimetric irregularities is an aspect that must be taken into account.

Moreover, the occurrence of the out-of-plane overturning mechanism is strongly reduced in case of good constraint conditions, and it can be further reduced in case of a good connection with the floors. For example, Figure 1.3 and Figure 1.4 show how it is possible to limit the occurrence of mode I mechanisms even with local interventions, such as the placement of steel tie-rods. In this case, the vulnerability is reduced by transferring the façade's out-of-plane response to the spine walls' in-plane response.

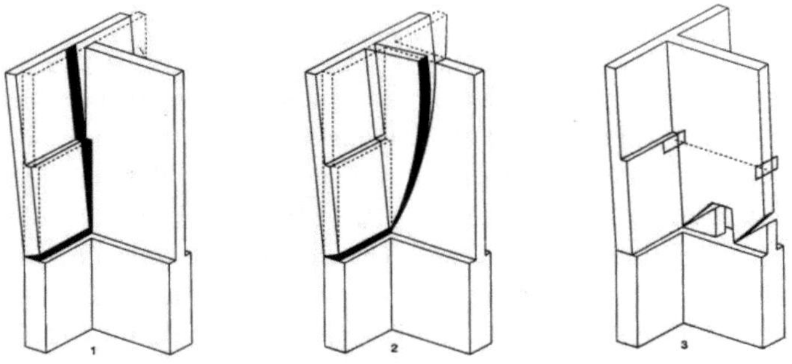

Figure 1.3: Façade's out-of-plane overturning mechanisms: (1) without any connection with the spine wall, (2) connection with the spine wall, (3) presence of a steel tie-rod [1]

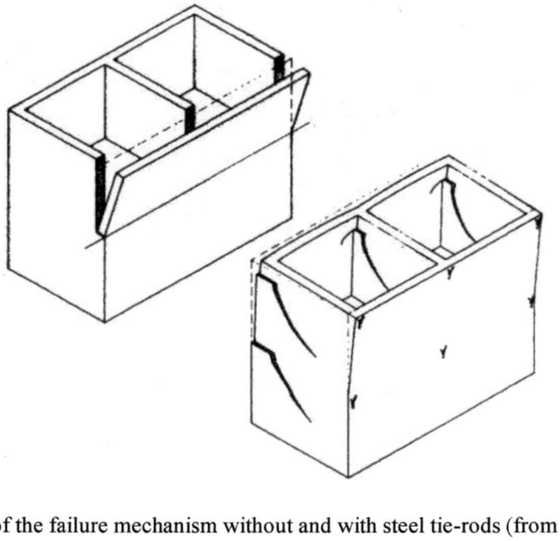

Figure 1.4: Changing of the failure mechanism without and with steel tie-rods (from Mode I to Mode II), [1]

The box-like behaviour allows defining a building's global vulnerability, depending on the entire structural seismic response. The latter is governed by the walls' in-plane response, connection degree, and floors' stiffness. Therefore, it is important to analyze the walls' in-plane response, examine the possible failure mechanisms, and use adequate calculation tools to investigate the building's three-dimensional behaviour.

Concerning the horizontal actions, it is possible to consider the masonry wall as an assembly of panels. Observing the damage induced by real earthquakes and analysing experimental data have shown that the damage is typically concentrated in well-defined portions of a wall: the vertical panels (masonry piers) and the coupling masonry beams (spandrels). On the other hand, the connection areas between the piers and spandrels are generally not cracked (Figure 1.5).

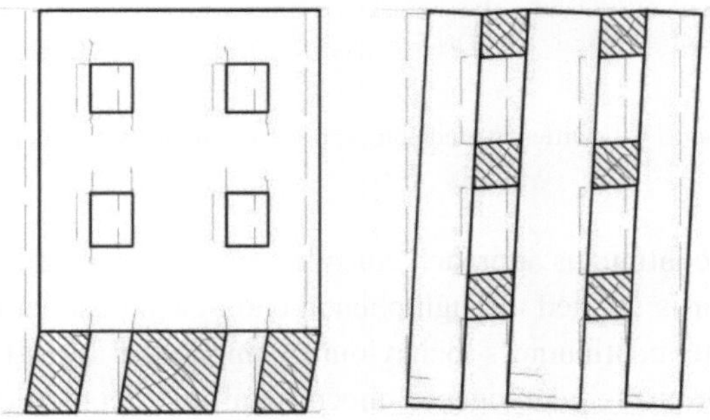

Figure 1.5: Examples of in-plane failure mechanisms of masonry walls [3]

These observations allow introducing the so-called equivalent frame modelling for masonry buildings, discussed in detail in § 1.5. Finally, any possible local failure related to the Mode I mechanism should be considered even in the case of box-like behaviour, as some local failures can occur in unconstrained portions of walls or if the in-plane local ductility demands exceed those available [4].

1.2 Modelling strategies of masonry structures

Several modelling strategies for the masonry structures are available in the literature depending on the model's scale. In this context, a distinction can be made between the finite element method (f.e.m.) and the macro-modelling. In the first category, the masonry behaviour is reproduced by adopting appropriate nonlinear constitutive material laws, and two different finite element modelling approaches are possible: discrete and continuous (Figure 1.6).

The discrete approach consists of the micro-modelling of masonry. The mortar joints and blocks are considered distinct units, in which their interface can be modelled through discontinuous elements or condensed into a single interface element. However, this type of modelling requires a broad knowledge of the individual components' mechanical properties, and a considerable computational effort is necessary. Therefore, the micro-modelling is suitable only for small structures subjected to heterogeneous stress and deformation states.

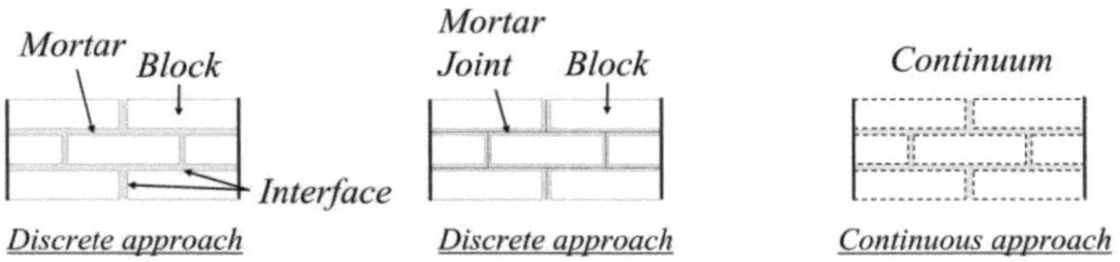

Figure 1.6: Different modelling approaches of masonry structures

On the contrary, the continuous approach considers masonry a continuous material whose mechanical behaviour is defined through phenomenological observations or homogenisation procedures. The continuum's behaviour is modelled by adopting macroscopic quantities obtained from its components' mechanical and geometrical properties. Compared to the micro-modelling, this approach reduces the computational effort and aims to analyze even large complex structures where the local stress states can be considered almost homogeneous.

It is worth noting that, due to the high degree of accuracy, both modelling techniques are commonly adopted in the research. However, the particularly high computational burden limits its application in the three-dimensional analysis of entire buildings. Moreover, the results' processing according to the most common design formulations is also rather onerous and, in many cases, problematic (e.g. the definition of collapse conditions).

Therefore, the macro-modelling technique is particularly interesting, as it allows reproducing the behaviour of simple macro-elements (panels) and the global response of complex structures with sufficient accuracy at the same time.

The first macro-modelling approaches were based on the principles of the limit analysis. It was assumed that the masonry had an infinite compressive strength, and its tensile strength was neglected. Thanks to this hypothesis, reducing the masonry wall to a kinematic chain of rigid bodies is possible (Figure 1.7). The configuration of the system is a function only of the displacement of a point. The static horizontal collapse load multiplier is calculated according to the assumed kinematic motion [5], [6].

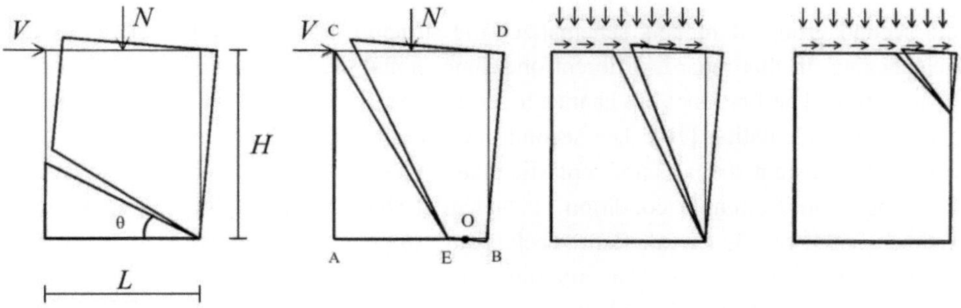

Figure 1.7: In-plane failure kinematic mechanisms of masonry piers [7]

This method leads to a good assessment of the structural vulnerability in the case of failure mode I [1], particularly after some structural improvement works, such as the placement of steel tie-rods or FRP (fibre reinforced polymer) strips. On the contrary, it is excessively cautious in the case of mode II [5], [6], as it allows calculating only the lower limit of residual resistance of the wall associated with the overturning mechanism.

An alternative to this calculation method is the approach that considers the masonry's elastic deformations eventually followed by inelastic deformations. Within this range of models, two main groups can be identified. The first one provides for the use of one-dimensional elements, commonly based on a strut idealization. In [8], [9], it is proposed to model the reacting portion of a masonry panel through a compressed strut, whose inclination and stiffness reproduce the panel's behaviour (Figure 1.8).

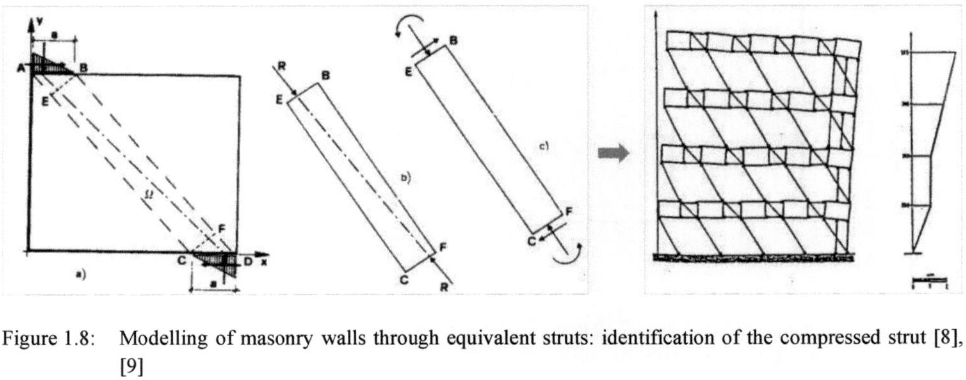

Figure 1.8: Modelling of masonry walls through equivalent struts: identification of the compressed strut [8], [9]

When the partialization increases, the equivalent strut's geometric properties, i.e. inclination, and dimensions of the section, change consequently. Therefore, this approach may also be classified as a "method with variable geometry". Each panel's failure occurs by reaching an equilibrium limit configuration or the diagonal strut's compressive failure.

The second group of models schematizes the structure as an assembly of deformable shear beams. In this regard, different one-dimensional elements have been proposed in the literature. The first ones are characterized by a variable stiffness based on the partialized section calculation [10]. The second ones are elements characterized by an elastic phase with constant stiffness and a plastic phase. The nonlinear behaviour is activated by achieving a limit strength condition [11]–[13]. It is worth specifying that most of the methods based on the "weak floor mechanism" (including the so-called POR method) adopt this type of element. An important distinction between these latter approaches consists of the number of possible mechanisms during the inelastic phase, particularly the failure mechanisms of the panel and the complete wall.

It is important to stress that the POR method [11] was largely adopted in Italy after the 1980 Irpinia earthquake [14]. It considered the masonry piers the only site of deformations and failures, without evaluating whether other elements, such as the spandrels, could participate. This assumption corresponds to the idealization of a "strong spandrels-weak piers" (SSWP) model (Figure 1.9), in which the piers crack first and prevent, in this way, the spandrels' failure. Therefore, the latter may be assumed as infinitely stiff portions that ensure a perfect coupling between the piers. After this assumption, the pier extremities cannot rotate, and only a "storey mechanism" can be activated.

Furthermore, in the first version of the method, the only possible collapse mechanism of a masonry wall was the piers' diagonal cracking. Other failures modes, e.g. rocking or shear-sliding, were neglected.

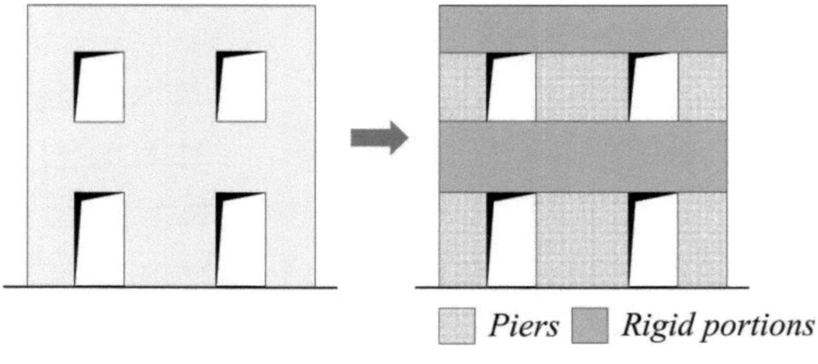

Figure 1.9: Modelling of a masonry wall through the POR method: identification of the piers and rigid portions

Therefore, additional strength criteria were introduced [12], [13], improving the method and taking into account other possible failure modes for piers. However, it has been impossible to overcome the intrinsic limit of the method, i.e. considering only the in-plane collapse. The model was based on this hypothesis and consisted of performing a nonlinear shear-displacement analysis separately for each interstorey (Figure 1.10). This approach greatly simplified the calculations, but it could not consider the determination of the internal forces and moments of the spandrels. Moreover, the interstorey shear-displacement analysis requires assumptions on the degree of constraint at each end of the piers, which depends on the stiffness and resistance of the horizontal coupling elements (i.e. masonry spandrels).

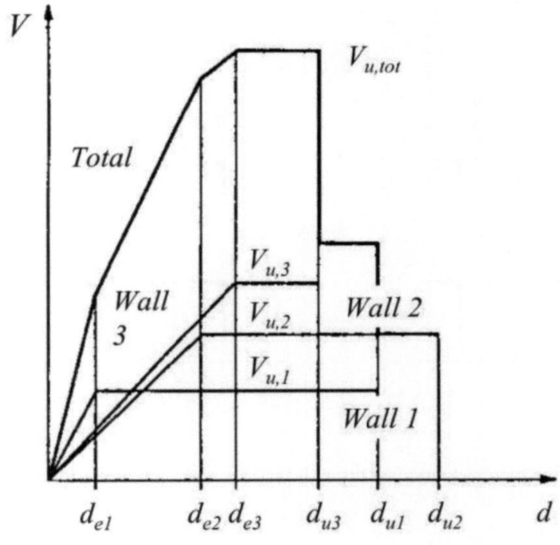

Figure 1.10: Capacity curve of a masonry building obtained with the POR method as the sum of that of each wall [15]

Moreover, the spandrels can be increasingly stressed if the horizontal seismic forces increase, and they could crack or even break. Therefore, the only possibility of accurately considering these phenomena is the global analysis of the multistorey wall or building. The latter is also the only way to calculate the local and global equilibriums correctly. In this regard, the interstorey analysis cannot account for the variations of normal force acting in masonry walls when the seismic forces increase, affecting their stiffness and resistance [4].

Another simplified modelling approach is the so-called “weak spandrels-strong piers” (WSSP in Figure 1.11), in which the piers are assumed uncoupled (cantilever idealization), and the spandrels have both null strength and null stiffness. However, it is important to stress that, in most cases, it is correct to consider the horizontal displacement of the vertical structural elements, at least coupled at the floor levels by the horizontal diaphragms. Thus, the definition of the piers’ effective height and boundary conditions is fundamental to assess the overall capacity of the wall since only pier elements are modelled. Furthermore, preliminary evaluations of spandrels’ effectiveness are necessary to orientate the choice between these two extreme schematizations correctly. FEMA guidelines expressly suggest both SSWP and WSSP models. On the contrary, the SSWP hypothesis (POR method) is no more allowed in the Italian Building Code [16] to assess multistorey masonry buildings.

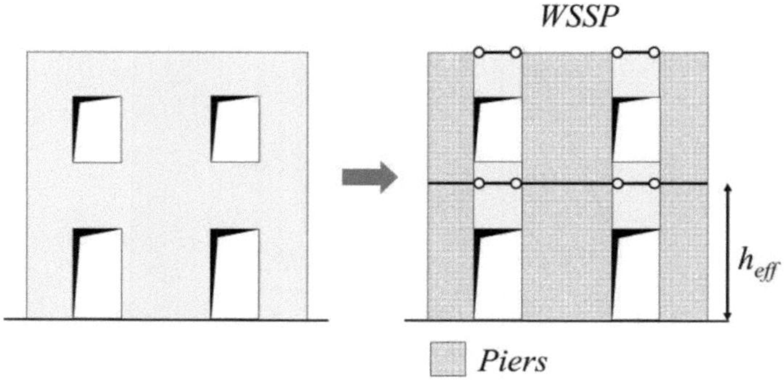

Figure 1.11: Modelling of a masonry wall through WSSP (weak spandrels-strong piers) simplification

It is important to specify these simplified models are inappropriate for the walls that may show both types of response in different regions or another behaviour during the nonlinear response. Moreover, the structural damages observed after past earthquakes have demonstrated that the masonry buildings’ global seismic behaviour first depends on some constructional details that can prevent local failure mechanisms, e.g. good connections between walls, bond-beams, etc. In this case, the seismic structural behaviour can mainly depend on the in-plane strength of the load-bearing walls positioned in the same direction of the horizontal action.

Since the in-plane seismic damage of a wall is generally concentrated in masonry portions between the openings, the following types of macroelements may be defined:

- Piers: main vertical structural elements that can carry both vertical and horizontal loads;
- Spandrels: secondary horizontal elements with the principal function to transfer the loads from the floor slabs to the piers;
- Nodes: masonry portions assumed rigid since the damage induced from an earthquake is generally localized in the piers and spandrels. They mainly have the function to connect these latter.

According to this type of discretization, each load-bearing wall can be considered an idealized frame, in which the structural elements, i.e. piers and spandrels, have only one node at each end (Figure 1.12).

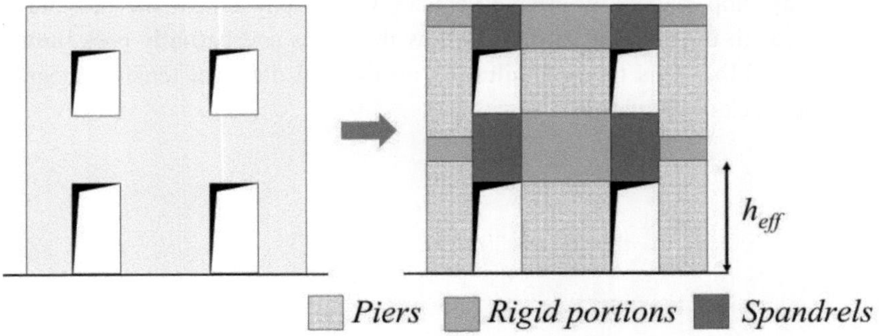

Figure 1.12: Modelling of a masonry wall through the Equivalent frame method

This modelling strategy is also called the "Equivalent frame method". It allows creating numerical models with much lower d.o.f than those obtained with finite element modelling. Therefore, a lower computational burden is provided, and the seismic response of masonry buildings can be predicted in a very realistic way at the same time. Moreover, it is also possible to model other structural elements with the masonry ones, such as reinforced concrete beams or columns, which are quite common in existing buildings. For this reason, technical standards, such as Eurocode 8 [17], Italian Building Code NTC 2018 [16], refer to macroelement-based models.

In this work, the equivalent frame model implemented in the TREMURI computer program has been adopted. A detailed description of the method is given in § 1.5.

1.3 In-plane strength of masonry piers

The damage observed on complex masonry walls after an earthquake and the experimental results of several laboratory tests [18] have shown that masonry piers subjected to in-plane loading may present a shear, flexural or mixed behaviour.

The first one is associated with the sliding shear and diagonal cracking failure. If the mortar joints' quality is poor and the vertical compressive stresses are low, the seismic forces may cause the sliding of a part of the panel along one of the bed joints (Figure 1.13a). This type of failure usually occurs in the upper storeys of a masonry building below a stiff roof structure, where the compressive stresses are low and the seismic accelerations are high.

On the contrary, this phenomenon rarely happens in the buildings' bottom storeys. The piers usually develop diagonally oriented cracks, which may follow the bed- and head-joints, pass through the units or partially follow the joints and partially pass through the bricks (Figure 1.13b). This type of failure is also called diagonal tension shear failure because of the cracks' orientation.

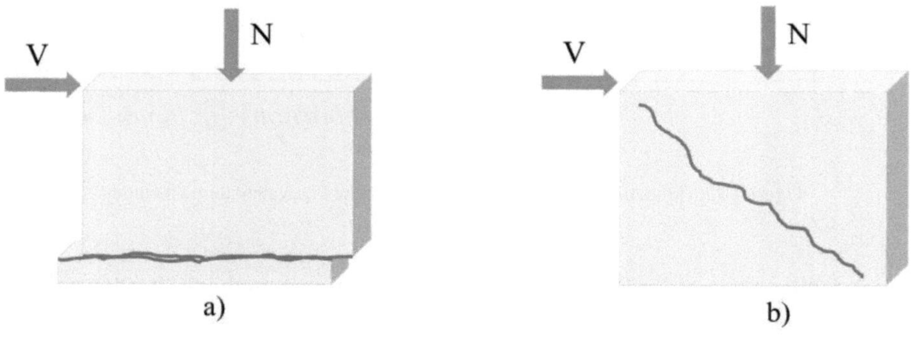

Figure 1.13: Scheme of shear failure mechanisms in masonry piers: a) sliding on a bed joint, b) diagonal cracking

Examples of diagonal shear cracking in load-bearing walls after an earthquake are depicted in Figure 1.14.

It is important to stress that masonry heterogeneity plays a dominant role in diagonal cracking. The failure generally results from some interacting factors, such as the quality of mortar and bricks. In this respect, two main types of simplified approaches are often used to predict the shear strength associated with diagonal cracking.

Figure 1.14: Diagonal shear cracking in brick masonry piers of a three-storey building (on the left), in stone-masonry piers of a historic building (on the right) after an earthquake [19]

The first one has been developed based on experimental tests performed on URM piers with doubly fixed boundary conditions in Ljubljana (Slovenia). It assumes that the masonry panel's behaviour is ideal elastic, homogeneous and isotropic up to the failure. The diagonal cracking starts at the centre of the pier and propagates afterwards to the corners. It occurs when the principal stress at the panel's centre σ_ξ reaches the masonry reference tensile strength $f_{m,t}$ (Figure 1.15). The equation for calculating the shear resistance V_t is derived based on the elasticity theory [20].

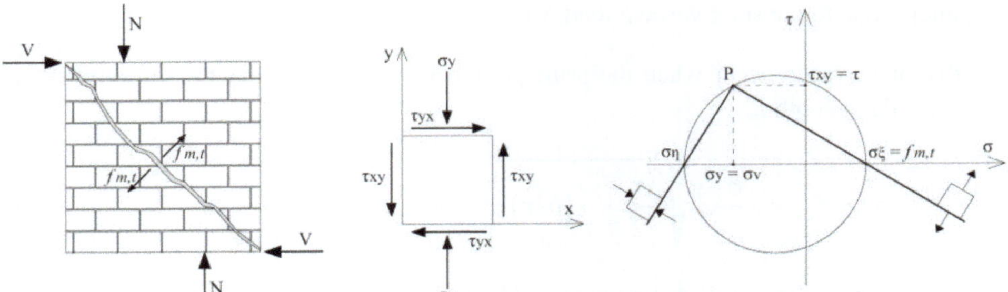

Figure 1.15: Diagonal shear cracking of a masonry pier: Mohr's circle and principal stresses

Thus, if the panel is sufficiently slender, it is possible to apply the Jourawski shear theory. In the generic cross-section, the shear stress varies with a parabolic law that assumes zero value at the edges and maximum value at the centre of gravity. On the contrary, if the panel is squat, the shear stress distribution is almost constant with a value equal to the average. Therefore, the maximum shear stress τ_{max} can be expressed as follows:

$$\tau_{\max} = b \cdot \tau = b \cdot \frac{V}{l \cdot t} \tag{1.1}$$

Where b is a factor that considers the shear stress distribution at the centre of the pier, τ is the mean shear stress acting in the horizontal middle section due to the shear force V, l and t are the length and thickness of the pier. The b factor depends on the panel's dimensions, and the values proposed in [21] are commonly adopted (Eq. (1.2)):

$$b = \begin{cases} 1.5 & \text{if } \frac{h}{l} > 1.5 \\ \frac{h}{l} & \text{if } 1 \leq \frac{h}{l} \leq 1.5 \\ 1.0 & \text{if } \frac{h}{l} < 1.0 \end{cases} \tag{1.2}$$

Therefore, the principal compressive and tensile stresses that develop in the middle section of the pier due to the vertical and shear load are equal to:

$$\sigma_{\eta,\xi} = \pm\frac{\sigma_v}{2} + \sqrt{\left(\frac{\sigma_v}{2}\right)^2 + (b \cdot \tau)^2} \tag{1.3}$$

Where σ_v is the average compressive stress in the horizontal cross-section at the centre of the panel A due to constant vertical load N ($\sigma_v = N/A$).

The diagonal cracks occur when the principal tensile stress σ_ξ attains the masonry referential tensile strength $f_{m,t}$:

$$\sigma_\xi = f_{m,t} = -\frac{\sigma_v}{2} + \sqrt{\left(\frac{\sigma_v}{2}\right)^2 + (b \cdot \tau)^2} \tag{1.4}$$

Thus, the shear resistance V_t can be obtained by rearranging Eq. (1.4),:

$$V_t = \frac{l \cdot t}{b} \cdot f_{m,t} \cdot \sqrt{1 + \frac{\sigma_v}{f_{m,t}}} \tag{1.5}$$

It is worth noting that this formulation is implemented in the explanatory Circular of the Italian Building Code NTC 2018 [22] to assess existing masonry panels with irregular patterns and in the seismic standards of other countries, such as the former Yugoslavia [19].

The second shear strength criterion is formulated according to the Mohr-Coulomb theory, which may be physically associated with the diagonal cracking through the mortar bed and head joints. Therefore, it is suitable to predict the shear strength of masonry panels with a regular pattern, and it is usually so formulated:

$$\tau = c + \mu \cdot \sigma_v \tag{1.6}$$

Where τ is the failure shear stress, c and μ are the mortar joints' cohesion and friction coefficient, σ_v is the average vertical stress acting on the bed joints.

By adopting Eq. (1.6), different ways are possible to evaluate the ultimate shear force V_t of a URM panel depending on the verification section's position.

The first approach is to consider the horizontal section at the top or bottom of the wall, taking into account also the flexural cracking. In this case, the average failure shear stress τ is referred to the uncracked section length l_c.

$$V_t = \tau \cdot l_c \cdot t = (c + \mu \cdot \sigma_v) \cdot l_c \cdot t = \left(c + \mu \cdot \frac{N}{l_c \cdot t} \right) \cdot l_c \cdot t \tag{1.7}$$

It is worth noting that Eq. (1.7) seems more suitable for describing the shear sliding rather than the diagonal cracking since it provides for the failure in the more partialized sections. Moreover, the quantities c and μ are assumed uniform along with the compressed length l_c. Thus, they should be considered "local" parameters that characterize the joint.

The effective uncracked section's length is generally calculated assuming a simplified distribution of compression stresses (constant or linear) and neglecting the masonry's tensile strength. If a linear distribution is adopted and the eccentricity e of the axial load N exceeds 1/6 of the panel's length, l_c can be calculated as:

$$l_c = 3 \cdot \left(\frac{l}{2} - e \right) \tag{1.8}$$

It is worth pointing out that Eurocode 8 adopts this approach to assess and retrofit masonry buildings. The expressions suggested by Annex C of DIN-EN-1998-3 [17] are the following:

$$V_{Rd} = \frac{f_v}{CF \cdot \gamma_m} \cdot l_c \cdot t \tag{1.9}$$

$$f_v = f_{vm0} + 0.4 \cdot \frac{N}{l_c \cdot t} \leq 0.065 f_{m,u} \tag{1.10}$$

Where f_{vm0} and $f_{m,u}$ are the average values of the masonry initial shear strength and compressive strength, CF and γ_m are the confidence and partial safety factors, respectively. Moreover, the upper limit of the shear strength in Eq. (1.10) considers the diagonal cracking failure with cracks passing through the bricks in a simplified way. Therefore, since the vertical and seismic loads should be known to calculate the compressed part of the wall's length, this approach is only useful in traditional safety verifications, where each structural element's resistance capacity is compared with the design action. Concerning the nonlinear pushover analyses, iterations would be required due to the changes in lateral load distribution in the nonlinear range [19].

Furthermore, the second possible approach to calculate the shear resistance of a URM panel is to consider the horizontal section at the centre of the pier, in which the average vertical stress σ_v is uniform:

$$V_t = \tau \cdot l \cdot t = (c + \mu \cdot \sigma_v) \cdot l \cdot t = \left(c + \mu \cdot \frac{N}{l \cdot t} \right) \cdot l \cdot t \qquad (1.11)$$

In this case, the strength parameters c and μ are intended as global material properties. They cannot be related to the bed joints' local cohesion and friction coefficient since the real normal stress distribution is non-uniform. Compared to Eq. (1.7), this approach can better interpret the failure mechanism that involves the mortar joints but with diagonal cracks. The results of various experimental campaigns and the seismic damage observed on the structures have shown the prevalence of diagonal cracking in most cases. On the contrary, the sliding along the horizontal base joint is confined to a limited number of cases. It can generally occur in rather squat panels subjected to low values of the normal compression.

Thus, the second approach should be more representative of a "stepped-stairs" failure, where the τ provided by Eq. (1.11) is assumed as the average shear stress of the panel's cross-section. Therefore, the resistance V_t must be obtained by multiplying this value by the entire cross-section.

Moreover, it is important to highlight that the masonry pattern plays a fundamental role in defining the inclined sliding plane of the diagonal cracking failure. In this respect, the approach proposed by Mann and Müller [23], which is aimed to determine the failure shear stress τ, is illustrated below, and it is based on the following hypotheses:

- The units are much stiffer than the mortar joints,
- The mechanical properties of head joints are negligible.

The second assumption is reasonable since several factors cannot guarantee the transmission of shear stresses, such as the bricks' smooth header faces and the usually poor mortar's quality of the head joints.

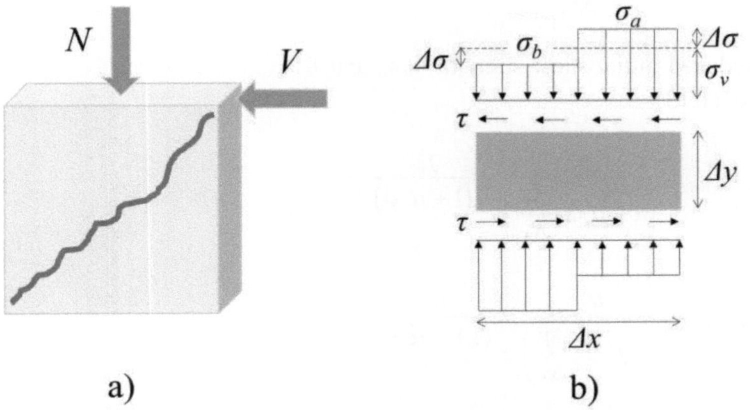

Figure 1.16: a) URM pier under in-plane shear masonry wall, b) single block

Therefore, considering a single masonry block (Figure 1.16 (b)), the shear stresses τ acting in the upper and lower mortar joint generate a torsional moment. The latter can only be equilibrated by an opposing and equally large torsional moment due to the normal stresses. As shown in Figure 1.16 (b), the caused additional stress $\Delta\sigma$ is added or subtracted to the ones generated from the dead and accidental loads, σ_v, increasing, in this way, the vertical stresses in one half of the block and reducing these in the other half. Therefore, an approximately stepped normal distribution is created, and the single brick's equilibrium is only guaranteed if the vertical loads are present. Otherwise, the shear stresses must consequently be transferred through the head joints, as no tensile stresses can be carried by to the bed joints.

The rotational equilibrium of the single block illustrated in Figure 1.16 (b) leads to Eq. (1.12):

$$2 \cdot \Delta\sigma \cdot \frac{\Delta x}{2} \cdot \frac{\Delta x}{4} = \tau \cdot \Delta x \cdot \frac{\Delta y}{2} \Rightarrow \Delta\sigma = 2\tau \cdot \frac{\Delta y}{\Delta x} \tag{1.12}$$

Since the Mohr-Coulomb criterion well interprets the shear failure along mortar joints, the shear stress τ can be calculated in the less compressed mortar joint part as follows:

$$\tau = f_{v0} + \mu \cdot \sigma_b = f_{v0} + \mu \cdot (\sigma_v - \Delta\sigma) = f_{v0} + \mu \cdot \left(\sigma_v - 2\tau \cdot \frac{\Delta y}{\Delta x} \right) \Rightarrow$$

$$\tau = \frac{f_{v0} + \mu \cdot \sigma_v}{\left(1 + \mu \cdot 2 \cdot \frac{\Delta y}{\Delta x}\right)} = \frac{f_{v0}}{\left(1 + \mu \cdot 2 \cdot \frac{\Delta y}{\Delta x}\right)} + \frac{\mu \cdot \sigma_v}{\left(1 + \mu \cdot 2 \cdot \frac{\Delta y}{\Delta x}\right)} = f_{v0,red} + \mu_{red} \cdot \sigma_v \tag{1.13}$$

Thus, the reduced initial shear strength $f_{v0,red}$ and friction coefficient μ_{red} can be defined based on Eq. (1.13):

$$f_{v0,red} = \frac{f_{v0}}{\left(1 + \mu \cdot 2 \cdot \frac{\Delta y}{\Delta x}\right)} = \frac{f_{v0}}{(1 + \mu \cdot \phi)} \tag{1.14}$$

$$\mu_{red} = \frac{\mu}{\left(1 + \mu \cdot 2 \cdot \frac{\Delta y}{\Delta x}\right)} = \frac{\mu}{(1 + \mu \cdot \phi)} \tag{1.15}$$

The product $2 \cdot \Delta y / \Delta x$ corresponds to the interlocking parameter ϕ of a regular masonry pattern with blocks' overlap length of two successive courses equal to $\Delta x/2$ and represents the tangent of the average inclination angle of the diagonal "stepped-stair" cracks. For example, common values of the ratio $\Delta y/\Delta x$ and friction coefficient μ for regular brick masonry are 0.5 and 0.577. In this case, the reduction factor $1/(1+\mu \cdot 2 \cdot \Delta y/\Delta x)$ is equal to 0.634. Thus, thanks to the reduced cohesion and friction coefficient, the masonry shear strength can be calculated assuming a uniform distribution of normal stresses σ_v without considering any reduction or increase.

However, under high values of vertical loads, the mortar joint's frictional resistance can become so great that it is not reached, and the block cracks by attaining the main tensile stress. In this case, the masonry panel can be considered a homogeneous shell loaded with shear and normal stresses.

As shown in Figure 1.17, the masonry unit must absorb the shear stresses of two courses since the head joints' load-bearing capacity is neglected. For simplicity, only the normal stress acting in the vertical direction σ_v is taken into account ($\sigma_x = 0$). In this case, the stepped normal stress distribution causes an increased moment and shear load in the single block so that the shear stress in the brick is about 2.3 times higher [24]. Therefore, the principal tensile stress σ_ξ can be written as follows:

$$\sigma_\xi = -\frac{\sigma_v}{2} + \sqrt{\left(\frac{\sigma_v}{2}\right)^2 + (2.3\tau)^2} \tag{1.16}$$

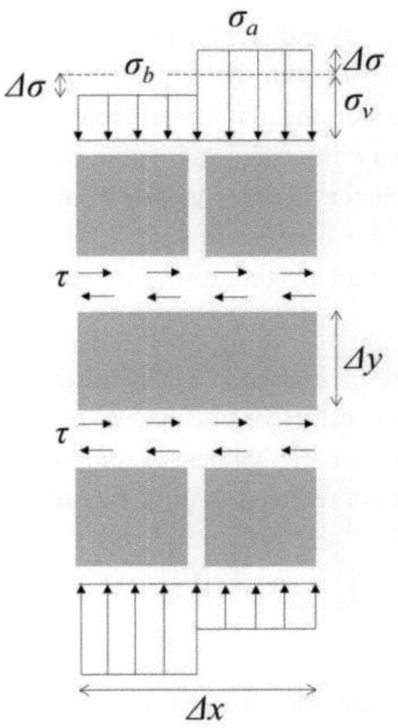

Figure 1.17: Tensile shear failure of masonry units

The masonry unit cracks when its tensile strength f_{bt} is attained. Thus, assuming σ_ξ equal to f_{bt}, the failure shear stress τ can be obtained by rearranging Eq. (1.16):

$$\tau = \frac{f_{bt}}{2,3}\sqrt{1 + \frac{\sigma_v}{f_{bt}}} \tag{1.17}$$

The Italian Building Code's explanatory circular [22] and the technical standard CNR-DT 212/2013 "Guide for the Probabilistic Assessment of the Seismic Safety of Existing Buildings" [25] adopt this approach to assess the in-plane load-bearing capacity of existing masonry panels with regular patterns. The following formulations are suggested, where V_t and $V_{t,lim}$ are the ultimate shear resistance of a URM panel with diagonal cracks passing through the mortar joints or masonry units:

$$V_t = \frac{l \cdot t}{b} \cdot \left(\frac{f_{vm0}}{1+\mu \cdot \phi} + \frac{\mu}{1+\mu \cdot \phi} \cdot \sigma_v \right) \leq V_{t,\lim} \tag{1.18}$$

$$V_{t,\lim} = \frac{l \cdot t}{b} \cdot \frac{f_{bt}}{2.3} \cdot \sqrt{1 + \frac{\sigma_v}{f_{bt}}} \tag{1.19}$$

It is worth noting that Eqs. (1.18) and (1.19) are obtained by multiplying the failure shear stress τ in Eqs. (1.13) and (1.17) with the entire panel's cross-section ($l \cdot t$). Then, the b factor is introduced to consider the parabolic shear stress distribution related to the panel's slenderness. In this way, the maximum value of shear stress τ_{max} can be correlated to the average one τ. The b factor assumes values from 1.0 to 1.5 (Eq.(1.2)), and several authors proposed the use of this coefficient [26], [27]. Furthermore, numerical analyses confirmed that the diagonal cracking failure occurs starting from the centre of the panel [15].

The German National Annex of Eurocode 6 [28] also adopt the approach proposed by Mann and Müller [23], but the effective uncracked section's length l_c is considered (Eq. (1.20)). Therefore, the characteristic values of initial shear strength f_{vk0} and friction coefficient (assumed 0.4) are not reduced through the interlocking parameter.

$$f_{vlt1} = f_{vk0} + 0.4\sigma_v = f_{vk0} + 0.4 \cdot \left(\frac{N}{l_c \cdot t} \right) \tag{1.20}$$

The brick's tensile failure is taken into account through Eq. (1.21), which is similar to Eq. (1.17):

$$f_{vlt2} = 0.45 \cdot f_{bt,cal} \cdot \sqrt{1 + \frac{\sigma_v}{f_{bt,cal}}} = 0.45 \cdot f_{bt,cal} \cdot \sqrt{1 + \frac{N}{f_{bt,cal} \cdot l_c \cdot t}} \tag{1.21}$$

Where $f_{bt,cal}$ is the characteristic value of the brick's tensile strength.

Thus, the German National Annex of Eurocode 6 [28] proposes to calculate the ultimate shear stress f_{vlt} as follows:

$$f_{vlt} = \min\left(f_{vlt,1}; f_{vlt,2}\right) \tag{1.22}$$

Furthermore, other possible in-plane failure modes of a URM pier are flexural rocking and crushing. In the first case, the panel behaves as a nearly rigid body and rotates about the toe (Figure 1.18). In the second one, it crushes with sub-vertical cracks oriented towards the compressed corners. The rocking phenomenon can occur in slender piers with low vertical stresses. In contrast, the crushing generally occurs under very high normal stress values when the masonry attains its compressive strength. Regarding the rocking behaviour, it is possible to calculate the ultimate bending moment M_{Rd} of a URM pier by neglecting the masonry tensile strength and assuming an appropriate nonlinear distribution of the compressive stresses at one end section.

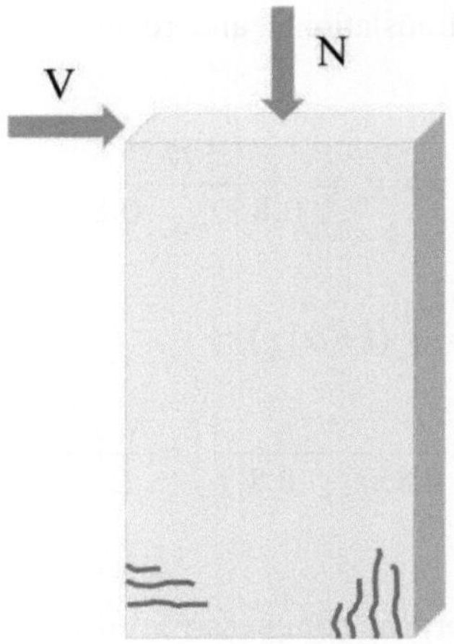

Figure 1.18: In-plane flexural failure (rocking) of a URM pier

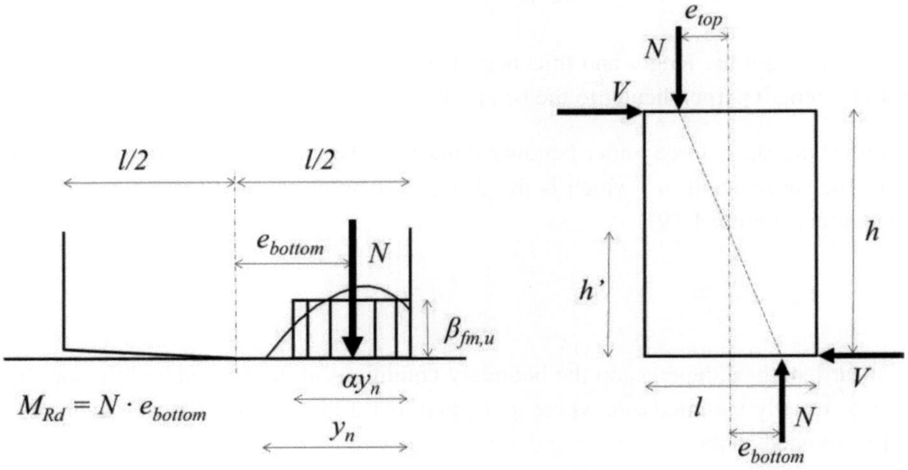

Figure 1.19: Stress distribution at the compressed toe of a URM pier under in-plane flexural failure

Therefore, considering the boundary conditions of Figure 1.19, in which the eccentricity of the normal force at the base section of the pier e_{bottom} is higher or equal to e_{top}, the ultimate bending moment is defined by the crushing failure at the compressed toe of the panel (the self-weight of the pier is neglected).

Adopting the stress-block formulation, in which the coefficients α and β may be assumed equal to 0.80 and 0.85, the translational and rotational equilibrium equations can be written as follows:

$$N = \beta f_{m,u} \cdot \alpha y_n \cdot t \Rightarrow y_n = \frac{N}{0.85 f_{m,u} \cdot 0.8 \cdot t} \tag{1.23}$$

$$\left.\begin{aligned} M_{Rd} &= N \cdot e_{bottom} = N \cdot \frac{1}{2}(l - \alpha y_n) \Rightarrow \\ M_{Rd} &= N \cdot \frac{1}{2}\left(l - 0.8 \frac{N}{0.85 f_{m,u} \cdot 0.8 \cdot t}\right) = \frac{N \cdot l}{2} - \frac{N^2}{2 \cdot 0.85 f_{m,u} \cdot t} \end{aligned}\right\} \tag{1.24}$$

Defining σ_v as the average vertical compressive stress referred to the total area of the section, $\sigma_v = N/l \cdot t$, Eq. (1.25) can be obtained:

$$M_{Rd} = \frac{l^2 \cdot t}{2} \cdot \sigma_v \cdot \left(1 - \frac{\sigma_v}{0.85 \cdot f_{m,u}}\right) \tag{1.25}$$

Where l and t are the length and thickness of the pier, $f_{m,u}$ is the masonry ultimate compressive strength perpendicular to the bed joints.

Therefore, the shear force under bending failure can be calculated through the introduction of the shear length h', which is the distance between the end section and that with zero moment (Figure 1.19):

$$V_{Rd}^F = \frac{M_{Rd}}{h'} \tag{1.26}$$

The shear length h' depends on the boundary conditions at the bottom and the top of the pier. It is usually assumed $\alpha \cdot h$, where α is equal to 1.0 in the case of a cantilever and 0.5 in a fixed-ended panel.

Finally, it is worth noting that several technical standards, such as Eurocode 8 [17], Italian Building Code NTC 2018 [16] and CNR-DT 212/2013 [25], adopt this calculation approach.

1.4 In-plane strength of masonry spandrels

The past strong earthquakes have shown that URM buildings can suffer severe damage to spandrels, especially if local out-of-plane mechanisms do not occur and the walls are mainly subjected to in-plane seismic forces [29]. Furthermore, they are often the first elements to crack or fail. For example, Figure 1.20 shows a URM building in L'Aquila (Italy) after the earthquake on 6 April 2009 ($M_w = 6.3$).

a) b)

Figure 1.20: Shear failure of spandrels supported by shallow masonry arches in an old URM building after the earthquake on 6 April 2009 in L'Aquila (Italy): (a) Entire building, (b) detail of a spandrel [30]

All the spandrels between the window openings have failed with diagonal shear cracks. They were restrained by relatively wide piers and horizontal steel tie bars positioned along the facade and embedded in the masonry [30].

a) b)

Figure 1.21: Flexural failure of spandrels in an old URM building after the earthquake on 6 April 2009 in L'Aquila (Italy): (a) Entire structure, (b) detail of a spandrel [30]

Furthermore, Figure 1.21 also shows a URM building located in L'Aquila (Italy). In this case, the steel tie bars were not present, and the piers were very slender. Therefore, they could provide a minor restraint to the axial elongation of the spandrels, which failed with wide flexural vertical cracks at both ends. As illustrated in Figure 1.21 (b), one spandrel completely collapsed because of the out-of-plane accelerations. This building was pulled down shortly after the earthquake. [30]

Therefore, both observations of the seismic damage to complex masonry walls and experimental laboratory tests have shown that the main failure modes of spandrels are the flexural and diagonal cracking ones [29]–[31]. In this regard, Benedetti et al. carried out a broad research program in which 24 half-scale building models were tested on the shaking table [32]. Some models were made of brick masonry with spandrels supported by wooden lintels, while other models were made of stone masonry with regular arches above openings. The experimental results showed that the energy dissipation of URM buildings can be maximized if the damage develops within spandrels rather than piers. Furthermore, Bothara et al. carried out a shaking table test on a half-scale URM building model with wooden lintels and flat masonry arches [33]. The test confirmed that spandrels can fail, not only showing diagonal shear cracking but also vertical flexural cracks that develop at the end sections of the element.

Moreover, since the spandrels are generally characterized by low axial load values, the crushing failure, represented by widespread sub-vertical cracks, is a very rare instance. The sliding shear can also not occur because of the interlocking phenomena that can develop at the interface between the element's end sections and the contiguous masonry [34].

Different types of spandrels can be identified in existing URM buildings. The lintel above the opening and the presence or absence of tensile resistant elements, such as steel tie bar or concrete beam, define the spandrel type. In ancient URM buildings, the spandrels are not designed for earthquake actions and have generally stone, masonry, wooden, or steel lintels above the openings. These elements only allow transferring the gravity loads to the adjacent piers. Furthermore, the steel lintels were usually realized with two I-beams transversely connected by steel ties (Figure 1.22) to avoid potential torsional-flexural buckling. The lateral external space along the beams was filled with mortar to provide flat surfaces [29]. The spandrels of historical URM buildings are generally supported by regular or flat arches (Figure 1.22).

On the contrary, the modern existing URM buildings, typical in Germany and Switzerland, are usually characterized by reinforced concrete lintels and flat slabs. In other countries, such as Italy, the r.c. ring beams are quite common.

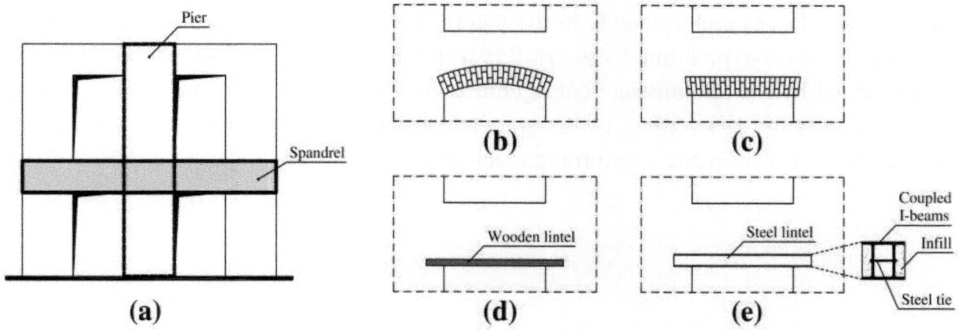

Figure 1.22: Typical masonry spandrels in existing old URM building [29]

Therefore, three different types of modern spandrels may be distinguished. As shown in Figure 1.23, the first one is supported by an r.c. lintel (a). This type belongs to older buildings, while the most recent constructions usually have large window openings. In the latter case, no lintel (Figure 1.23 (b)) or even no masonry spandrel above the slab (Figure 1.23 (c)) can be provided. The opening reaches over the entire storey height, and the coupling of the masonry piers results only from the r.c. slab.

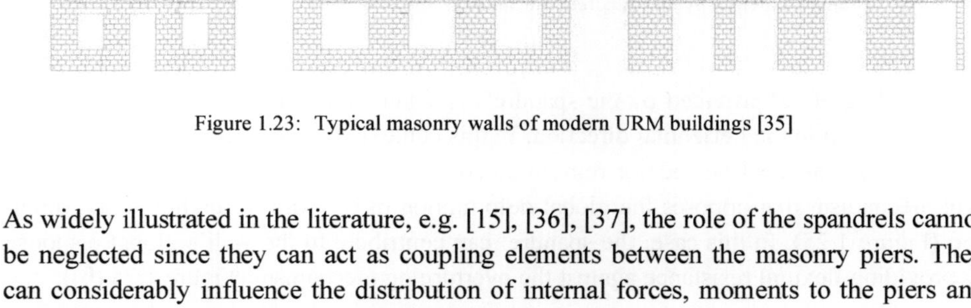

Figure 1.23: Typical masonry walls of modern URM buildings [35]

As widely illustrated in the literature, e.g. [15], [36], [37], the role of the spandrels cannot be neglected since they can act as coupling elements between the masonry piers. They can considerably influence the distribution of internal forces, moments to the piers and the failure mechanisms that may be activated.

In this respect, Figure 1.24 illustrates the behaviour of a multistorey masonry wall subjected to horizontal seismic forces. Cases *a)* and *c)* represent the extreme cases in which the spandrel elements are idealized as infinitely flexible or stiff. The depicted diagrams of internal shear forces and moments in the piers refer to the static scheme of a cantilever (case *a)*) or a shear-type frame (case *c)*). It is worth noting that the most realistic situation is represented by the intermediate configuration *b)*. Regarding the failure modes, case *a)* is the worst, as the slenderness of the completely decoupled masonry piers accentuates the formation of an in-plane overturning mechanism.

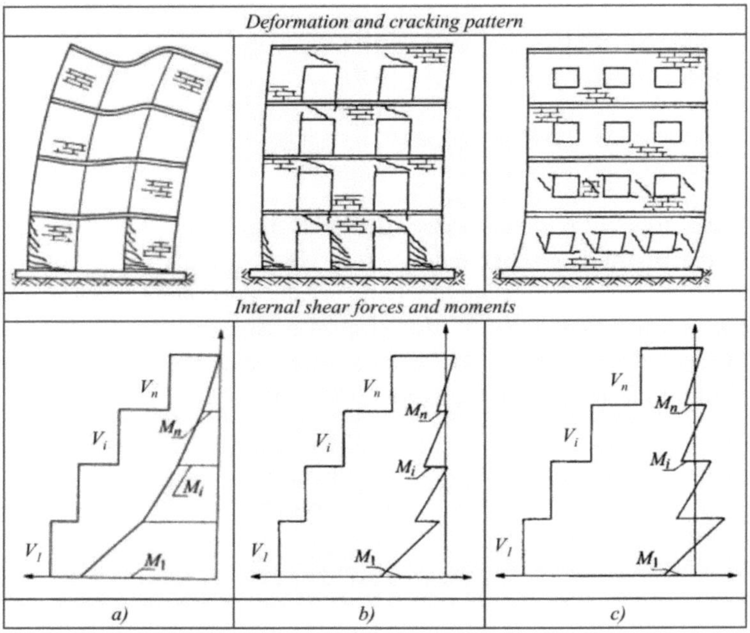

Figure 1.24: Examples of deformation, internal forces and moments in masonry piers with different degrees of coupling provided by the spandrel elements [15]

The coupling effect provided by the spandrels is a function of their stiffness and degree of compression in the horizontal direction. In this context, the presence of tensile resistant elements such as steel tie-rods or reinforced concrete beams induces the activation of a strut mechanism that opposes the global deformation of the wall in the horizontal direction (Figure 1.25). In this case, the spandrel can contribute to the wall's global response by providing flexural resistance against the overturning mechanism (Figure 1.25 (b)).

Reinforced concrete slabs and ring beams can also act as horizontal coupling elements between the masonry piers, increasing the buildings' load-bearing capacity by transforming a cantilever wall system into a frame system. As shown in Figure 1.23, the openings of very modern masonry buildings may span the entire story height (typical in Germany or Switzerland). In this case, the slab may be treated as a coupling beam with an effective width [38].

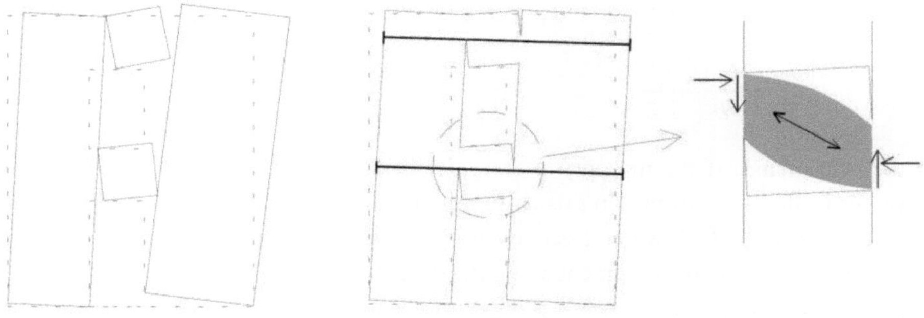

Figure 1.25: Behaviour of masonry spandrels under seismic actions without any tensile resistant element (a) and coupled with steel tie-rods [37]

Concerning the strength verification of the spandrel, a common practice is to adapt the formulations proposed for the piers to these elements. For example, suppose the axial force N acting in the spandrel (i.e. in the direction parallel to the bed joints) is known from the analysis. In that case, the element's behaviour is assumed as a pier rotated to 90°. Moreover, if the masonry is characterized by a regular pattern, the diagonal cracking may be "stepped" through the mortar joints or masonry units. Therefore, the criterion proposed by Mann and Muller [23] can be adopted as follows:

$$V_t = \frac{h \cdot t}{b} \cdot \left(\frac{f_{v0}}{1+\mu \cdot \phi} + \frac{\mu}{1+\mu \cdot \phi} \cdot \sigma_0 \right) \leq V_{t,\lim} \tag{1.27}$$

$$V_{t,\lim} = \frac{h \cdot t}{b} \cdot \frac{f_{bt}}{2.3} \cdot \sqrt{1 + \frac{\sigma_0}{f_{bt}}} \tag{1.28}$$

Where h and t are the spandrel's height and thickness, σ_0 is the normal stress. If the axial force N is reliably known from the analysis, σ_0 can be assumed as the greater between the horizontal and vertical stress. The latter can be eventually evaluated from the loads transmitted by the floors and from the vertical stresses in the adjacent piers.

It is worth noting that the shear strength is mainly a function of the compression normal to the bed joints and, in a lesser degree, of the compression in the direction parallel to the bed joints. Since the first one is essentially negligible between the openings, the shear strength may be very low. Therefore, neglecting the normal stress σ_0, only the contributions of the mortar joint's cohesion and the brick tensile strength are considered:

$$V_t = \frac{h \cdot t}{b} \cdot \left(\frac{f_{v0}}{1+\mu \cdot \phi} \right) \leq V_{t,\lim} \tag{1.29}$$

$$V_{t,\lim} = \frac{h \cdot t}{b} \cdot \frac{f_{bt}}{2.3} \tag{1.30}$$

On the other hand, if the masonry is irregular and not well organised, the mechanical behaviour is almost isotropic. In this case, it is reasonable to adopt the model of Turnšek and Čačovič (Eq. (1.31)), which assumes masonry as an equivalent isotropic material and considers the cracking development along the principal stress direction.

$$V_t = h \cdot t \cdot \frac{1.5\tau}{b} \sqrt{1+\frac{\sigma_0}{1.5\tau}} = h \cdot t \cdot \frac{f_{m,t}}{b} \sqrt{1+\frac{\sigma_0}{f_{m,t}}} \tag{1.31}$$

The average compressive stress σ_0 in Eq. (1.31) can eventually be neglected for the same reasons mentioned above.

Concerning the spandrels' flexural failure, the most common practice is to adopt the same model developed for piers if the acting axial load N is known from the analysis. In this case, the spandrel is assumed as a pier rotated to 90°, and the rocking failure occurs when the masonry attains its compression strength at the compressed toe. The ultimate bending moment M_{Rd} is usually calculated after the beam theory, neglecting the tensile strength of the masonry and assuming an appropriate normal stress distribution at the compressed corner (Eq. (1.32)).

$$M_{Rd} = \frac{h^2 \cdot t}{2} \cdot \sigma_n \cdot \left(1 - \frac{\sigma_n}{0.85 \cdot f_{m,u}} \right) \tag{1.32}$$

Where σ_n is the horizontal stress acting in the spandrel and $f_{m,u}$ is the masonry ultimate compressive strength perpendicular to the bed joints.

It is worth noting that Eq. (1.32) leads to very precautionary predictions of the flexural resistance since moderate values of the axial load generally characterize the spandrels. As a result, the rocking failure tends to prevail over the diagonal cracking much more frequently than observed in experimental campaigns or existing buildings after an earthquake.

Moreover, it is important to specify that the flexure due to seismic actions causes tensile stresses normal to the horizontal mortar joints in the piers and tensile stresses normal to the vertical mortar joints in the spandrels. Since masonry is an anisotropic material, the response of these structural elements is not the same. Moreover, the spandrels may be moderately confined by the floors and the deformation of the contiguous masonry regions (rigid nodes in the "equivalent frame" idealisation).

Therefore, the authors Cattari and Lagomarsino [34] propose a formulation based on the assumption that the spandrel's response may occur as an "equivalent strut" because of the interlocking phenomena that develop at the interface between the spandrel's end sections and the contiguous masonry. Thus, an "equivalent" tensile strength f_{tu}, which properly characterizes the element and not the masonry material, can be defined.

The formulation proposed by Cattari and Lagomarsino [34] is based on the following hypotheses:

- Uniform distribution of the tensile stresses perpendicular to vertical mortar joints;
- Uniform shear stresses along the horizontal mortar joint;
- Negligible mechanical properties of the vertical joints.

The hypothesis of neglecting the vertical joints is reasonable since their cohesive contribution is modest compared to the other factors. Moreover, it may also not exist because of the cracking phenomena.

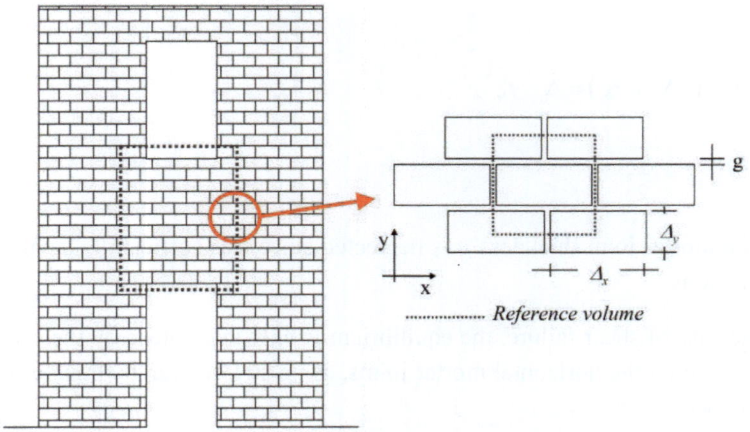

Figure 1.26: Masonry spandrel and Reference volume [34]

Therefore, a reference volume is considered at the end sections of a spandrel (Figure 1.26). It is identified by the height $2(\Delta_y+g)$ and width (Δ_x+g), where Δ_y, Δ_x are the height and width of the block, g is the mortar joint thickness.

Two main failure mechanisms are possible:

1. Tensile failure of the block;

2. Shear failure of the horizontal mortar joints.

In the case of tensile failure of the block, the masonry unit's tensile strength f_{bt} is attained. As shown in Figure 1.27, the horizontal equilibrium leads to:

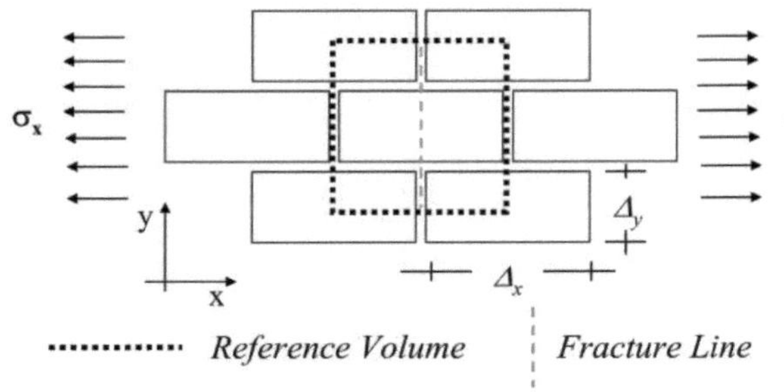

Figure 1.27: Reference volume and scheme of the tensile failure of the block [15]

$$\sigma_x \cdot (2\Delta_y + 2g) = \Delta_y \cdot f_{bt} \quad (1.33)$$

$$f_{mt,u1} = \frac{f_{tb}}{2} \quad (1.34)$$

In which the mortar joint thickness g is neglected since it is sufficiently smaller than the block dimensions.

Then, in the case of shear failure, the equilibrium is only guaranteed by the shear stresses that can develop on the horizontal mortar joints, as no stresses can be transferred through the vertical ones.

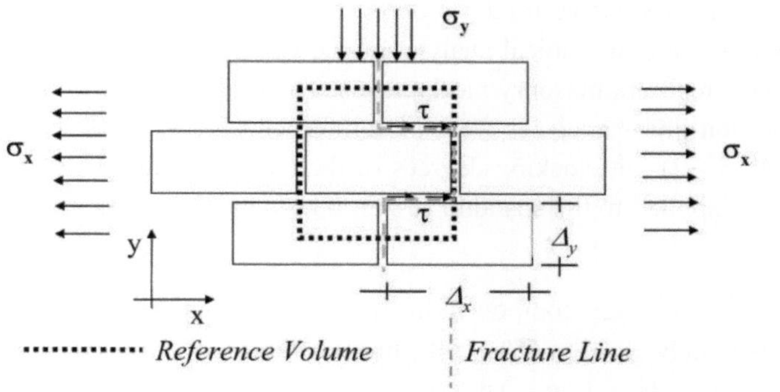

Figure 1.28: Reference volume and scheme of the shear failure of the horizontal mortar joints [15]

If the Mohr-Coulomb type criterion is adopted and the cohesive contribution of the vertical mortar joints is neglected, the shear stresses τ only depend on the stress σ_y normal to the bed joint (Figure 1.28). Therefore the equivalent tensile strength $f_{mt,u2}$ associated with this failure mechanism may be expressed as follows:

$$\sigma_x \cdot (2\Delta_y) = \tau \cdot 2 \cdot \left(\frac{\Delta_x}{2}\right) \tag{1.35}$$

$$f_{mt,u2} = \frac{\Delta x}{2\Delta y} \cdot \mu \cdot \sigma_y \tag{1.36}$$

Where μ is the friction coefficient, and the ratio $\Delta_x/2\Delta_y$ corresponds to the interlocking parameter ϕ.

Therefore, the spandrel equivalent tensile strength $f_{mt,u}$ is equal to the minimum value obtained between $f_{mt,u1}$ and $f_{mt,u2}$. It is worth noting that the block's tensile failure is brittle, and it usually occurs in masonry characterized by very weak bricks. On the contrary, the shear failure may be classified as ductile. In this case, the entity of the compressive stress σ_y, acting at the spandrel's end sections, assumes a decisive role. For this reason, a masonry spandrel located at mid-storeys and characterized by a regular pattern can show a more significant strength from this failure mechanism [34].

Moreover, the determination of the compressive stress σ_y in the modelling with the equivalent frame approach needs some specifications. Approximate criteria based on the vertical load acting in the adjacent piers may be adopted to consider the stress distribution that develops at the end sections interface. In this regard, it is worth noting that the normal stress due to the seismic actions is variable in the piers. Therefore, Cattari and Lagomarsino [34] performed several non-linear analyses with the finite element method and the non-linear constitutive model proposed by Calderini and Lagomarsino [39]. It was based on the micromechanical analysis of the composite continuum with both friction and cohesive resistant masonry mechanisms. The presence of a wood lintel was also modelled. The non-linear analyses were performed considering different spandrel slenderness (λ = 1.35 - 2), interlocking degrees of the masonry pattern (Δ_x/Δ_y = 2 - 4) and axial load values applied to the spandrel (0 ÷ 100 kN) and the adjacent piers (37.5 ÷ 225 kN), [34].

The analysis of the σ_y stress component at the spandrel's interface end-sections showed that it is approximately equal to 65% of the mean compressive stresses acting on the piers in the first response phase (the elastic one). Therefore, this factor was assumed in Eq. (1.36) for computing $f_{mt,u2}$, and the inelastic response was analyzed. The results showed only an increase of the σ_x stress component at the tensioned toe of the spandrel's section until the attainment of the peak shear force. It is important to specify that the maximum value of σ_x was only 15-20% greater than the analytical one deduced from Eq. (1.35). Thus, the authors could conclude that, even if the σ_y stress distribution varies, a precautionary evaluation of f_{tu} can be obtained based on the axial load acting on piers in static conditions [34].

Moreover, assuming the behaviour of the masonry spandrel elastic-perfectly plastic with limited ductility in tension μ_t and compression μ_c (Figure 1.29 (a)), the spandrel's failure domain can be determined (Figure 1.30).

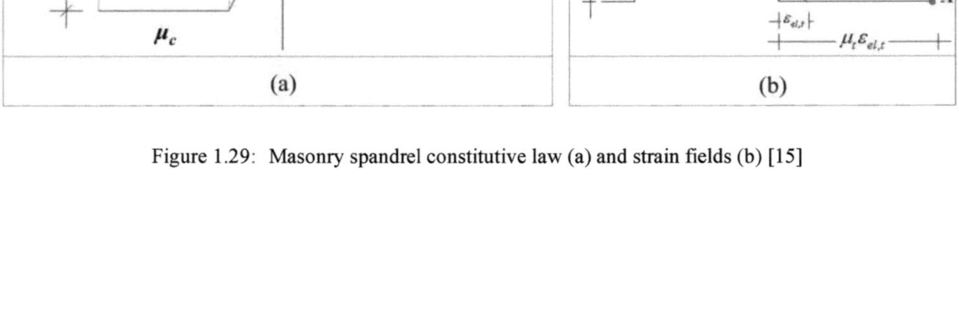

Figure 1.29: Masonry spandrel constitutive law (a) and strain fields (b) [15]

Thus, the ultimate bending moment $M_{Rd} = f(N, f_{mh,u}, f_{mt,u}, \mu_c, \mu_t)$ [1] can be obtained by idealizing the masonry as a homogenous continuum, assuming that the section remains plane, and solving a system of translation and rotation equilibrium equations [15].

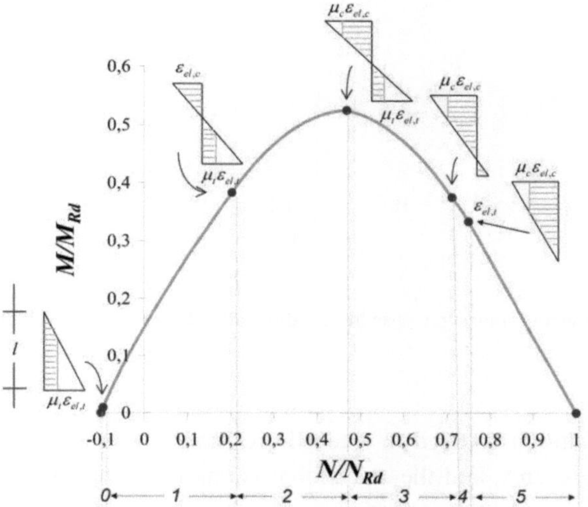

Figure 1.30: Masonry spandrel flexural strength domain and strain fields [15]

To this aim, six strain fields can be identified. They start from the uniform compression of the spandrel's cross-section (field (0)) and finish with the constant tension (field (5)). Figure 1.29 (b) shows the strain fields from (1) to (4), in which the cross-section is partialized.

Furthermore, Figure 1.31 illustrates the strength domain for different ratios between $f_{mt,u}$ and $f_{mh,u}$ (indicated as η). Cattari and Lagomarsino [34] have observed that for rather low values of assigned ductility in tension μ_t, the resulting domain differs not significantly from the case of infinite ductility. Since the shear failure along the horizontal mortar joints is a ductile mechanism, the latter assumption is also considered reasonable. Moreover, compared to the flexural resistance obtained with Eq. (1.32), where the spandrel is treated like a pier rotated of 90°, the bending capacity obtained with this approach can be truly increased, in particular for low values of N, which usually characterize these elements. As shown in Figure 1.31, this effect is beneficial even for very moderate values of η (for example, 0.01).

[1] $f_{mh,u}$ is the masonry compressive strength in the horizontal direction.

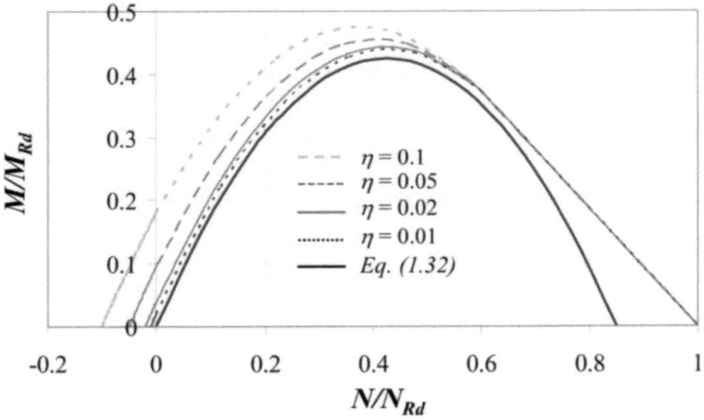

Figure 1.31: Failure domain of a masonry spandrel for different values of η (μ_c = 1.25, μ_t = at infinite), from [34]

Finally, if the spandrel is coupled to a tensile resistant element, such as steel tie-rods, concrete slabs, ring beams, and the axial force is unknown from the analysis, formulations based on the equivalent strut model can also be adopted. In this regard, Figure 1.32 illustrates a masonry portion between two openings supported by a concrete slab. In this case, the spandrel can act as a compression strut.

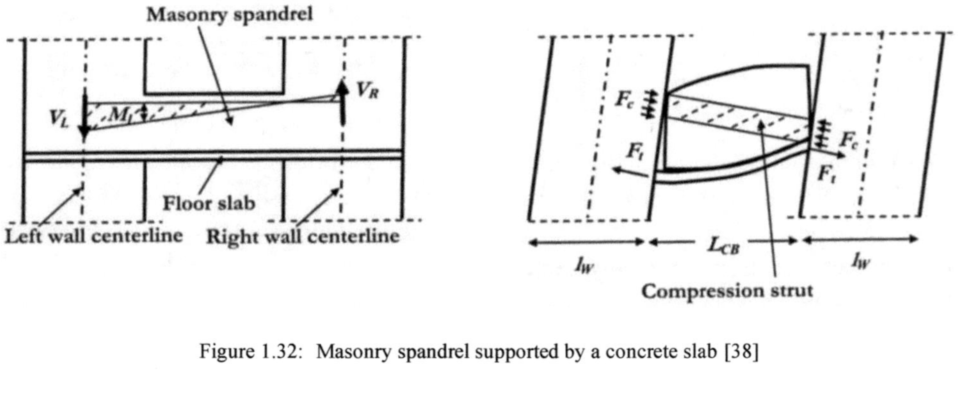

Figure 1.32: Masonry spandrel supported by a concrete slab [38]

A bending moment close to zero results at the bottom right side of the element, while a maximum bending moment results at the left side. The moment can be approximately calculated as the tensile force developed in the tensioned element multiplied by the lever arm, equal to the depth of the spandrel minus one half of the compression depth. The so calculated bending moment is limited by the masonry compressive strength in the horizontal direction, which can be significantly lower than the vertical one, and by the maximum tensile strength that can develop in the slab.

In this context, the Italian Building Code NTC 2018 [16] proposes this approach to evaluate the ultimate bending moment of a spandrel with a tensile resistant element if the normal force N is unknown (which is the case of floors assumed as infinitely stiff). Thus, Eq. (1.37) is suggested, which is consistent with the compression strut model shown in Figure 1.32:

$$M_{Rd} = H_p \cdot \frac{h}{2} \cdot \left[1 - \frac{H_p}{\left(0.85 \cdot f_{mh,u} \cdot h \cdot t\right)} \right] \tag{1.37}$$

H_p is the minimum between the tensile resistance of the stretched interposed element inside the spandrel and $0.4 \cdot f_{mh,u} \cdot h \cdot t$, where $f_{mh,u}$ is the masonry compressive strength in the horizontal direction, h and t are the height and thickness of the spandrel.

1.5 Equivalent frame method

The analysis of the modelling strategies available for masonry structures presented in § 1.2 has shown that only a few fundamental aspects of the seismic response can be reproduced through simplified approaches. On the contrary, the equivalent frame modelling can reproduce the three-dimensional behaviour of masonry buildings, taking into account the interaction of the various parts and their non-linearities. In this work, the method implemented in the TREMURI computer program has been adopted. The software was originally developed in 2001, gradually improved at the University of Genoa (Italy), and also implemented in the commercial version 3Muri [14].

The equivalent frame modelling is based on the assumption that the masonry wall can be considered a set of vertical piers and horizontal spandrels. This simplification is justified by observing that the seismic damage is usually concentrated only in these parts while the connection areas remain undamaged. In this regard, it is reasonable to suppose that the deformation state of the connection areas always remains within the elastic limits. Therefore, it can be neglected in evaluating the wall's behaviour, as the latter is limited by the non-linear strains of its structural elements.

The masonry piers and spandrels can be represented by a single finite element identified by a limited number of degrees of freedom. On this basis, the entire wall can be modelled by assembling non-linear elements mutually connected by rigid parts. In this way, the models have a limited number of degrees of freedom. The response of a wall subject to static (monotonous or cyclic) and dynamic actions can be represented with a modest computational burden. It is important to stress that the reliability of this modelling approach is linked to the ability to describe the damage phenomena that occur in the masonry. Therefore, the single nonlinear finite element must correctly reproduce the main in-plane failure modes of the masonry, i.e. shear and bending (rocking) failure, and must also capture the strength degradation in the nonlinear phase.

Furthermore, the possibility of nonlinear modelling of other structural elements, such as reinforced concrete (r.c.), steel or wooden beams, is very useful for analysing new and existing buildings. It is worth noting that many mixed masonry r.c. structures have been built from the beginning of the twentieth century because of the spreading of r.c. technology (e.g. buildings with internal r.c. frames and outer masonry walls).

Moreover, structural interventions carried out on existing masonry structures have caused the birth of mixed buildings (e.g. additional r.c. storeys, replacement of internal masonry walls by r.c. frames or r.c. walls inserted to support lifts and staircases). Since these structural modifications may increase the existing structure's seismic vulnerability, it is important to consider them in the modelling.

Therefore, the strategies to idealize the masonry wall in an equivalent frame, the nonlinear beam element with lumped inelasticity and piecewise behaviour implemented in the TREMURI program are illustrated in § 1.5.1 and § 1.5.2. Finally, the assembling method of three-dimensional models is presented in § 1.5.3.

1.5.1 Modelling of the wall: definition of piers and spandrels

The masonry wall's modelling as an equivalent frame requires identifying its main structural components: piers and spandrels. Conventional criteria based on the damage observed after earthquakes and experimental campaigns are often assumed in the literature since rigorous formulations are not defined. Moreover, a systematic parametrical analysis, either numerical or experimental, has never been performed [14].

The height of the masonry piers is usually defined as a function of adjacent openings' height. A common criterion is to assume a maximum inclination of the cracks starting from the opening corners equal to 30°. This assumption is the same initial hypothesis proposed in [40] to define the masonry panels' equivalent height. It was based on the storey mechanism and provided an increased height for the external piers. In [41], it is defined as the height over which a compression strut is likely to develop at the steepest possible angle. Under this assumption, the cracks can develop either horizontally or at 45°.

In this context, Figure 1.33 shows an example of the frame idealization procedure implemented in the TREMURI software in the case of a regularly perforated masonry wall. The spandrel elements are identified based on the vertical alignment and overlap of openings. In the case of full alignment, the spandrel's length and height are assumed equal to the distance and width of the adjacent openings.

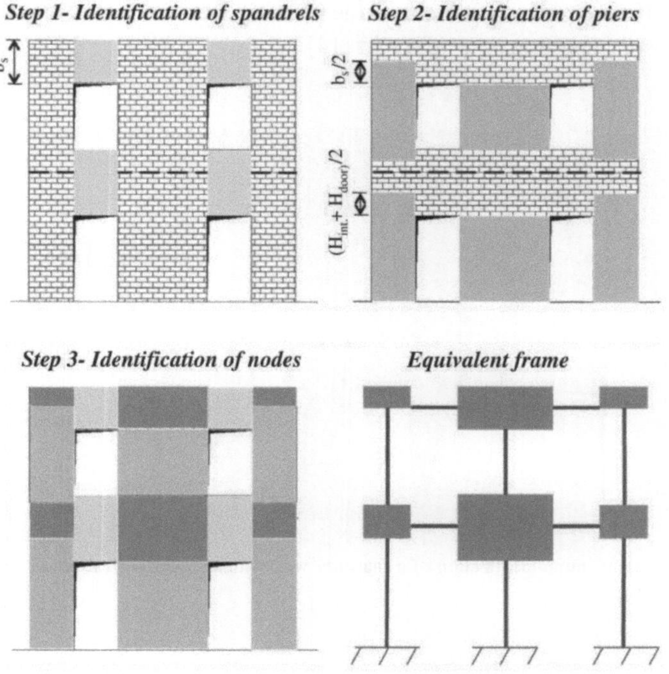

Figure 1.33: Example of equivalent frame idealization in TREMURI for a perforated wall with regularly distributed openings [14]

Then, the pier elements are defined starting from the height of adjacent openings. The height of the internal piers is assumed equal to that of openings when these latter are perfectly aligned. Concerning the external piers' height, the possible development of inclined cracks from the opening corners and/or from the lintel edges must be considered. An approximate way is to define the pier's height as the average of the adjacent opening's and interstorey height. Finally, the rigid nodes are directly obtained from the previously defined elements connected to them.

This calculation is done separately for each storey and wall to obtain the complete frame idealization. It is important to stress that if the cone diffusion angle is not limited, the effective in-plane aspect ratio of the external piers may be significantly overestimated if adjacent openings with a limited height and close to the wall edge are present.

Although the masonry piers and spandrels' identification may result rather easily in walls with regularly distributed openings, it can become more difficult when they are irregularly arranged. In this case, a possible solution is to assume the spandrel' height conventionally as a mean value based on the overlapping part between the openings at the two levels (Figure 1.34). If no overlap is present or the opening lacks, it seems more appropriate to assume the masonry portion as a rigid part [14].

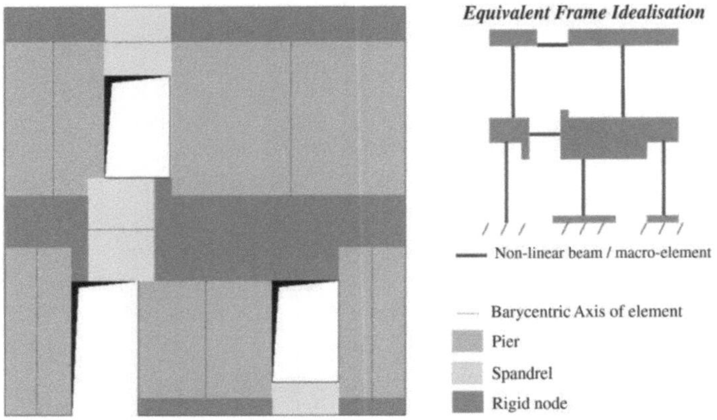

Figure 1.34: Equivalent frame idealization of a masonry wall with irregularly distributed openings in TREMURI [14]

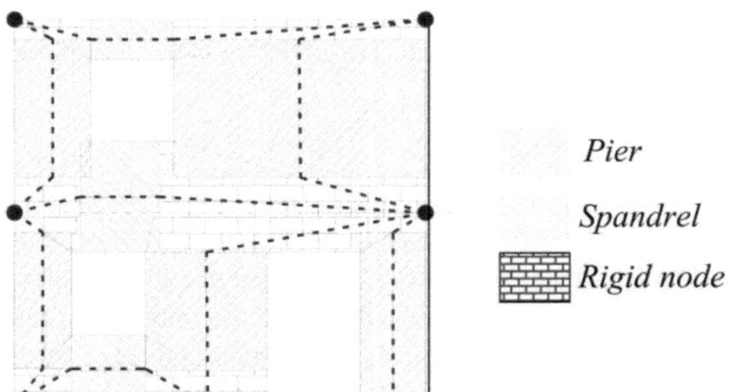

Figure 1.35: Idealisation of the masonry wall as an equivalent frame in TREMURI [15]

Furthermore, as shown in Figure 1.35, the barycentric axes of the deformable elements may not coincide with the model's nodes. Thus, an eccentricity is caused, which may be considered by applying a rigid end matrix to the stiffness matrix of the element itself. In this regard, Figure 1.36 illustrates the masonry element with two nodes, *i* and *j*, one for each extremity. The degrees of freedom at each node, (u_i, w_i, φ_i) and (u_j, w_j, φ_j), are the displacements in x-, y-direction and the rotation about the z-axis, positive if anticlockwise. Therefore, considering that the barycenters of the rigid nodes 1 and 2 have eccentricities with respect to the nodes *i* and *j* (Figure 1.36), the following kinematic relations can be written under the hypothesis of small displacements:

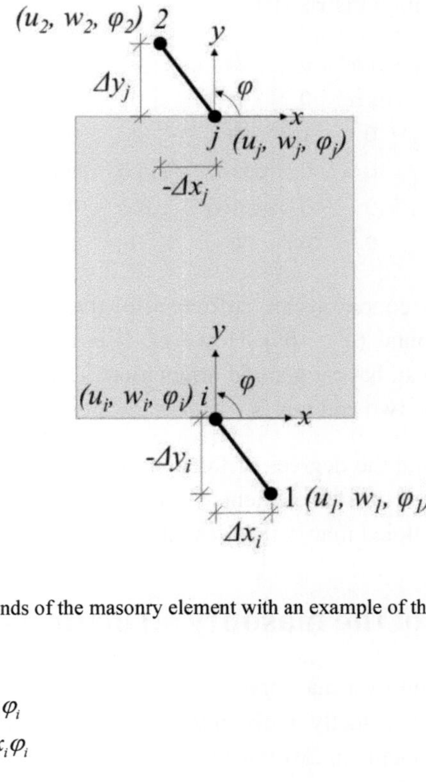

Figure 1.36: Rigid ends of the masonry element with an example of their possible eccentricity

$$\begin{cases} u_1 = u_i - \Delta y_i \varphi_i \\ w_1 = w_i + \Delta x_i \varphi_i \\ \varphi_1 = \varphi_i \\ u_2 = u_j - \Delta y_j \varphi_j \\ w_2 = w_j + \Delta x_j \varphi_j \\ \varphi_2 = \varphi_j \end{cases} \tag{1.38}$$

Thus, the rigid end matrix $[D_e]$ can be written as follows:

$$[D_e] = \begin{bmatrix} 1 & 0 & -\Delta y_i & 0 & 0 & 0 \\ 0 & 1 & \Delta x_i & 0 & 0 & 0 \\ 0 & 0 & 1 & 0 & 0 & 0 \\ 0 & 0 & 0 & 1 & 0 & -\Delta y_j \\ 0 & 0 & 0 & 0 & 1 & \Delta x_j \\ 0 & 0 & 0 & 0 & 0 & 1 \end{bmatrix} \tag{1.39}$$

In this way, the elastic stiffness matrix of the element $[K_e]$ and the vector of nodal actions $\{q\}$ are modified through the rigid end matrix [15].

Furthermore, the modelling of the spandrels requires the rotation of the element, which is performed through the rotation matrix [R]:

$$[R] = \begin{bmatrix} \cos\alpha' & \sin\alpha' & 0 & 0 & 0 & 0 \\ -\sin\alpha' & \cos\alpha' & 0 & 0 & 0 & 0 \\ 0 & 0 & 1 & 0 & 0 & 0 \\ 0 & 0 & 0 & \cos\alpha' & \sin\alpha' & 0 \\ 0 & 0 & 0 & -\sin\alpha' & \cos\alpha' & 0 \\ 0 & 0 & 0 & 0 & 0 & 1 \end{bmatrix} \tag{1.40}$$

It is worth noting that two conditions are sufficient for the piers and spandrels, i.e. vertical (α' = 90°) and horizontal (α' = 0°). However, it is often necessary to model also beams or tie-rods, which can have a generic orientation. They are identified in the wall's plane by the position of the two end nodes.

Finally, once the system and the degrees of freedom have been univocally numbered, the stiffness matrixes of the structural elements (masonry piers/spandrel, beams, tie-rods, etc.) are assembled into a global matrix that describes the entire wall's stiffness [4].

1.5.2 Modelling of the masonry structural elements

After the frame idealization of a masonry wall, it is important to interpret its structural elements' seismic response correctly. Only in this way, a reliable prediction of the wall's overall behaviour can be obtained. Several formulations are proposed in the literature to calculate the in-plane lateral strength of masonry panels (§ 1.3 and § 1.4). They may be used for the EF modelling adopting a nonlinear beam element with lumped inelasticity as finite element.

The mechanical behaviour may be described in terms of global stiffness, strength, and ultimate displacement capacity by assuming a proper force-displacement piecewise-linear relationship and defining appropriate drift limits. This type of 2D element is implemented in the TREMURI program [42], and it is indicated in the following as ML-BEAM. It allows describing the nonlinear response of a masonry panel until very severe damage levels (from DL1 from DL5) by assigning progressive strength degradations at defined values of drift. The kinematic variables and generalized forces aimed to describe the element are indicated in Figure 1.37. It is worth pointing out that loads are applied only on nodes and not along with the element.

Two sets of parameters can be identified to describe the backbone and the ML-BEAM element's hysteretic response. The first one includes parameters necessary to define the initial stiffness k_{el} and its progressive degradation, the maximum strength V_{Rd}, and the nonlinear response of the panel for increasing damage levels (d_i, drift value at the attainment of the ith DL, $\beta_{E,i}$, residual shear strength with respect to V_{Rd}).

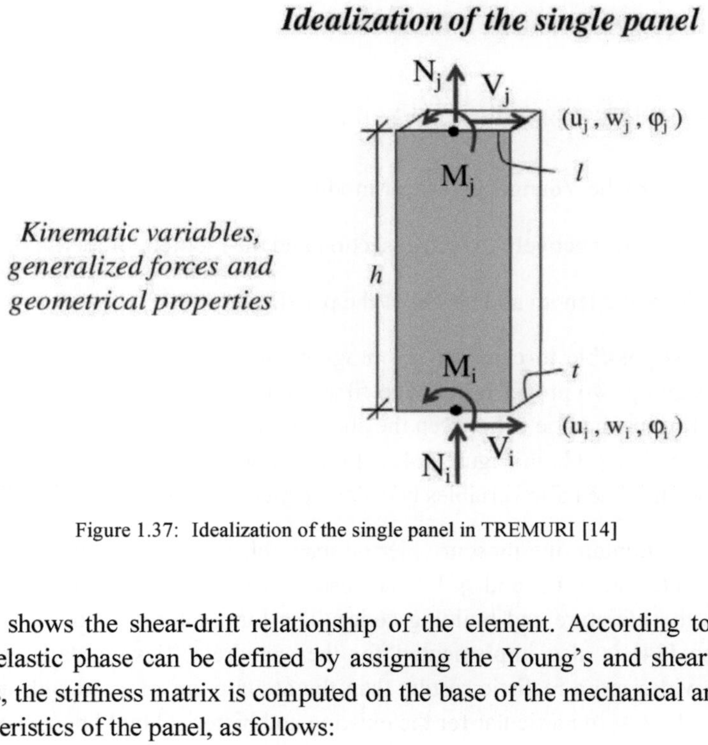

Figure 1.37: Idealization of the single panel in TREMURI [14]

Figure 1.41 shows the shear-drift relationship of the element. According to the beam theory, the elastic phase can be defined by assigning the Young's and shear moduli, E and G. Thus, the stiffness matrix is computed on the base of the mechanical and geometrical characteristics of the panel, as follows:

$$\begin{Bmatrix} V_i \\ N_i \\ M_i \\ V_j \\ N_j \\ M_j \end{Bmatrix} = \begin{bmatrix} \frac{12EI}{h^3(1+\psi)} & 0 & -\frac{6EI}{h^2(1+\psi)} & -\frac{12EI}{h^3(1+\psi)} & 0 & -\frac{6EI}{h^2(1+\psi)} \\ 0 & \frac{EA}{h} & 0 & 0 & -\frac{EA}{h} & 0 \\ -\frac{6EI}{h^2(1+\psi)} & 0 & \frac{EI(4+\psi)}{h(1+\psi)} & \frac{6EI}{h^2(1+\psi)} & 0 & \frac{EI(2-\psi)}{h(1+\psi)} \\ -\frac{12EI}{h^3(1+\psi)} & 0 & \frac{6EI}{h^2(1+\psi)} & \frac{12EI}{h^3(1+\psi)} & 0 & \frac{6EI}{h^2(1+\psi)} \\ 0 & -\frac{EA}{h} & 0 & 0 & \frac{EA}{h} & 0 \\ -\frac{6EI}{h^2(1+\psi)} & 0 & \frac{EI(2-\psi)}{h(1+\psi)} & \frac{6EI}{h^2(1+\psi)} & 0 & \frac{EI(4+\psi)}{h(1+\psi)} \end{bmatrix} \begin{Bmatrix} u_i \\ v_i \\ \varphi_i \\ u_j \\ v_j \\ \varphi_j \end{Bmatrix} \tag{1.41}$$

Where:

- ψ is a coefficient equal to $\frac{1.2EI^2}{Gh^2}$;
- E and G are the Young's and shear moduli;
- A and I are respectively the cross-section and the moment of inertia of the panel;
- l and h are the length and height of the panel.

Moreover, it is possible to consider the progressive stiffness degradation in the elastic phase by assigning two proper ratios. The first one is between the initial secant stiffness, k_{el} and k_{sec}. The second one is between the shear at the end of the initial elastic phase and the shear strength V_{Rd} (k_0 in Figure 1.41). Furthermore, rigid end offsets are adopted to transfer static and kinematic variables between element ends and nodes.

The maximum strength of a masonry pier or spandrel, V_{Rd}, is computed according to the criteria illustrated in § 1.3 and § 1.4, consistent with the most common formulations proposed in the literature and building codes. The formulation proposed by Cattari and Lagomarsino [34] is also implemented for the spandrels' flexural failure mode. This criterion provides more realistic results than the assumption of a strength model analogous to that of piers, in particular for the existing and historical buildings where spandrels are often "weak" and are not coupled to other tensile resistant elements [14]. As discussed in [43], the comparison with experimental results have also confirmed that the adoption of this strength criterion can provide more reliable results.

The hysteretic response is formulated through a phenomenological approach, which is able to reproduce the different behaviours of piers and spandrels and the various possible failure modes (flexural, shear or even mixed type). The maximum strength of the panel, V_{Rd}, is equal to the minimum obtained from the failure criteria defined for the element and considering the current axial force acting at each step. Then, a procedure of nonlinear correction of the elastic strength prediction is carried out based on the comparison with the calculated limit resistance of the element. The redistribution of the internal forces is performed to ensure the element's equilibrium.

Different values of d_i and $\beta_{E,i}$ may be defined to describe the panel's flexural and shear response. In this respect, qualitative force-displacement curves of masonry piers under cyclic lateral shear load are illustrated in Figure 1.38.

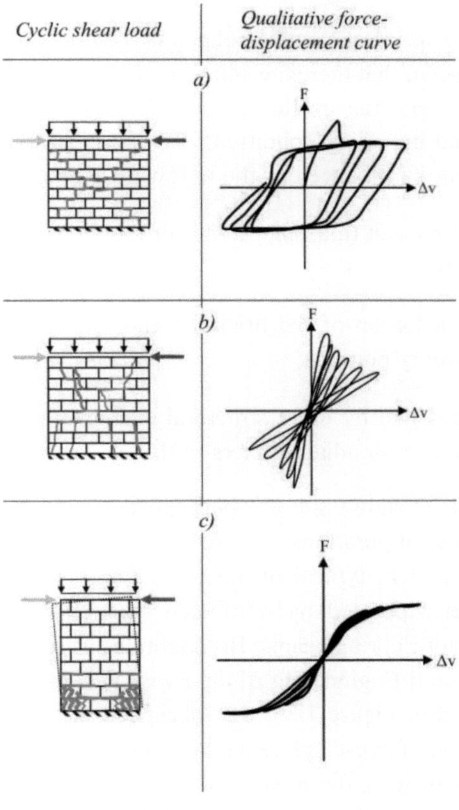

Figure 1.38: Qualitative force-displacement curves of masonry piers under cyclic shear load, from [44]

Case *a* shows the diagonal cracking developing by sliding along the mortar joints. In this case, the energy is dissipated by friction, and the hysteresis curves are full. The post-peak behaviour can be described as ductile and corresponds approximately to an elastoplastic material behaviour. On the contrary, the block tensile failure is brittle (case *b*). It is evident from the less full hysteresis loops together with the stiffness and load-bearing capacity decrease in each cycle after the peak.

Furthermore, flexural failure (case *c* in Figure 1.38) is more prevalent in slender masonry panels. The cyclic loads lead to a rocking movement that induces tensile and compressive stresses in the corner areas alternately. The deformations are large compared to the load that can be absorbed. The hysteresis curves are characterized by S-shaped loops with low dissipation of energy, where the slope of the curves decreases only very slightly. It is worth noting that the S-shaped loops do not mainly result from the material degradation but the reduction of the cross-section's compressed part with increasing deformation.

Concerning the masonry spandrel's cyclic behaviour, recent experimental campaigns, i.e. [29], [35], [45], have shown that there are some significant differences between piers and spandrels. In masonry piers, the cyclic response mainly depends on their mechanical properties, geometry and boundary conditions. On the contrary, the force-displacement response of the spandrels is influenced by the following aspects:

- type of lintels, if present (masonry arches or lintels in stone, timber, steel or reinforced concrete);
- interlocking phenomenon of the bricks at end-sections of the spandrel with the contiguous masonry portions;
- axial restraint provided by other structural elements such as reinforced concrete beams, steel tie-rods or adjacent piers [43].

In this regard, Figure 1.39 shows some experimental results of two full-scale spandrel elements with different configurations. The first is a solid clay brick masonry spandrel supported by a masonry arch, typical in older buildings. The second is a hollow clay brick masonry spandrel supported by reinforced concrete, typical in newer masonry buildings with r.c. slabs or r.c. ring beams. Both samples were tested under cyclic loading at the Institute of Structural Engineering of the Swiss Federal Institute of Technology in Zurich [35]. As depicted in Figure 1.39, the specimens failed by diagonal cracking but with different experimental force-displacement curves. In the first sample, after the linear elastic phase, the attainment of the maximum shear force is followed by a significant strength drop. On the contrary, the second specimen has shown only a minimum decrease in strength in the post-peak phase. Therefore, it is important to consider the different types of piers' and spandrels' behaviour in the EF modelling and the various in-plane response of the masonry spandrels based on their configuration. To this aim, different values of d_i and $\beta_{E,i}$ to describe the panel's flexural and shear response can be assigned. They can be differentiated in the case of spandrel and pier elements as well.

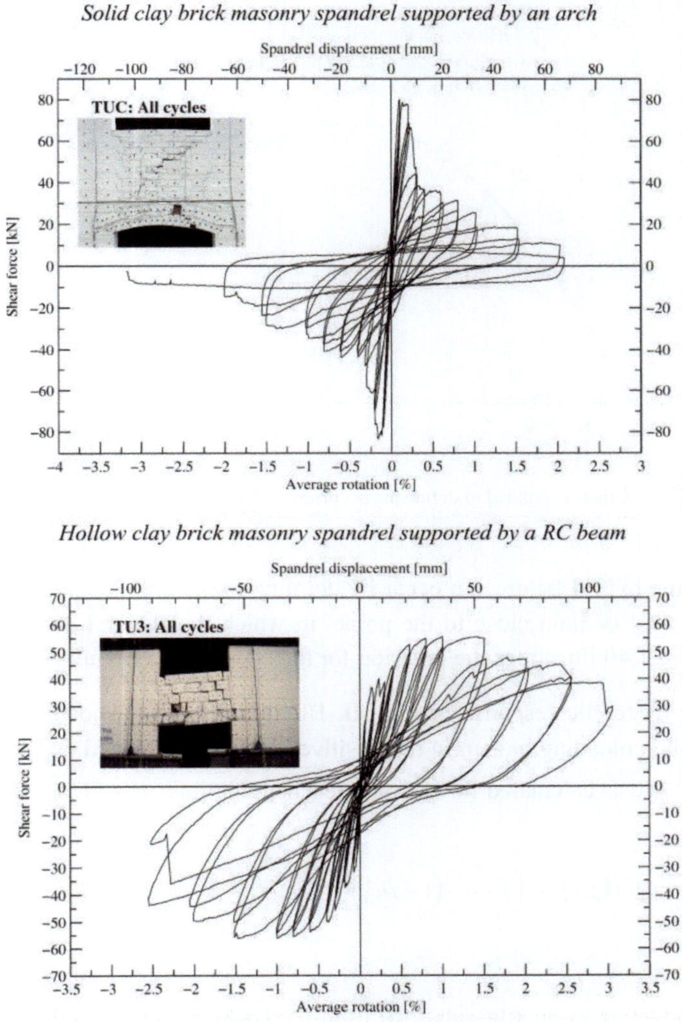

Figure 1.39: Experimental results of two full-scale masonry spandrels with different configurations [35]

Moreover, it happens frequently that the failure of a masonry panel is a combination of the shear and flexural one. In this case, the mixed failure mode is considered by the program calculating average values of d_i and $\beta_{E,i}$, starting from those assigned for the shear and flexural behaviour.

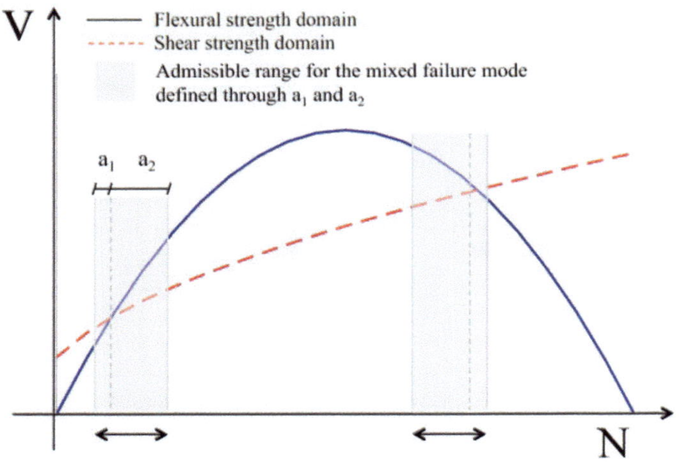

Figure 1.40: Criteria assumed to define the occurrence of a mixed failure in TREMURI [42]

Furthermore, the hybrid failure can occur by defining in the input an admissible range in the panel's V – N domain close to the points in which the flexural and shear domains intersect. Figure 1.40 illustrates the criterion for the occurrence of a mixed failure mode.

The complete hysteretic response of the ML-BEAM element is shown in Figure 1.41. The slope of the unloading branch in the positive quadrant is defined by the stiffness k_U^+ (from A^+ to C^+). It is calculated through the reached ductility μ^+ and strength decay $\beta_{E,i}$, as follows:

$$k_U^+ = k_{sec} \cdot (\mu^+)^{-c_1} \cdot \left[1 - c_2 \cdot (1 - \beta_{E,i})\right] \qquad (1.42)$$

where:

- μ^+ is the value of ductility attained from the backbone curve in the positive quadrant;
- c_1 aims to degrade the stiffness k_U^+ with respect to the secant one k_{sec}. It may assume values from 0 (elastoplastic law) to 1 (secant stiffness);

- c_2 is a parameter that can further degrade the value of k_U^+ through the progressive strength decay $\beta_{E,i}$ reached on the backbone. It may assume values from 0 to 1.

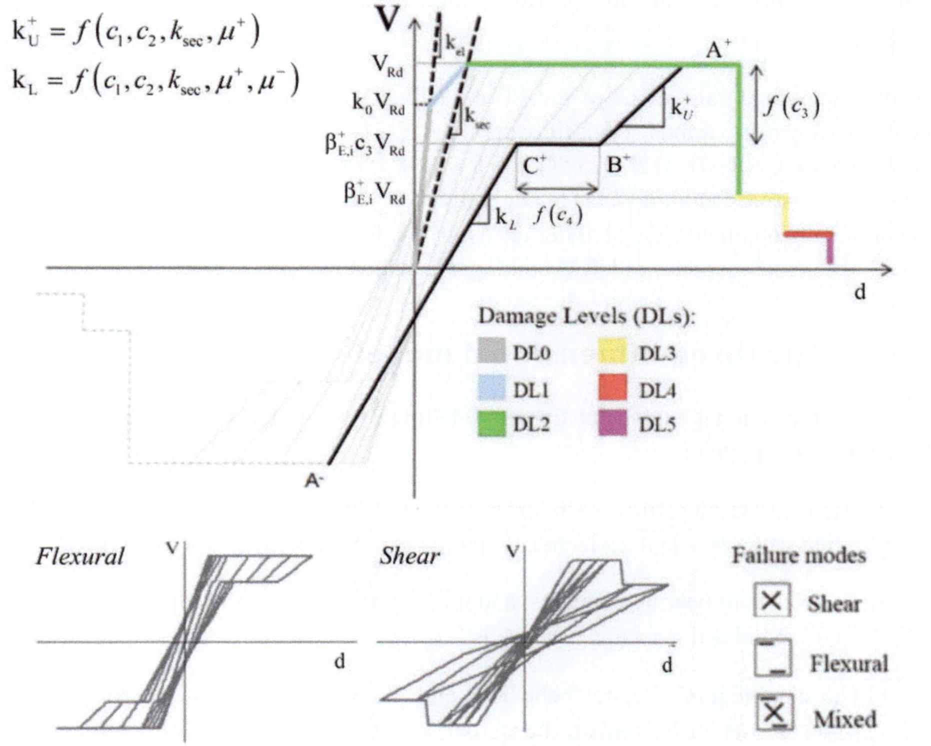

Figure 1.41: Multilinear constitutive law and hysteretic response of the ML-BEAM element implemented in TREMURI [46]

In the same way, the expression of k_U^- for the negative quadrant is defined:

$$k_U^- = k_{sec} \cdot (\mu^-)^{-c_1} \cdot \left[1 - c_2 \cdot (1 - \beta_{E,i})\right] \tag{1.43}$$

As shown in Figure 1.41, the unloading curve may also present a horizontal branch, which is typical of bending failure. In that case:

- the coefficient c_3 fixes the point B^+. It can vary from 0 (A^+ - B^+ branch until the abscissa axis) to 1 (non-linear elastic condition);

- the coefficient c_4 defines the extension of the B$^+$ - C$^+$ branch. The suggested range of values is 0 -1 [42].

Then, the loading branch from C$^+$ to A$^-$ is ruled by the stiffness k_L. It is computed considering the stiffness k_U^+ and the ductility value reached in both positive and negative quadrants (μ^+, μ^-).

Finally, numerical validations of the EF modelling with the described ML-BEAM element are shown in various scientific works, e.g. [43], [47], [48], and in the Italian technical standard CNR-DT 212/2013 "Guide for the Probabilistic Assessment of the Seismic Safety of Existing Buildings" [25]. The validation has been performed through experimental results obtained from shaking tables, cyclic lateral shear tests on single spandrels or with the actual response of URM buildings damaged by seismic events.

1.5.3 The three-dimensional model

The three-dimensional equivalent frame modelling of whole URM buildings is based on the following hypotheses:

- The load-bearing structure, referred to the vertical and horizontal actions, is identified with walls and horizontal diaphragms (i.e. roofs, floors or vaults);
- The walls are bearing elements, and the diaphragms have the function to share the horizontal actions among the walls;
- The diaphragms' flexural behaviour and the wall out-of-plane response are neglected, as their in-plane behaviour mainly governs the global structural response.

It is worth noting that the global seismic behaviour is possible only if vertical and horizontal elements are properly connected. If necessary, "local" out-of-plane mechanisms may be verified separately through suitable analytical methods [14].

Therefore, the first step to assemble the 3D model is to define a global Cartesian coordinate system (X, Y, Z). Then, the coordinates of one point and the angle θ formed with the global X-axis identify the vertical wall planes (Figure 1.42). Thus, each wall can be modelled as a plane frame in the local coordinate system. The internal nodes are two-dimensional nodes with three degrees of freedom (d.o.f.). The nodes at corners and intersections of two or more walls are three-dimensional nodes characterized by five d.o.f. in the global coordinate system (u_x, u_y, u_z, φ_x, φ_y). It is worth noting that the rotational d.o.f. around the vertical Z-axis can be neglected, as the membrane behaviour is adopted for walls and floors.

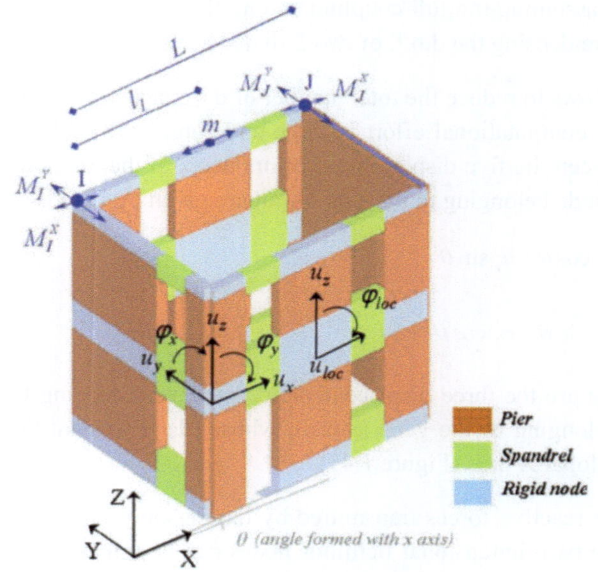

Figure 1.42: Three-dimensional assembling of masonry walls in TREMURI: classification of 3D and 2D rigid nodes and mass sharing [14]

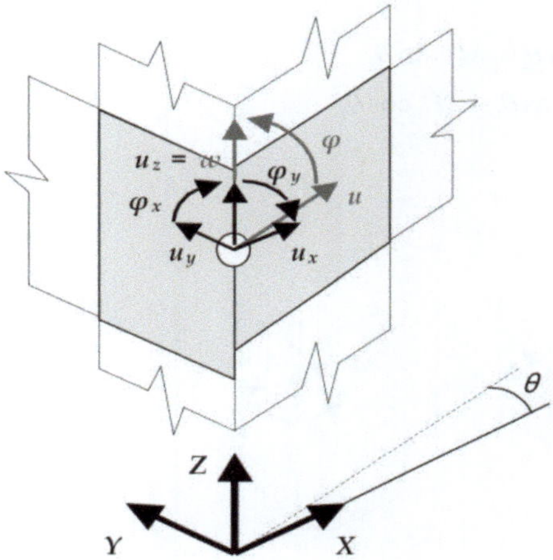

Figure 1.43: D.o.f. of the three-dimensional node [15]

Furthermore, the three-dimensional nodes can be obtained by assembling 2D virtual rigid nodes acting in each wall plane and projecting the local d.o.f. along global axes (Figure

1.43). Therefore, assuming the full coupling among the connected walls, the assemblage is performed by condensing the d.o.f. of two 2-dimensional nodes.

This procedure allows to reduce the total number of d.o.f. and perform nonlinear analyses with a reasonable computational effort in large and complex building models [14]. The relationships between the five displacement components of the 3D node and the three of the fictitious 2D node belonging to the single wall are given by Eq. (1.44):

$$\begin{cases} u = u_x \cos\theta + u_y \sin\theta \\ w = u_z \\ \varphi = \varphi_x \sin\theta - \varphi_y \cos\theta \end{cases} \tag{1.44}$$

Where u, w and φ are the three displacement components according to the d.o.f. of the fictitious node belonging to the generic wall, whose plane is identified by the angle θ formed with the global X-axis (Figure 1.43).

Consequently, the reactive forces transmitted by the masonry elements belonging to the single walls to the two-dimensional fictitious nodes can be referred to the global coordinate system:

$$\begin{cases} F_x = F_h^1 \cos\theta_1 + F_h^2 \cos\theta_2 \\ F_y = F_h^1 \sin\theta_1 + F_h^2 \sin\theta_2 \\ F_z = F_v^1 + F_v^2 \\ M_x = M^1 \sin\theta_1 + M^2 \sin\theta_2 \\ M_y = -M^1 \cos\theta_1 - M^2 \cos\theta_2 \end{cases} \tag{1.45}$$

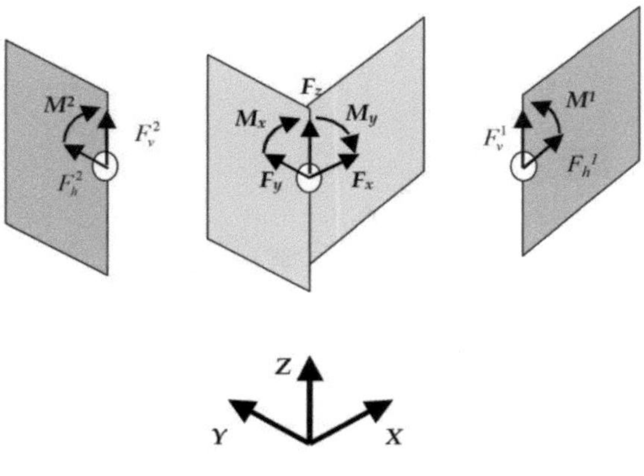

Figure 1.44: Forces acting on the node with five d.o.f. and on the corresponding virtual nodes with three d.o.f. [15]

As shown in Figure 1.44, the terms with superscripts 1 and 2 correspond to the forces and moments acting on the virtual nodes in walls 1 and 2, to which the three-dimensional node belongs.

In this way, the wall's in-plane modelling can be performed, and the nodes belonging to a single wall remain two-dimensional with only three d.o.f. instead of five.

Moreover, the vertical loads' contribution is computed as nodal mass added to all nodes based on the areas of influence of each node and considering the floor's spanning direction. In this regard, the 2D nodes have no d.o.f. in the direction perpendicular to the wall plane. Therefore, the nodal mass component related to the out-of-plane d.o.f is shared to the corresponding d.o.f. of the two nearest 3D nodes of the same wall and storey after the following relations [14]:

$$\left.\begin{aligned} M_I^X &= M_I^X + m\left(1-\left|\cos\theta\right|\right)\frac{L-l_I}{L} \\ M_I^Y &= M_I^Y + m\left(1-\left|\sin\theta\right|\right)\frac{L-l_I}{L} \end{aligned}\right\} \qquad (1.46)$$

Figure 1.42 shows the terms of Eq. (1.46).

Furthermore, it is important to highlight that a proper assumption on the diaphragm stiffness may significantly influence the global response. If the floors are modelled as "infinitely" flexible, they cannot transfer the seismic action from heavily damaged walls to the still efficient ones. On the contrary, if the diaphragms are considered "infinitely" stiff, their contribution could be overestimated. Despite this, the floor behaviour in the 3D modelling is frequently assumed completely rigid. It is worth noting that this consideration may also be unrealistic in historic masonry buildings, where ancient constructive technologies, such as timber floors/roofs, structural brick or stone vaults, were commonly adopted.

Therefore, specific floor elements are used in the TREMURI also to simulate the presence of flexible diaphragms. They are characterized by the Young's modulus E_1 in the principal direction (i.e. the floor spanning orientation), the Young's modulus E_2 along the perpendicular direction, the Poisson ratio ν and the shear modulus G_{12}. It is worth noting that both moduli of elasticity, E_1 and E_2, provide for the connection's degree between walls and horizontal diaphragm. They allow the formation of a link between the in-plane horizontal displacements of the nodes belonging to the same wall-to-floor intersection, influencing in this way the axial force computed in the spandrels. Furthermore, the shear modulus G_{12} affects the diaphragm's tangential stiffness and the horizontal force transfer among the walls in the linear and nonlinear phases [14]. The reference finite element is a three-node orthotropic membrane element (plane stress) with two d.o.f. at each node (u_x, u_y) in the global coordinate system (Figure 1.45).

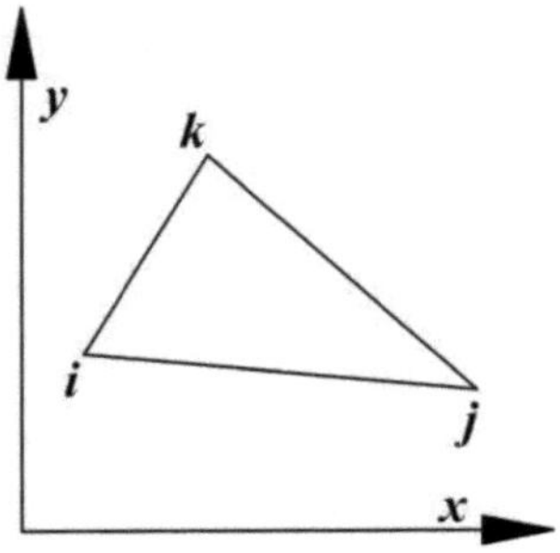

Figure 1.45: The three-node element [4]

Therefore, in the case of orthotropy and plane stress state, the relation between stresses and deformations is given from Eq. (1.47):

$$\left[\hat{D}\right]\cdot\{e\}^T = \{s\}^T \Rightarrow \begin{bmatrix} \frac{E_1}{1-m_{12}\nu^2} & \frac{m_{12}\nu E_1}{1-m_{12}\nu^2} & 0 \\ \frac{m_{12}\upsilon E_1}{1-m_{12}\nu^2} & \frac{E_1}{1-m_{12}\nu^2} & 0 \\ 0 & 0 & G_{12} \end{bmatrix} \cdot \{\varepsilon_1\ \varepsilon_2\ \gamma_{12}\} = \{\sigma_1\ \sigma_2\ \tau_{12}\} \tag{1.47}$$

Where v is the Poisson ratio, and m_{12} is equal to the ratio E_2/E_1.

Assuming α^* as the angle formed by the main load-bearing direction of the floor with the global X-axis, it is possible to rewrite the matrix [$\hat{D}$] in the rotated configuration to take into account the actual orientation of the diaphragm:

$$[D] = [R]^T \left[\hat{D}\right] \cdot [R] \tag{1.48}$$

Where $[R]$ is the rotation matrix defined as follows:

$$[R] = \begin{bmatrix} \cos\alpha^* & \sin\alpha^* & 0 \\ -\sin\alpha^* & \cos\alpha^* & 0 \\ 0 & 0 & 1 \end{bmatrix} \tag{1.49}$$

Then, the matrix $[B_i]$ can be defined for each node i of the three-node element adopting linear shape functions:

$$[B_i] = \frac{1}{2A_t}\begin{pmatrix} y_j - y_k & 0 \\ 0 & x_k - y_j \\ x_k - y_j & y_j - y_k \end{pmatrix} \tag{1.50}$$

Where x_j, y_j and x_k, y_k are the coordinates of the nodes j and k, while A_t is the triangle area.

Therefore, starting from the definition of the matrices $[B_i]$ and $[D]$, the stiffness matrix of the 3-nodes membrane element $[K^e]$ can be assembled:

$$\left[K^e\right] = \begin{bmatrix} \left[k_{ii}^e\right] & \left[k_{ij}^e\right] & \left[k_{ik}^e\right] \\ \left[k_{ji}^e\right] & \left[k_{jj}^e\right] & \left[k_{jk}^e\right] \\ \left[k_{ki}^e\right] & \left[k_{kj}^e\right] & \left[k_{kk}^e\right] \end{bmatrix} \tag{1.51}$$

with

$$\left[k_{ij}^e\right] = [B_i]^T \cdot [D] \cdot \left[B_j\right] \cdot A_t \cdot s \tag{1.52}$$

Where s is the equivalent thickness assumed for the membrane element.

Finally, in the case of 4-nodes elements (Figure 1.46), the average contribution of the two possible meshes of 3-node elements is computed to obtain the stiffness matrix. In this way, an irregular quadrilateral floor area can be modelled with a single element.

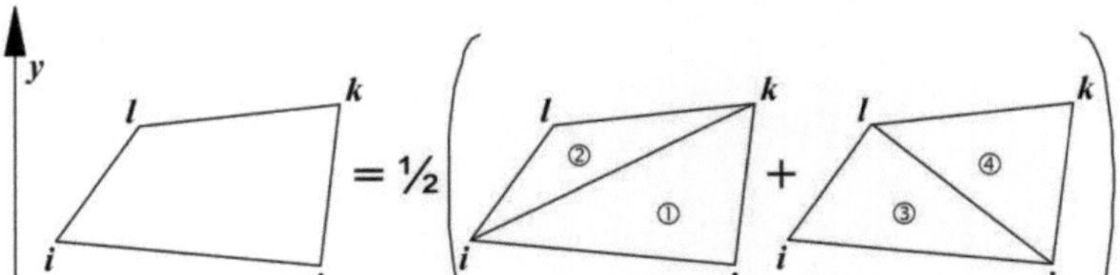

Figure 1.46: The four-node element [4]

Chapter 2

Strengthening of masonry structures with externally bonded textile reinforced mortar

Externally bonded textile reinforced mortar is a recently emerged technology for repairing and strengthening masonry structures (Figure 2.1). It consists mainly of two components, i.e. matrix and grid. The former has the function to cover and protect the fibre reinforcement that is embedded inside, ensuring, at the same time, the stress transfer between the masonry substrate and the textile component. It is generally made of fine-grained mortar with Portland cement (cement-based) or natural hydraulic lime (lime-based). Furthermore, dry organic polymers may also be added to improve the workability, setting time and bond, but less than 5% by weight since they decrease the matrix's vapour permeability and fire resistance.

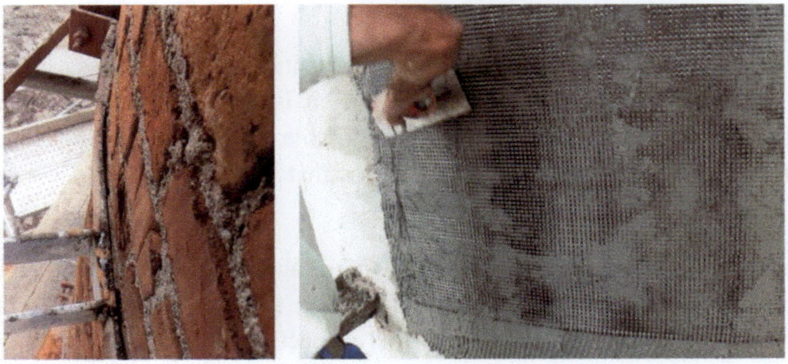

a) *b)*

Figure 2.1: Strengthening of a masonry chimney with externally bonded textile reinforced mortar: *a)* original surface, *b)* application of the reinforcing system [49]

The textile component usually consists of an open mesh of yarns in alkali-resistant (AR) glass, carbon, aramid, basalt, or polyparaphenylene benzobisoxazole (PBO) continuous fibres. They can be either dry, coated, or preimpregnated and are generally arranged in two or more directions. Examples of textiles' types are shown in Figure 2.2.

Different terminologies are diffused for this new class of composites, such as textile-reinforced mortar (TRM) or fabric-reinforced cementitious matrix (FRCM). For the sake of simplicity, they are named only FRCM in this work.

It is worth noting that the composite system's mechanical effectiveness strongly depends on the matrix's ability to saturate the fibre threads. The spacing of the yarns should allow the inorganic matrix to penetrate the mesh openings. For this reason, it shouldn't be a generic mortar. It should be specifically formulated to be coupled with the textile component and the masonry substrate.

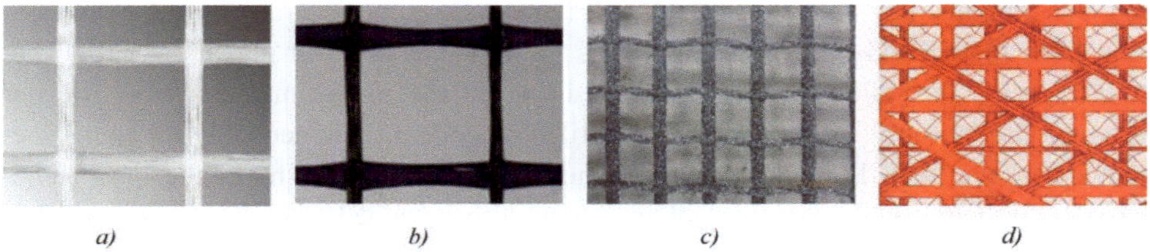

Figure 2.2: Four examples of textile grids: *a)* AR-glass; *b)* basalt; *c)* carbon; *d)* multi-axial hybrid AR-glass and polypropylene (courtesy of Dr Günther Kast GmbH & Co.)

Furthermore, these systems offer important advantages in thickness, weight, tensile strength, and no concrete cover against corrosion is required. They can improve the structural members' ultimate capacity with negligible mass increase thanks to their high strength-to-weight ratio. It is important to stress that this aspect is very important since the seismic actions are proportional to the mass.

Moreover, they are usually applied to the outer surface of structural members. Thus, the application can fit the contour of the strengthened element, making, in this way, the installation process fast and versatile. In this respect, the intervention's reversibility is generally possible for the lime-based mortars, making the FRCM also suitable for repairing historic architectural heritage. Therefore, compared to conventional techniques, e.g. section enlargement, steel tie rods, reinforced concrete overlays, the FRCM is a competitive technology for repair, seismic retrofit, and rehabilitation of existing masonry structures.

2.1 Historical background

The first structural cement-based textile composite was the ferrocement patented in 1855 by Lambot in France [50]. It consisted of a steel reinforcement grid embedded in a cementitious matrix of about 50 mm in thickness. It is worth remembering that ferrocement technology was almost absent in the construction industry until the 1960s. In those years, this technique was mainly used to build slender shell structures such as water tanks or roofs and, only in the 1990s it was also adopted as a repair solution. In this context, ferrocement laminates were used to repair damaged reinforced concrete elements. Although various research studies confirmed the increase in the ultimate capacity of strengthened reinforced concrete beams and columns, this technique was mainly used in new construction. Therefore, the application of the ferrocement as a repair technology remained limited [50].

Then, the composite materials based on the polymeric matrix were developed in the 1960s and 1970s. They consisted of one or more layers of high-strength fibre sheets made of carbon, glass, basalt, or aramid embedded in a polymer-based matrix. Widely known as fibre-reinforced polymers (FRP), this technology's application was initially limited to the aerospace and defence industries. Only in the early 1990s it was also available for the construction industry, as FRP manufacturing cost was reduced. The main application fields of the FRPs in the construction sector were repairing or strengthening concrete and masonry structures. Furthermore, various studies confirmed the effectiveness of externally bonded FRP systems in improving the axial, shear, and flexural resistance of structural elements [50]. However, despite these advantages, the FRP composites are characterized by some limitations, such as:

- Heat sensitivity of the organic matrix;
- Problematic application on wet surfaces;
- Impermeability of the polymeric matrix: it can trap moisture within the masonry, leading to the loss of bond with the substrate;
- Surface preparation necessary for the application of FRP laminates on uneven or jagged surfaces;
- Hazardous materials: the polymers need special handling processes before, during, and after their use.

Therefore, mortar-based composites have been developed to overcome all these problems. The organic matrix of the FRP has been replaced with an inorganic one generally based on cement or lime mortar. Thus, compared to the FRPs, the FRCM systems have the following advantages:

- Resistance of the inorganic matrix to elevated temperatures and ultraviolet radiation;
- Application on irregular or wet substrates with minimal or no surface preparation;
- Vapour permeability, physical/chemical compatibility of the lime-based mortars with masonry substrates: this aspect is fundamental for applications to historic structures.

Finally, it is worth noting that the FRCM composites also present many analogies with the textile-reinforced concrete (TRC). The latter is a thin precast structural element made of alkali-resistant glass or carbon fabrics embedded in high performance finely grained cement concrete [51], which is commonly adopted as a stand-alone load-bearing element for new constructions. On the contrary, the FRCM is specifically designed for the external strengthening of existing structures.

2.2 Characterization methods of the textile reinforced mortar systems (FRCM)

Two different test methods have been recently proposed in the scientific literature about the mechanical characterization of the FRCM composites. The first one combines the results of direct tensile tests on bare textile specimens with those obtained from single-lap shear bond tests. This procedure has been developed within the RILEM Technical Committee (TC) 250-CSM, "Composites for the Sustainable Strengthening of Masonry". It is also adopted by the Italian technical standard CNR-DT 215/2018 "Guide for the Design and Construction of Externally Bonded Fibre Reinforced Inorganic Matrix Systems for Strengthening Existing Structures" [52].

The second characterization method is based on performing direct tensile tests on composite specimens. In this case, the samples are clamped between the wedges of a universal testing machine without applying any lateral pressure through clevis-type grips. This test procedure is defined in the acceptance criteria AC434 of the International Code Council–Evaluation Service in the US [53]. It is adopted by the American technical standard ACI 549.4R-13 "Guide to Design and Construction of Externally Bonded Fabric-Reinforced Cementitious Matrix (FRCM) Systems for Repair and Strengthening Concrete and Masonry Structures" [54]. Finally, the test method parameters have also been validated by Arboleda, who conducted an extensive study on testing FRCM composites with clevis grips [55].

2.2.1 Characterization method of FRCM composite tested with clamping grips

The characterization method presented in this paragraph combines the results of tensile tests performed with clamping grips on bare textile samples and single-lap shear bond tests. The latter is commonly carried out to investigate the load transfer capacity between the externally bonded strengthening system and the structural member, i.e. the substrate. It is important to stress that in many structural applications, such as strengthening masonry walls, arches or vaults, the performance of the FRCM usually depends on the bond behaviour. Therefore, it needs to be experimentally investigated for the design and assessment of the reinforced structural element. Furthermore, recent experimental studies, e.g. [56]–[59], have demonstrated that the shear bond failure may occur not only within the substrate but also within the thickness of the system (differently from FRPs). In this regard, the grid-to-matrix adhesion or interlocking, the fibre-to-fibre and matrix to substrate bond are factors that may influence the load transfer mechanism. In more detail, the following aspects generally affect the load transfer capacity between substrate and system:

- textile architecture;
- coating or pre-impregnation of the fibre;
- thickness of the matrix,
- characteristics and mechanical properties of the substrate (roughness, moisture content, strength and Young's modulus)
- quality of the application and curing conditions.

It is worth noting that no norm exists on the single-lap shear-bond tests. Only a recommendation of the RILEM Technical Committee (TC) 250-CSM [60] has been recently published. Figure 2.3 illustrates the test setup proposed in [60], where the specimen consists of an FRCM strip applied onto a substrate prism. The unbonded textile end of the sample is clamped in the wedges of the testing machine, and it is subjected to a tensile load parallel to the composite-to-substrate interface. In this way, the load transfer capacity between the externally bonded strengthening system and the structural member (i.e. the substrate) can be evaluated, and the following possible failure modes can occur:

- Mode A: complete debonding of the system from the masonry support (cohesive failure);
- Mode B: debonding of the system at the matrix-to-substrate interface (adhesive failure);

- Mode C: debonding of the system at the matrix-to-textile interface;
- Mode D: slippage of the textile within the matrix;
- Mode E: tensile break of the textile fibres in the unbonded part;
- Mode F: tensile rupture of the textile fibres within the mortar matrix [61].

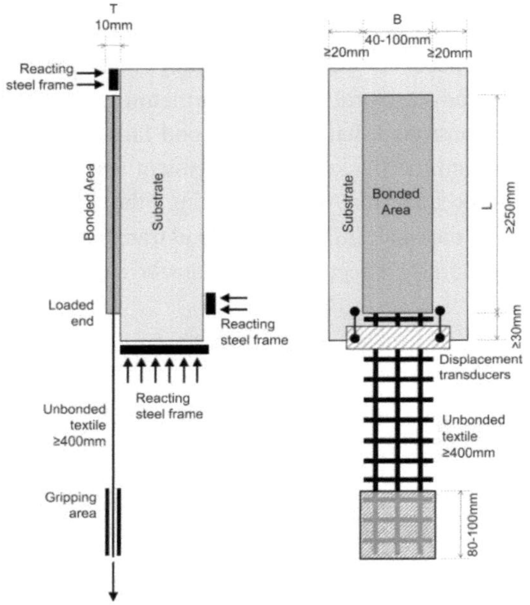

Figure 2.3: Test setup for the single-lap shear bond tests proposed in [60]

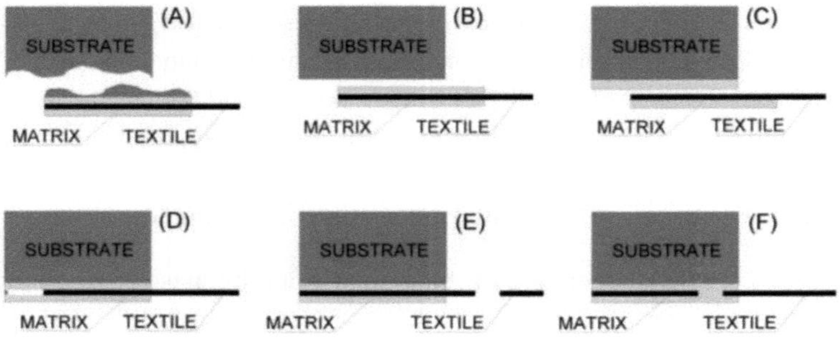

Figure 2.4: Possible shear bond failure modes of the FRCM systems when applied to the masonry substrate [61]

The experimental results are usually expressed in terms of axial stress-slip relationship. The stresses σ are conventionally referred to the equivalent cross-sectional area of the dry textile A_f, as follows:

$$\sigma = \frac{F}{A_f} \tag{2.1}$$

Where F is the axial force recorded during the tests, and A_f [mm²] is calculated with Eq. (2.2):

$$A_f = \frac{T_x \cdot n_y}{\rho_{fib} \cdot 1000} \cdot b_f \tag{2.2}$$

T_x is the yarn count expressed in Tex [g/km], n_y is the number of yarns per unit width expressed in [n°/cm], ρ_{fib} is the fibre density [g/cm^2], b_f is the width of the textile [mm].

In this way, any possible variation of mortar thickness cannot influence the equivalent cross-sectional area of the FRCM. Furthermore, the equivalent fibre thickness [mm] of the dry textile can be defined from A_f as follows:

$$t_f = \frac{A_f}{b_f} \tag{2.3}$$

It is important to stress that if the textile mesh consists of the same type and number of yarns in its main directions, the equivalent cross-sectional area and thickness is the same. Otherwise, they depend on the considered direction, i.e. vertical, horizontal and eventually diagonal, if present.

Furthermore, the typical force-slip curves that may result from single-lap shear-bond tests are depicted in Figure 2.5. The failure modes A, B, and C are quite brittle. The curve generally shows a nearly flat branch until the debonding of the system occurs. On the contrary, mode D displays a diagram with a soft load decrease after the peak due to the progressive slippage of the textile mesh within the mortar. Moreover, the response curve of the failure mode E is characterized by a sudden strength reduction due to the tensile breaking of the fibre threads. In this case, the curve's flat branch is commonly not reached. Finally, mode F shows a strength drop at the telescopic failure of the fibre wires embedded in the matrix. Then, a further load decrease usually occurs since the first portion of the textile out of the mortar can slip [61].

Furthermore, the boundary conditions, the stress state and the cracking pattern of the applied FRCM system may differ from those reproduced through the tensile tests on FRCM coupons with clamping grips. Finally, some doubts about the repeatability of the tests and the method's robustness have been raised, especially for the FRCM systems, whose conventional stress falls in the uncracked or crack development phases of the tensile response.

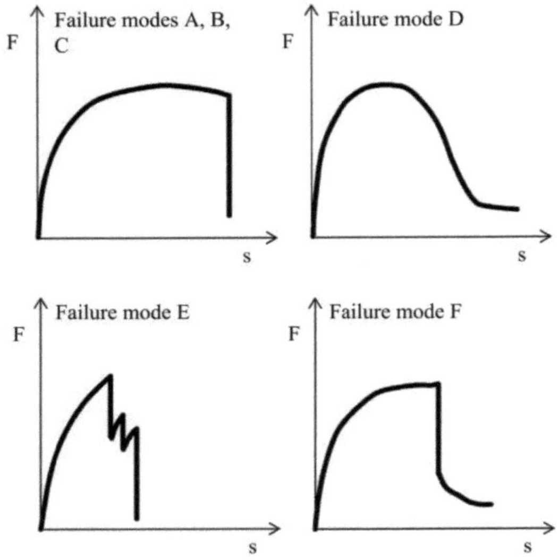

Figure 2.5: Typical axial force-slip curves obtained from the single-lap shear bond tests [61]

Therefore, the acceptance criteria proposed by Ascione in [61] may be particularly sensitive to these aspects. The acceptance strain and tensile modulus of elasticity can be derived from direct tensile tests on bare textile samples and not on FRCM specimens. The latter are performed by clamping grid strips in the testing machine's wedges with sufficient lateral pressure. Only in this way can the slippage of the textile strip in the load introduction areas be avoided, and the tensile break of the fibres can be achieved.

This choice was also taken from the Italian standardization board set within the National Research Council (CNR) to develop qualification guidelines for FRCM composites [50]. In this regard, the Italian technical standard CNR-DT 215/2018 defines the limit conventional FRCM tensile strength $\sigma_{lim,conv}$ for design and assessment purposes. As shown in Figure 2.6, it is equal to the characteristic value of the maximum axial force obtained from the single-lap shear-bond tests divided by the equivalent cross-sectional area A_f. Then, the corresponding modulus of elasticity E_f and ultimate strain $\varepsilon_{lim,conv}$ are derived from the clamping-grip tensile tests on bare textile specimens. It is worth noting that if the grid doesn't consist of the same type and number of yarns in its main directions, the tensile properties of the FRCM system depend on the considered direction. Otherwise, they are the same. Finally, the single-lap shear-bond test results should be valid only for the tested FRCM system and substrate. They cannot be directly extended or extrapolated to different composites or substrate materials [60].

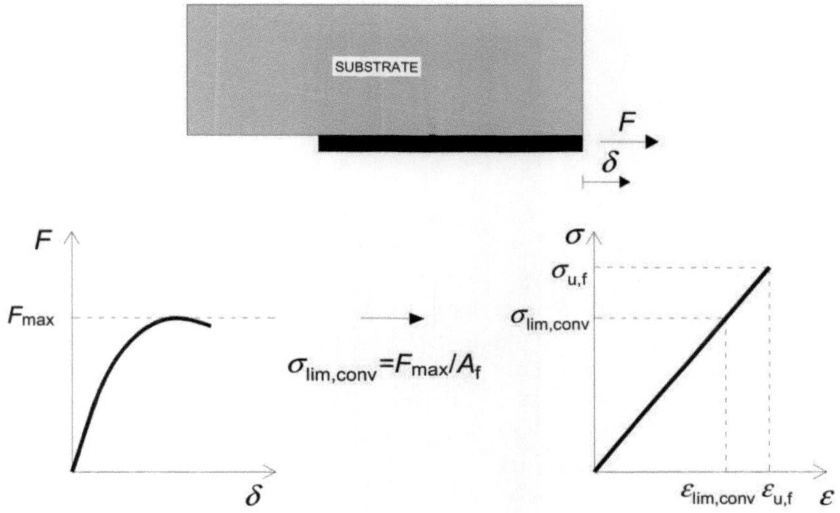

Figure 2.6: Determination of $\sigma_{lim,conv}$ and $\varepsilon_{lim,conv}$ [52]

2.2.2 Characterization method of FRCM composites tested with clevis-type grips

The second characterization method of an FRCM system is based on direct tensile tests on composite coupons (Figure 2.7). Adopting the test method suggested by the AC434 guidelines, the axial load is transferred from the frame of a universal testing machine to the composite specimen through clevis-type grips. Two metal tabs, preferably aluminium, are glued at each end of the sample. In this way, the composite coupon is connected to the gripping mechanism.

This test setup allows reproducing the application case where the composite is not mechanically anchored. Thus, the system's tensile response depends mainly on the grid-matrix interface bond. The break of the textile fibres or the grid's slippage within the matrix are possible failures. It is worth specifying that the other shear bond failure modes illustrated in Figure 2.4 are not considered from this test setup. The reason for this different acceptance method is to be found in the main foreseen applications. In particular, the American approach has been developed considering the application of FRCM systems on concrete structures or masonry with concrete units. On the contrary, the method developed within RILEM and adopted by CNR DT-215/2018 is mainly based on their application on clay brick or stone masonry [50]. Figure 2.8 shows the typical stress-strain diagram of an FRCM system obtained from uniaxial tensile tests with clevis-type grips.

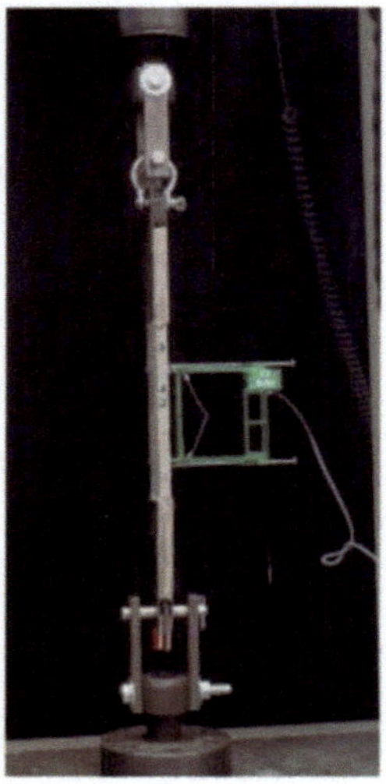

Figure 2.7: Uniaxial tensile tests with clevis-type grips on FRCM composite specimens [54]

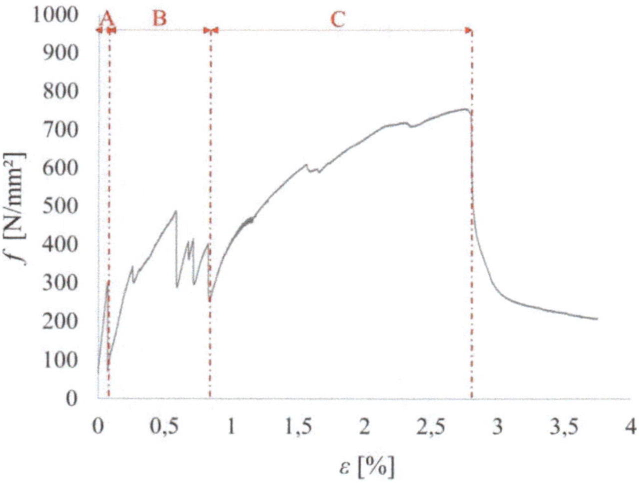

Figure 2.8: Example of uniaxial tensile stress-strain diagram of FRCM composite specimens obtained from direct tensile tests with clevis-type grips

Three consecutive branches can be usually identified:

- Stage A: uncracked phase
- Stage B: crack development
- Stage C: fully cracked phase.

It is worth noting that the width of these stages generally depends on several factors, such as mechanical properties and nature of the matrix (cementitious or hydraulic lime), layers' thickness and interlocking between grid and mortar.

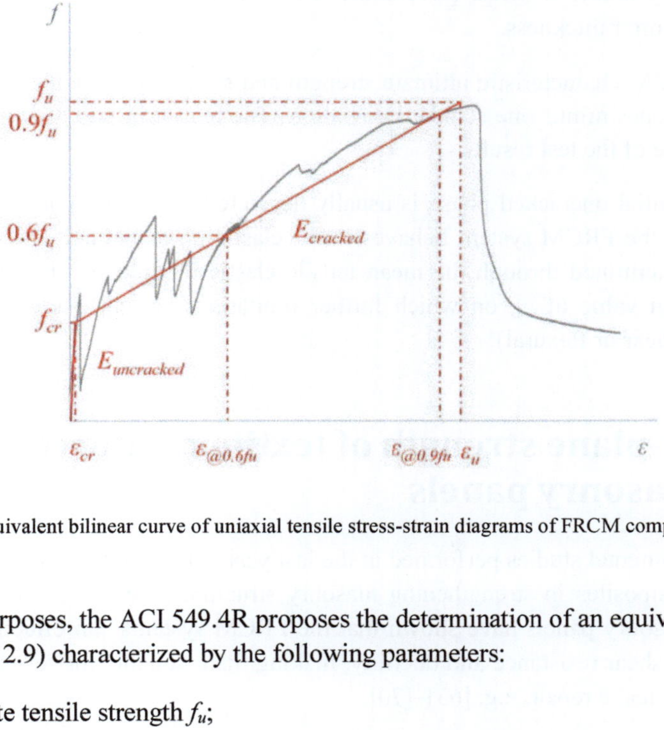

Figure 2.9: Equivalent bilinear curve of uniaxial tensile stress-strain diagrams of FRCM composite specimens

For design purposes, the ACI 549.4R proposes the determination of an equivalent bilinear curve (Figure 2.9) characterized by the following parameters:

- ultimate tensile strength f_u;
- tensile modulus of elasticity of the cracked specimen $E_{cracked}$, calculated with Eq. (2.4):

$$E_{cracked} = \frac{0.3 f_u}{\left(\varepsilon_{@0.9 f_u} - \varepsilon_{@0.6 f_u}\right)} \tag{2.4}$$

Where $\varepsilon_{@0.9fu}$ and $\varepsilon_{@0.6fu}$ are the strains at $0.9f_u$ and $0.6f_u$;

- ultimate tensile strain ε_u calculated with Eq. (2.5):

$$\varepsilon_u = \varepsilon_{@0.6f_u} + \frac{0.4f_u}{E_{cracked}} \quad (2.5)$$

- stress and strain of the transition point between the uncracked and cracked phases;
- tensile elastic modulus of the uncracked specimen. It is defined on the initial linear segment of the response bilinear. Two points that connect the results in a line close to the trend and slope of the response curve identify the initial line.

Also in this case, the stresses are conventionally referred to the equivalent cross-sectional area of the dry textile A_f (Eqs. (2.1) and (2.2)) to avoid the influence of any possible variation of mortar thickness.

Then, the FRCM characteristic ultimate strength and strain are calculated as the average experimental ones minus one standard deviation. The tensile elastic modulus is equal to the mean value of the test results.

Finally, the initial uncracked phase is usually neglected for design purposes. After ACI 549.4R-2013, the FRCM system behaves linear elastic up to failure. The design tensile strength is determined through the mean tensile elastic modulus of the cracked FRCM and the design value of ε_u, on which further limitations are set based on the type of application (shear or flexural).

2.3 In-plane strength of textile reinforced masonry panels

Several experimental studies performed in the last years have confirmed the effectiveness of FRCM composites in strengthening masonry structures. Tests conducted on textile-reinforced masonry panels have shown that the FRCM systems can effectively improve their in-plane shear resistance and ductility, making them suitable for seismic retrofitting and post-earthquake repair, e.g. [63]–[70].

A common approach to calculate the in-plane shear strength of a textile-reinforced panel is to add the increase of resistance due to the FRCM system to the shear strength of the URM panel. In this context, both technical standards CNR-DT 215/2018 and ACI 549.4R-13 propose design formulations.

Therefore, starting from the calculation method of CNR-DT 215/2018, the in-plane shear resistance of a strengthened masonry panel, $V_{Rd,s}$, can be determined as follows:

$$V_{Rd,s} = V_{URM} + V_{t,f} \tag{2.6}$$

V_{URM} is the shear resistance of the URM panel under diagonal cracking, and $V_{t,f}$ is the increase of strength due to the reinforcing system calculated through Eq. (2.7).

$$V_{t,f} = \frac{1}{\gamma_{Rd}} \cdot n_f \cdot t_{f,s} \cdot l_f \cdot \alpha_t \cdot \varepsilon_{fd} \cdot E_f \tag{2.7}$$

Where:

- γ_{Rd} is a model safety factor equal to 2, according to current knowledge on the FRCM systems;
- n_f is the total number of applied textile layers;
- $t_{f,s}$ is the equivalent thickness of one textile layer with fibres arranged parallel to the shear force;
- l_f is the design length of the applied reinforcement. It is measured orthogonally to the shear force, and in any case, it cannot be assumed to be greater than the dimension H of the panel (Figure 2.10). The limitation $l_f \leq H$ is introduced, since the product $n_f \cdot t_{f,s} \cdot l_f$ represents the equivalent cross-section area of reinforcement effective in shear, which intersects a diagonal crack inclined at 45°;
- α_t is a coefficient that considers the reduced tensile strength of the fibres under shear actions. It may be assumed equal to 0.8 in the case of no experimental testing.
- ε_{fd} is the design strain of the yarns arranged parallel to the shear force. It is derived from the conventional strain limit $\varepsilon_{lim,conv}$ using the partial safety factors indicated for design situations;
- E_f is the average Young's modulus of the bare textile.

Moreover, CNR-DT 215/2018 limits the in-plane shear strength of the textile-reinforced panel by the masonry crushing capacity, $V_{t,c}$. In this case, the shear force V_{Ed} must not exceed the diagonal crushing resistance defined by Eq. (2.8):

$$V_{t,c} = 0.25 \cdot f_{m,u} \cdot t \cdot d_f \tag{2.8}$$

Where:

- $f_{m,u}$ is the ultimate masonry compressive strength;
- t is the thickness of the panel;

- d_f is the distance between the compressed edge of the panel and the fibre of the reinforcement that attains the highest tensile strain (Figure 2.10).

Concerning the American design approach, it is important to stress that the limit-state design principles adopted by ACI are different from the European ones. For this reason, the design formulations proposed by ACI 549.4R-13 vary from those suggested by CNR-DT 215/2018, although both standards have a similar calculation method.

Therefore, the shear strength of a masonry panel reinforced with FRCM can be determined after ACI 549.4R-13 as follows:

$$\varphi_v V_n = \varphi_v \left(V_m + V_f \right) \tag{2.9}$$

Where φ_v is the shear strength reduction factor equal to 0.75, and V_n is the nominal shear strength. V_m and V_f are the masonry and FRCM strength contribution, respectively. V_m is calculated according to the Building Code Requirements and Specification for Masonry Structures [71], while V_f is defined in Eq. (2.10):

$$V_f = 2n_f \cdot A_f \cdot l \cdot f_{fv} \tag{2.10}$$

Where n_f is the number of applied reinforcement layers; A_f is the equivalent cross-sectional area of the grid by unit width effective in shear; l is the wall's length in the direction parallel to the shear force. It is worth noting that ACI 549.4R-13 doesn't explain the reason for factor 2 in Eq. (2.10). It only suggests applying the composite system preferably on both sides of the wall for symmetry and effectiveness. Furthermore, no indications are given for the one side application. Factor 2 is probably introduced to consider only the double-sided FRCM application. Thus, n_f should indicate the number of reinforcement plies per side and not the total number of applied layers.

Regarding the design tensile strength of the FRCM reinforcement, f_{fv}, it is calculated through Eq. (2.11):

$$f_{fv} = E_{cracked} \cdot \varepsilon_{fv} \tag{2.11}$$

In which $E_{cracked}$ is the tensile modulus of elasticity of the cracked FRCM and ε_{fv} is the tensile design strain in the FRCM shear reinforcement, defined as follows:

$$\varepsilon_{fv} = \min\left(\varepsilon_u ; 0.004\right) \tag{2.12}$$

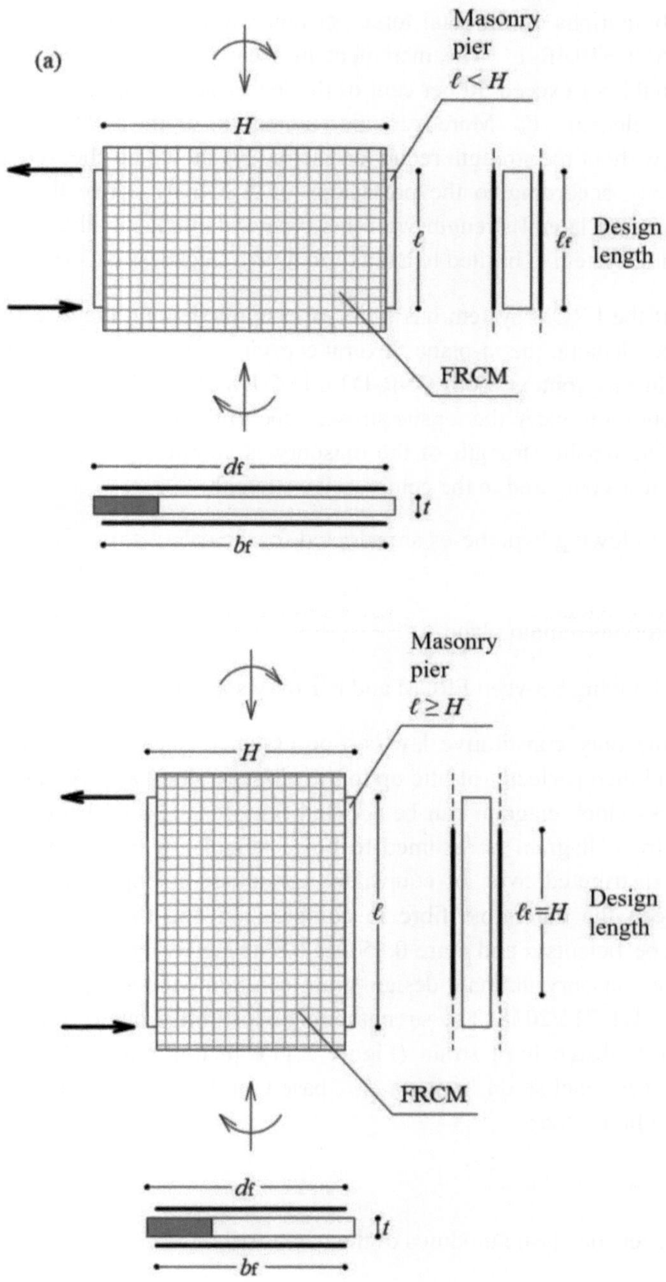

Figure 2.10: In-plane FRCM reinforcement of a masonry pier: definitions after CNR-DT 215/2018 [52]

Furthermore, limitations on the total force per unit width transferred to the masonry are given in the ACI 549.4R-13. The increment in shear strength provided by the FRCM system, V_f, should not exceed 50 per cent of the load-bearing capacity of the unstrengthened structural element, V_m. Moreover, the summation of the URM and FRCM shear contributions (without the strength reduction factor, φ_v) should be checked against the toe crushing capacity according to the provisions of ASCE 41, since the masonry panel attains its maximum lateral strength with this type of failure. Finally, the application of the strengthening system is limited to a maximum wall thickness of 305 mm.

Furthermore, if the FRCM system has fibre yarns applied along the direction of the axis of the structural element, the in-plane flexural capacity of the textile-reinforced panel can be increased. In this context, both CNR-DT 215/2018 and ACI 549.4R-13 consider the FRCM contribution to carry the tensile stresses due to the bending-rocking behaviour of the element. The tensile strength of the masonry is commonly neglected because it is generally very low compared to the compressive strength.

Therefore, the following hypotheses are adopted for the calculation of the design bending moment:

- Plane sections remain plane;
- Perfect bonding between FRCM and masonry support.

The uniaxial masonry constitutive law can be assumed linear elastic up to the design strength $f_{m,u}$ and then perfectly plastic up to the ultimate strain $\varepsilon_{m,u}$. As an alternative, the equivalent stress-block diagram can be adopted (Figure 2.11). In this case, the masonry compressive stress diagram is assumed to be rectangular with a uniform compressive strength $\alpha f_{m,u}$ distributed over an equivalent compressive length βy_n, where y_n is the distance between the outermost fibre in compression and the neutral axis. Common values of the coefficients α and β are 0.85 and 0.8, respectively. If experimental data are unavailable, the masonry ultimate design strain can be assumed equal to 3.5‰, as suggested by CNR DT 215/2018. The strengthening constitutive law of the FRCM is linear elastic up to the design limit strain (Figure 2.11). In this regard, CNR-DT 215/2018 defines the conventional strain limit, $\varepsilon_{lim,conv}$, based on the tensile strength obtained from single-lap shear bond tests:

$$\sigma_{\lim,conv} = E_f \cdot \varepsilon_{\lim,conv} \tag{2.13}$$

Where E_f is the tensile elastic modulus of the bare textile.

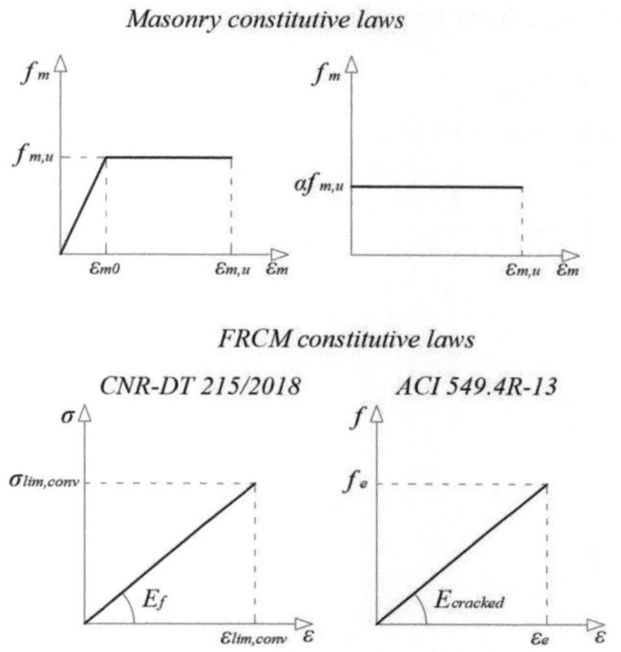

Figure 2.11: Uniaxial constitutive laws for masonry and FRCM system

On the contrary, ACI 549.4R-13 assumes that the effective tensile failure strain of the FRCM composite material, ε_e, is limited by the ultimate tensile strain of the FRCM system $\varepsilon_u \leq 1.2\%$. Then, the effective tensile stress level in the FRCM reinforcement attained at failure, f_e, is calculated through Eq. (2.14):

$$f_e = E_{cracked} \cdot \varepsilon_e \quad \text{with} \quad \varepsilon_e = \min(\varepsilon_u; 1.2\%) \tag{2.14}$$

Where $E_{cracked}$ is the tensile modulus of elasticity of the cracked FRCM composite material.

It is worth noting that if the neutral axis y_n intersects the reinforcing system, the compressed part of the FRCM is considered non-reactive. Only the masonry can carry compressive stresses since both standards assume that the FRCM system does not exhibit any stiffness or compressive strength. In this context, Figure 2.12 shows the flexural failure of a masonry pier strengthened with an FRCM applied on the entire surface of the element ($d_f = l$) and both sides. It is important to stress that, in this case, only the direction of the system parallel to the panel's axis can contribute to the in-plane flexural resistance, as shown by the induced strain and stress distributions in the cross-section.

Therefore, the following failure modes are possible:

- Masonry crushing;
- Tensile Failure of the FRCM system.

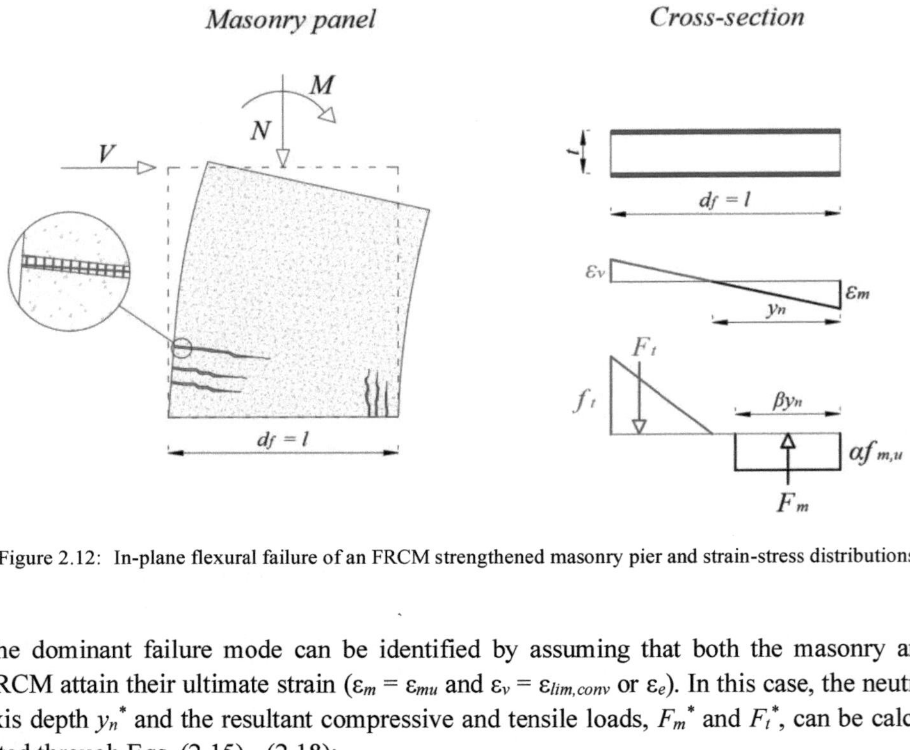

Figure 2.12: In-plane flexural failure of an FRCM strengthened masonry pier and strain-stress distributions

The dominant failure mode can be identified by assuming that both the masonry and FRCM attain their ultimate strain ($\varepsilon_m = \varepsilon_{mu}$ and $\varepsilon_v = \varepsilon_{lim,conv}$ or ε_e). In this case, the neutral axis depth y_n^* and the resultant compressive and tensile loads, F_m^* and F_t^*, can be calculated through Eqs. (2.15) - (2.18):

$$y_n^* = \frac{l \cdot \varepsilon_{m,u}}{\left(\varepsilon_{m,u} + \varepsilon_v\right)} \tag{2.15}$$

$$F_m^* = \alpha f_{m,u} \cdot t \cdot \beta y_n^* \tag{2.16}$$

$$F_t^* = 0.5 \cdot n_f \cdot \left(l - y_n^*\right) \cdot t_{f,b} \cdot E_f \cdot \varepsilon_{\text{lim},conv} \quad \text{(CNR-DT 215/2018)} \tag{2.17}$$

$$F_t^* = 0.5 \cdot n_f \cdot \left(l - y_n^*\right) \cdot t_{f,b} \cdot E_{cracked} \cdot \varepsilon_e \quad \text{(ACI 549.4R-13)} \tag{2.18}$$

Where $t_{f,b}$ is the equivalent fibre thickness of one textile layer in the direction parallel to the panel's axis, which is effective during the in-plane bending failure.

Thus, comparing the acting normal force N with the difference (F_m^* - F_t^*), the actual failure mode can be identified. In more detail, if $N > (F_m^* - F_t^*)$, masonry crushes by achieving the ultimate compressive strain ε_{mu} while the effective tensile strain in the FRCM is $\varepsilon_v = \varepsilon_{mu} \cdot (l - y_n)/y_n$.

On the contrary, if $N < (F_m^* - F_t^*)$, the FRCM attains the ultimate tensile strain and $\varepsilon_m < \varepsilon_{mu}$. In this case, it is worth noting that the masonry could be in the elastic range, and the hypothesis of a stress block distribution could be violated. However, this assumption appears reasonable since it only modifies the neutral axis depth y_n. The difference in the design flexural strength is negligible.

Therefore, once the failure mode is identified, y_n can be calculated by solving Eq. (2.19):

$$F_m - F_t = N \tag{2.19}$$

Then, the flexural capacity of the strengthened panel can be evaluated according to CNR-DT 215/2018 and ACI 549.4R-13 as follows:

$$M_{Rd,s} = F_m \cdot \frac{(l - \beta y_n)}{2} + F_t \cdot \frac{(l + 2y_n)}{6} \quad \text{(CNR-DT 215/2018)} \tag{2.20}$$

$$\phi_m M_n = \phi_m \left(F_m \cdot \frac{(l - \beta y_n)}{2} + F_t \cdot \frac{(l + 2y_n)}{6} \right) \quad \text{(ACI 549.4R-13)} \tag{2.21}$$

Where ϕ_m is the strength reduction factor for flexure equal to 0.6.

Moreover, the design lateral strength associated with the flexure-controlled failure mode can be determined through Eqs. (2.22) and (2.23):

$$V_{Rd,s}^F = \frac{M_{Rd,s}}{k \cdot h} \quad \text{(CNR-DT 215/2018)} \tag{2.22}$$

$$V_n^F = \frac{\phi_m M_n}{k \cdot h} \quad \text{(ACI 549.4R-13)} \tag{2.23}$$

Where k is a coefficient that accounts for the boundary condition of the panel, equal to 0.5 for a fixed-fixed wall and 1 for a fixed-free wall, h is the height of the panel.

Finally, only CNR DT 215/2018 gives indications to verify the strength of the masonry spandrels reinforced with FRCM. As shown in Figure 2.13, the element's behaviour is assumed as a pier rotated to 90°, and the axial force N in the direction parallel to the bed joints is neglected.

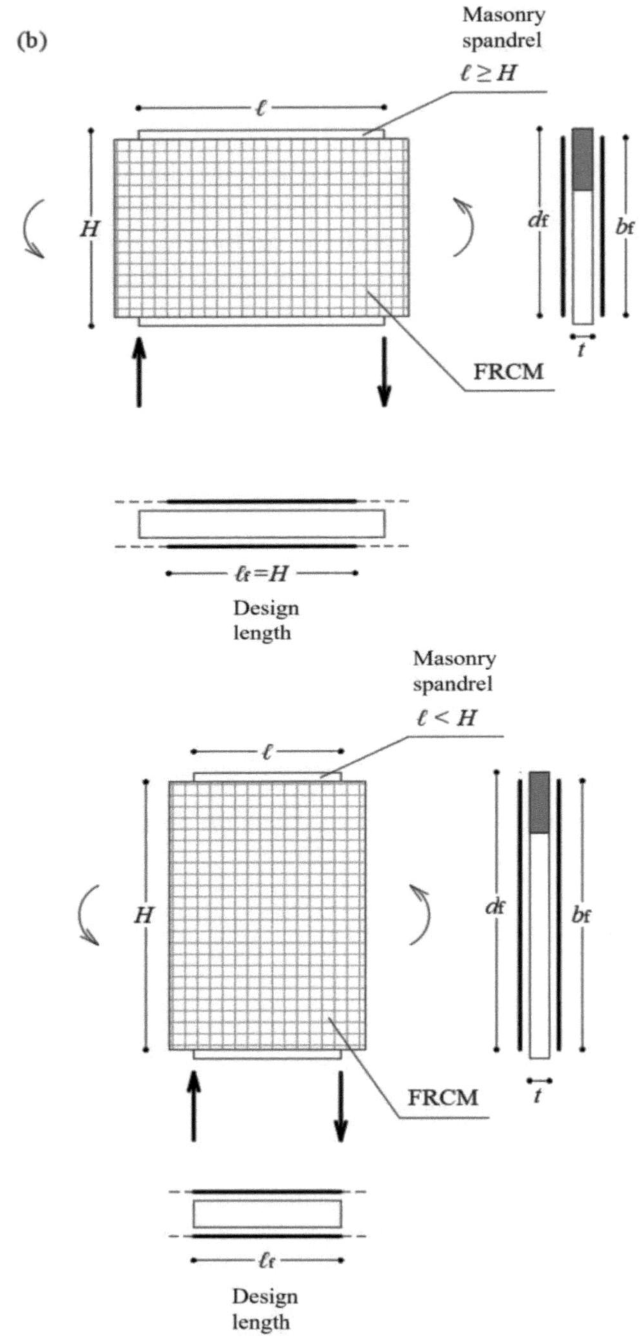

Figure 2.13: FRCM in-plane strengthening of a masonry spandrel: definitions after CNR-DT 215/2018 [52]

Therefore, the shear strength $V_{Rd,s}$ can be still calculated through Eqs. (2.6), (2.7) and (2.8), but the contribution of the normal force in the in-plane resistance of the URM panel under diagonal cracking should be not considered. Then, the cross-sectional analysis necessary to determine the improved flexural capacity should also be performed neglecting N.

2.4 Modelling of textile reinforced masonry structures with the Equivalent frame method

The Equivalent frame method described in § 1.5 is a modelling technique originally developed for URM buildings, which is also suitable for modelling textile-reinforced masonry structures. It assumes that a masonry wall can be schematized as a set of vertical piers and horizontal spandrels. Since the seismic damage is usually concentrated only in these parts, the connection areas generally remain undamaged.

Recent shaking table tests performed on masonry buildings retrofitted with FRCM, e.g. [72], [73], have shown that this phenomenon also occurs in the case of textile reinforced masonry structures. Thus, it can be assumed that the deformation state of connection areas always remains within the elastic limits. It is worth remembering that this assumption allows neglecting these parts in the evaluation of the wall behaviour. The latter is usually limited by the non-linear strains of its structural elements.

Therefore, the textile-reinforced masonry piers and spandrels can be still modelled through a single finite element with a limited number of degrees of freedom. The entire wall is obtained by assembling non-linear elements mutually connected by rigid parts. The numerical models are characterized by a limited number of degrees of freedom. Thus, the response of a wall subject to static (monotonous or cyclic) and dynamic actions can be represented with a modest computational burden.

Furthermore, recent experimental studies, e.g. [69], [74]–[76], have confirmed that the FRCM systems can increase the in-plane lateral strength and displacement capacity of masonry panels. Therefore, the ML-BEAM element implemented in the TREMURI software, described in § 1.5.2, can be suitable to reproduce the in-plane response of a textile strengthened masonry panel until very severe damage levels.

Moreover, as shown in § 2.2.2, the in-plane shear resistance of a textile reinforced masonry panel can be calculated as the sum of the shear strength of the URM panel and the increase of resistance due to the reinforcing system. It is worth remembering that the strength of the unreinforced ML-BEAM element implemented in TREMURI is computed according to the criteria illustrated in § 1.3 and § 1.4, consistent with the most common formulations proposed in the literature and building codes.

Therefore, the increase of in-plane load-bearing capacity can be obtained using the formulations described in § 2.2.2. In this way, no additional nodes or degrees of freedom are required to model the FRCM system, and a lower computational burden is still provided.

The kinematic variables and generalized forces of the strengthened ML-BEAM element are shown in Figure 2.14, which also depicts its monotonic behaviour. It is worth pointing out that if the strengthening system improves the panel's stiffness, the reinforced element's elastic properties can be defined in the input phase by assigning improved elastic and shear moduli (E and G in Eq. (1.41)). Moreover, the progressive stiffness degradation in the elastic phase can also be modelled by assigning two proper ratios, as described in § 1.5.2.

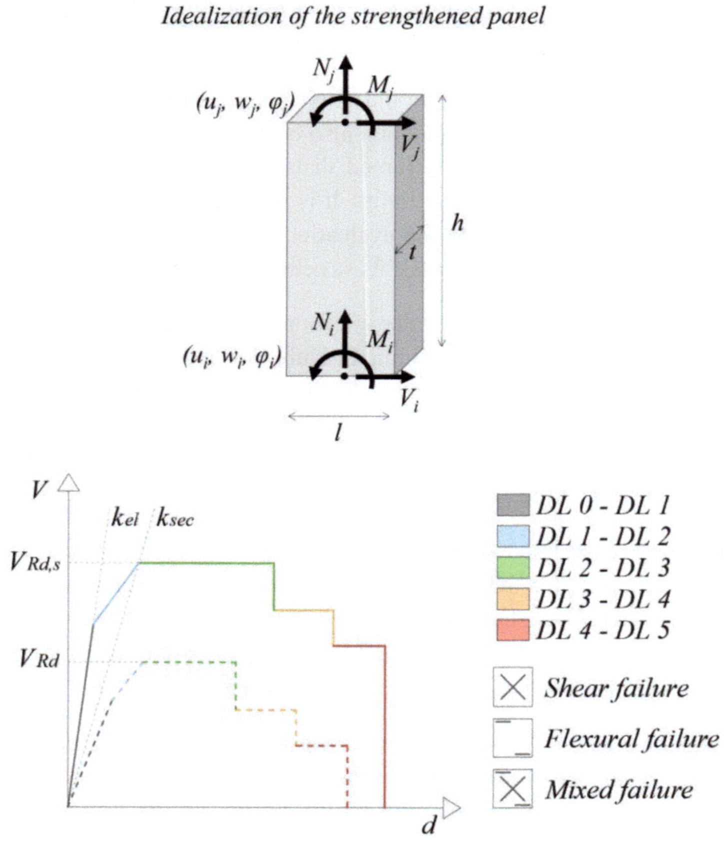

Figure 2.14: Idealization of the single panel and monotonic multilinear constitutive law of the ML-BEAM element strengthened with FRCM implemented in TREMURI

Then, the maximum strength of the panel, $V_{Rd,s}$, is computed as the minimum shear resistance obtained from the defined shear and flexural failure criteria, considering the current axial force acting at each analysis step. The TREMURI program calculates the increase of resistance due to the reinforcing system through the formulations proposed by the Italian technical standard CNR-DT 215/2018. A nonlinear correction of the elastic strength prediction is performed, comparing the latter with the element's limit resistance. Thus, the redistribution of the internal forces is carried out to ensure the element's equilibrium.

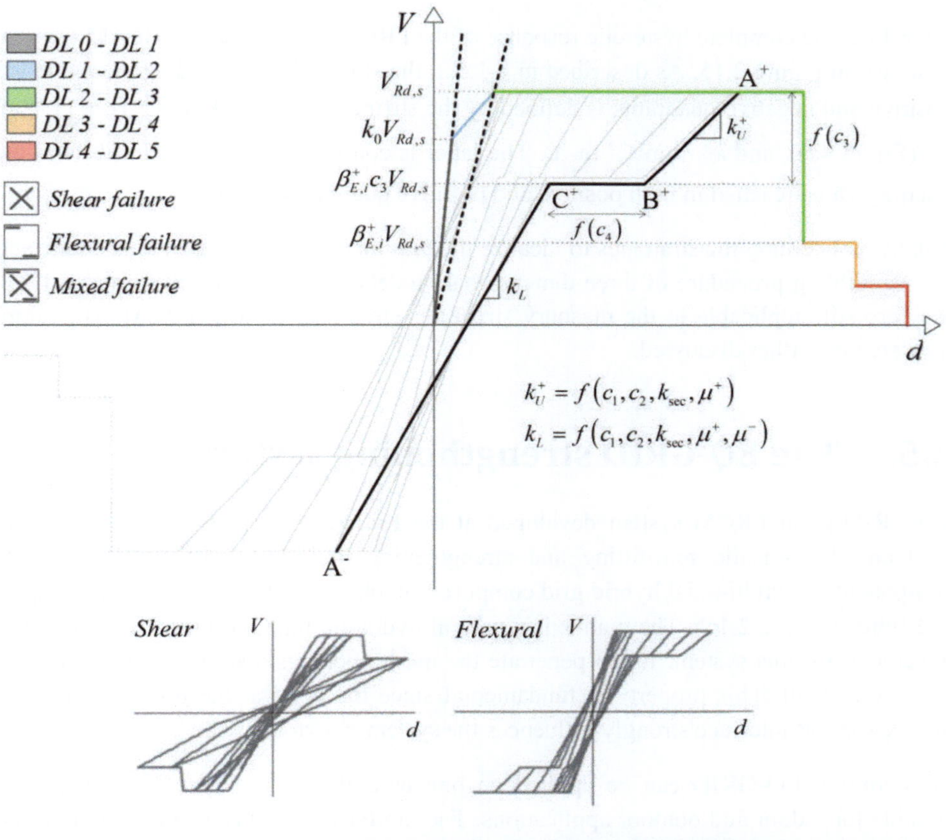

Figure 2.15: Multilinear constitutive law and hysteretic response of the ML-BEAM element strengthened with FRCM implemented in TREMURI

The modelling of the hysteretic behaviour is still based on a phenomenological approach, as for the unstrengthened element. It allows reproducing the possible failure modes of piers and spandrels reinforced with FRCM (flexural, shear and mixed type). As shown in

Figure 2.15, the nonlinear response of the strengthened panel can be modelled until very severe damage levels by assigning progressive strength degradations $\beta_{E,i}$ at defined drift values d_i. Different values of $\beta_{E,i}$ and d_i can be set to describe the panel's flexural and shear behaviour. They can also be differentiated for spandrels, piers and unstrengthened structural elements if they are present.

Furthermore, the hybrid failure is possible in a range of the panel's V – N domain close to the points in which the flexural and shear domains intersect by assigning in the input phase two factors, a_1 and a_2 (Figure 1.40). In this case, the program calculates average values of d_i and $\beta_{E,i}$ from those assigned for the shear and flexural behaviour.

Therefore, the complete hysteretic response of the FRCM reinforced ML-BEAM element is shown in Figure 2.15. As described in § 1.5.2, the slope of the unloading branch in the positive and negative quadrants is defined by the stiffness k_U^+ from A^+ to C^+ (Eq. (1.41)), k_U^- (Eq. (1.42)), and k_L from C^+ to A^-. The latter is computed by considering k_U^+ and the ductility value reached in both positive and negative quadrants (μ^+, μ^-).

Finally, concerning the strategies to idealize the masonry wall in an equivalent frame and the assembling procedure of three-dimensional models illustrated in § 1.5.1 and § 1.5.3, they are still applicable to the masonry structures strengthened with FRCM. Therefore, they are not further discussed.

2.5 The EQ-GRID strengthening system

EQ-GRID is an FRCM system developed at the Karlsruhe Institute of Technology in Germany for seismic retrofitting and strengthening masonry structures. The textile component is a multi-axial hybrid grid composed of alkali-resistant glass and polypropylene fibres (Figure 2.16). The matrix is a natural hydraulic lime mortar (NHL) explicitly developed for this system. It can penetrate the mesh openings and encapsulate the fibre yarns very well. This property is fundamental since the bond at the textile-matrix and matrix-support interface strongly influences the system's performance.

Furthermore, EQ-GRID can be applied on one or both sides of masonry walls. It is suitable for indoor and outdoor applications. The mortar is non-shrinkable, workable and quite viscous to be applied on vertical surfaces. It is very compatible with the masonry support since it is characterized by porosity and vapour permeability. Thus, the moisture cannot remain trapped within the wall. The humidity can migrate through the thickness, and no thermo-hygrometric barrier can occur.

Moreover, the grid's density is about 330 g/m^2, and the dry bulk density of the mortar is equal to about 1.42 kg/m^3. Since the total thickness of the applied system is only 8 mm, the mass addition to the original structure is negligible. Therefore, no increase in seismic action can be caused.

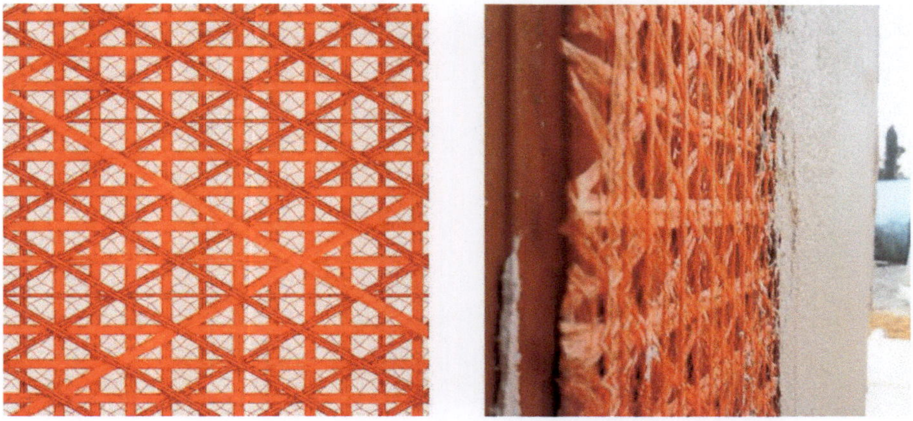

Figure 2.16: The EQ-GRID strengthening system (courtesy of Dr Günther Kast GmbH & Co.)

Furthermore, a broad experimental campaign was performed at the Karlsruhe Institute of Technology in Germany to identify the mechanical properties of the system and investigate the in-plane behaviour of masonry panels strengthened with EQ-GRID. The test results are presented and discussed in Chapter 3.

The experimental campaign has shown that the system can increase the in-plane resistance of masonry panels, especially for relatively poor masonry quality. The structural element reinforced with EQ-GRID is characterized by an improved performance in terms of strength and deformation thanks to the system's capacity to bear the tensile stresses induced by horizontal shear actions. This aspect is very important in the case of earthquake excitation and makes EQ-GRID suitable for repair, seismic retrofit, and rehabilitation of existing masonry structures.

Chapter 3

Experimental characterization of the EQ-GRID strengthening system

The test results presented in this chapter are obtained from a wide experimental campaign performed at the Karlsruhe Institute of Technology in Germany for the European research project "MULTITEXCO - High Performance Smart Multifunctional Technical Textiles for the Construction Sector". The main goal of the experimental testing was to characterize the EQ-GRID system by determining its mechanical properties and investigating the in-plane performance of strengthened masonry panels.

Therefore, several tensile tests were performed on bare grid samples and composite specimens for each main direction of the textile (vertical, horizontal and diagonal). Moreover, compression and bending tests were carried out to define the mechanical properties of the matrix. Then, the bonding behaviour of the system applied to standard support was investigated through double- and single-lap shear-bond tests. Finally, cyclic lateral shear tests were performed on masonry panels reinforced with EQ-GRID to evaluate the effectiveness of the strengthening system. The test results are depicted in terms of shear-drift curves, in which the increase of strength and displacement capacity is shown.

3.1 Tensile properties of the textile component of the EQ-GRID system

The EQ-GRID system's textile component is a hybrid multi-axial grid composed of AR glass and polypropylene wires. Since these two materials have very different mechanical properties, several tensile tests were carried out on textile strips and single yarns for each system's principal direction (vertical, horizontal, and diagonal). Thus, twenty-five grid specimens, eighteen AR glass yarns and ten polypropylene wires were tested after DIN EN 13473-2, which refers to the DIN EN ISO 527-4 about determining the tensile properties.

The dimensions of the textile strips were 50x440 mm for the vertical and horizontal direction and 85x440 mm for the diagonal one. In the adopted test setup, the specimens were clamped between wedge grips of the testing machine (Figure 3.1). Therefore, aluminium-sanded tabs were glued at each end with epoxy resin, and the samples were left at ambient laboratory conditions for the necessary curing time (two days). The distance between the tabs, i.e., the free length of the textile, was 200 mm.

Moreover, aluminium was used instead of steel because of the significantly lower Young's modulus. In this way, the overstress near the gripping area was reduced, and local failures could be avoided. The samples were tested under displacement control with a constant rate of 5 mm/min through an MTS electro-hydraulic universal testing machine equipped with a 100 kN load cell. The tensile load and the vertical displacement were recorded. The results are expressed in terms of stress-strain diagrams.

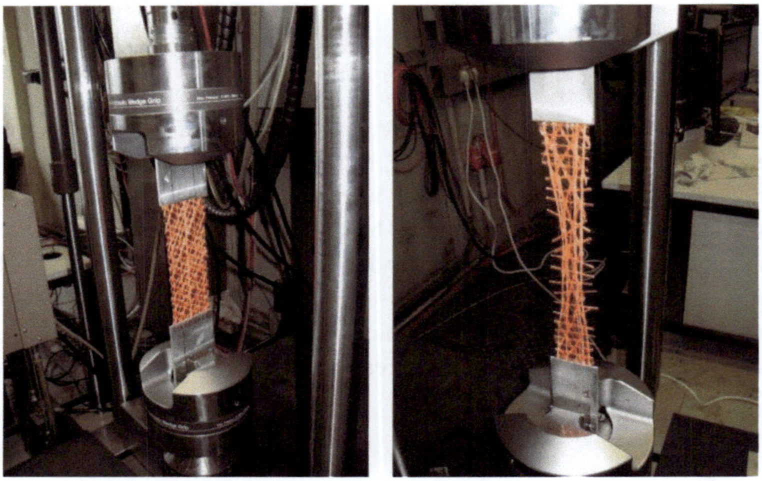

Figure 3.1: Direct tensile tests with clamping grips on bare EQ-GRID textile specimens

It is worth noting that, since the grid is hybrid and the textile strips were tested under displacement control, the AR glass and polypropylene wires were subjected to the same strain during the test, but not to the same stress because of the different elastic moduli that these materials have. Furthermore, the AR glass fibre has a linear elastic behaviour until failure. On the contrary, the polypropylene fibre exhibits a nonlinear behaviour before reaching the maximum tensile stress and can attain a very high ultimate strain (> 20%). However, its response can be assumed linear elastic in the strain range equal to that achieved by the glass fibre wires (Figure 3.2).

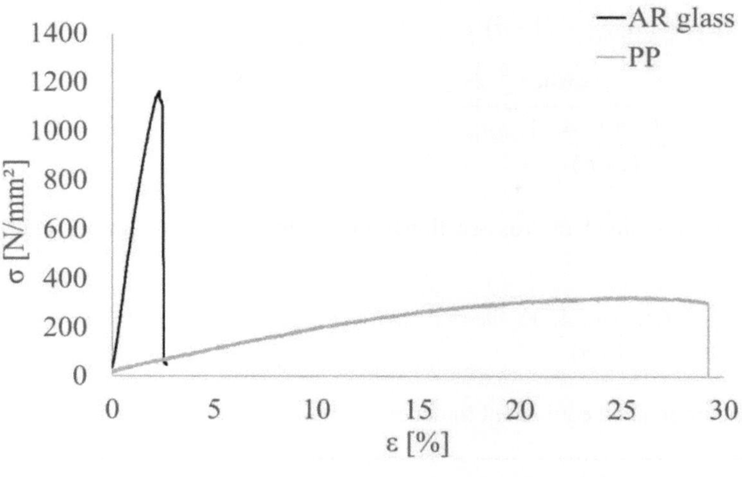

Figure 3.2: σ-ε diagram of one AR glass and one polypropylene wire

Under these assumptions, the homogenization coefficient n can be defined as the ratio between the elastic modulus of the polypropylene E_{PP} and the one of the AR glass E_{AR}:

$$\varepsilon_{AR} = \varepsilon_{PP} \Rightarrow \frac{\sigma_{AR}}{E_{AR}} = \frac{\sigma_{PP}}{E_{PP}} \Rightarrow \sigma_{PP} = \frac{E_{PP}}{E_{AR}} \cdot \sigma_{AR} = n \cdot \sigma_{AR} \tag{3.1}$$

Where ε_{AR}, σ_{AR} and ε_{PP}, σ_{PP} are respectively the strain and stress in the AR glass and polypropylene fibres.

Therefore, the measured tensile force F is equal to the sum of the forces occurring in the AR glass and polypropylene wires, F_{AR} and F_{PP}:

$$F = F_{AR} + F_{PP} = \sigma_{AR} \cdot A_{AR} + \sigma_{PP} \cdot A_{PP} \tag{3.2}$$

A_{AR} and A_{PP} are the equivalent cross-sectional areas of the AR glass and polypropylene wires, respectively.

Then, since σ_{PP} is equal to $n \cdot \sigma_{AR}$, the stress in the AR glass wires can be obtained from the total force F:

$$F = \sigma_{AR} \cdot A_{AR} + n \cdot \sigma_{AR} \cdot A_{PP} = \left(A_{AR} + n \cdot A_{PP}\right) \cdot \sigma_{AR} \Rightarrow$$
$$\sigma_{AR} = \frac{F}{\left(A_{AR} + n \cdot A_{PP}\right)} \tag{3.3}$$

Therefore, the total stress σ is equal to:

$$\sigma = \sigma_{AR} + \sigma_{PP} = (1+n)\cdot\sigma_{AR} = (1+n)\cdot\frac{F}{(A_{AR}+n\cdot A_{PP})} = \\ = \frac{F}{\frac{(A_{AR}+n\cdot A_{PP})}{(1+n)}} = \frac{F}{A_{eq}} \tag{3.4}$$

Where A_{eq} is the equivalent cross-sectional area of the dry textile [mm²] homogenized to the glass fibre:

$$A_{eq} = \frac{(A_{AR}+n\cdot A_{PP})}{(1+n)} \tag{3.5}$$

Thus, the homogenized equivalent thickness t_{eq} [mm] can be defined as follows:

$$t_{eq} = \frac{A_{eq}}{b_f} \tag{3.6}$$

Where b_f is the width of the grid [mm].

It is worth specifying that the coefficient n is defined for each direction of the system. Therefore, Table 1 shows three homogenization coefficients, n_v, n_h, n_d, determined through the elastic moduli of the tested AR glass and polypropylene wires.

Table 1: Homogenization coefficients n_v, n_h and n_d of the EQ-GRID equivalent cross-section

Vertical	Horizontal	Diagonal
n_v	n_h	n_d
0.047	0.053	0.051

Furthermore, the average Young's modulus E_f of the grid can be calculated between two assigned strain values, ε_I and ε_{II}:

$$E_f = \frac{\sigma_{II}-\sigma_I}{\varepsilon_{II}-\varepsilon_I} \tag{3.7}$$

Where σ_{II} and σ_I are the corresponding stress values.

According to DIN EN ISO 527-4, the following strain range should be adopted:

$$\left.\begin{aligned} \varepsilon_I &= 0.05\% \\ \varepsilon_{II} &= 0.25\% \end{aligned}\right\} \tag{3.8}$$

In this case, the linear approximation of the tensile response leads to an overestimation of the failure tensile strength σ_u in the vertical direction and an underestimation in the other two (Figure 3.3).

Therefore, the strain range suggested by CNR-DT 215/2018 for the characterization of the FRCM systems is also considered:

$$\left.\begin{aligned} &\varepsilon_I \text{ at } 0.1\sigma_u \\ &\varepsilon_{II} \text{ at } 0.5\sigma_u \end{aligned}\right\} \quad (3.9)$$

As shown in Figure 3.3, the failure strength is overestimated in all three directions by adopting Eq. (3.9) values. For this reason, another strain range needs to be chosen. In this context, a better linear approximation is obtained with the average secant elastic modulus calculated between zero and the maximum tensile stress. Figure 3.3 also shows the linear tensile response determined with the secant stiffness.

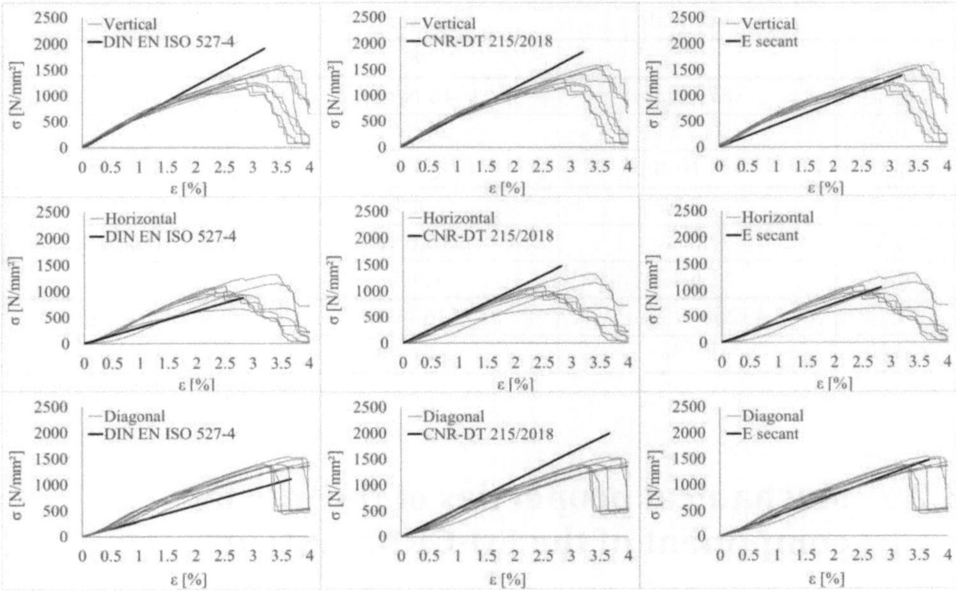

Figure 3.3: σ-ε diagrams obtained from the tensile tests on EQ-GRID textile strips and linear approximation of the tensile response

Thus, the mechanical properties of the multi-axial grid are summarized in Table 2. The strength $\sigma_{u,m}$ is calculated as the ratio between the average maximum tensile force and the equivalent cross-section A_{eq} of the dry textile homogenized to the glass fibre. The characteristic strength and strain values, $\sigma_{u,k}$ and $\varepsilon_{u,k}$, are obtained as the average ones minus k_n times the standard deviation *STD*, as recommended by Annex D of Eurocode 0. The factor k_n is assumed 2.0 since the number of the tested specimens is eight or even nine. Furthermore, the mean secant Young's modulus E_f is also indicated in Table 2.

It is worth noting that the coefficient of variation (Co.V.) assumes higher values for the specimens in the horizontal direction since the production process influences the tensile response of the horizontal AR glass wires. Therefore, both ultimate strength and strain values are lower than those obtained for the vertical direction, although the samples had a very similar threads' arrangement. Finally, the diagonal tensile strength is higher than the others. In this regard, it is important to remember that the stresses are referred to the homogenized grid's cross-section, and the specimens have a low glass fibre content in this direction.

Table 2: Mechanical properties of the hybrid multi-axial grid.

	Vertical	Horizontal	Diagonal
$n_{specimen}$	8	8	9
$\sigma_{u,m}$	1387 N/mm²	1046 N/mm²	1458 N/mm²
Co.V.	13.1 %	20.4 %	5.0 %
$\sigma_{u,k}$	1024 N/mm²	620 N/mm²	1311 N/mm²
$\varepsilon_{u,m}$	3.2 %	2.8 %	3.7 %
Co.V.	9.0 %	15.6 %	10.6 %
$\varepsilon_{u,k}$	2.6 %	1.9 %	2.9 %
E_f	43269 N/mm²	37516 N/mm²	40067 N/mm²
Co.V.	5.7 %	21.6 %	9.3 %

3.2 Mechanical properties of the mortar component of the EQ-GRID system

The matrix of the EQ-GRID system is a natural hydraulic lime-based (NHL) mortar whose mechanical properties were determined through bending and uniaxial compression tests.

The bending tests were performed on three specimens with dimensions equal to 40x40x160 mm after 28 days of maturation. Then, the six halves obtained by the previous samples were subjected to uniaxial compression. In this way, six compression tests on mortar prisms with dimensions 40x40x80 mm could be carried out. This test method is suggested by the DIN EN 196-1 and allows obtaining the average flexural tensile strength $f_{t,M}$ and compressive strength $f_{c,M}$ of the EQ-GRID matrix. The results are summarized in Table 3.

Table 3: Mechanical properties of the EQ-GRID system's matrix.

$f_{c,M}$ (28 days)	Co.V.	$f_{t,M}$ (28 days)	Co.V.	E_M	f_p
14.95 N/mm^2	3.6%	5.13 N/mm^2	0.9%	7.5 GPa	0.5 N/mm^2

Finally, Table 3 also reports the elastic modulus E_M and the pull-off strength f_p declared by the manufacturer [77].

3.3 Tensile properties of the EQ-GRID system

The mechanical properties of the EQ-GRID system were obtained from direct tensile tests performed on composite specimens. Two different test setups were adopted (Figure 3.4).

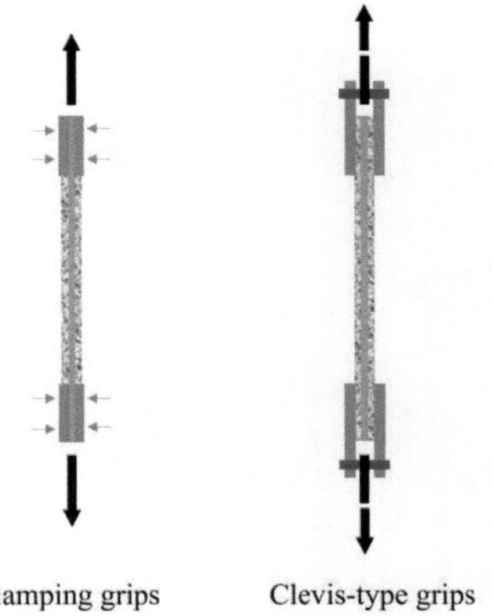

Figure 3.4: Direct tensile tests on composite coupons: scheme of the test setups

In the first one, the samples were clamped in the wedge grips of the testing machine. A lateral pressure of 5 bar was applied to avoid sliding the specimen in the load introduction area (Figure 3.4). On the contrary, the axial load was transferred by adhesion using the clevis-type grips (Figure 3.4). In this case, sanded aluminium tabs connected to the grips were glued with epoxy resin to the ends of the composite specimens. This gripping mechanism is suggested by Annex A of the American standard AC434 [53].

Therefore, thirty direct tensile tests were carried out monotonically under displacement control with a constant rate of 0.5 mm/min at ambient laboratory conditions. An MTS electro-hydraulic universal testing machine equipped with an MTS 100 kN load cell was used. The first fifteen tests were performed adopting the clamping grips and the second fifteen with the clevis-type. Therefore, ten samples were tested for each main direction of the grid (five per test setup). The vertical displacement and the tensile load were recorded.

3.3.1 Direct tensile tests with clamping grips

This test method aims to simulate a strengthening application in which the grid is mechanically anchored at its ends. The expected failure is the tensile breaking of the fibres. Some constructive details, such as connectors or turning the grid in the corners of the building (Figure 3.5), can reproduce a mechanical anchorage.

Figure 3.5: Examples of construction details for the application of the EQ-GRID system [73]

The preparation's stages of the coupons are illustrated in Figure 3.6. The first matrix layer (3 mm thick) was rolled out on the wood formwork. Then, grid strips with dimensions 50x440 mm and 85x440 mm were embedded in the still fresh mortar for a length of

300 mm, leaving free their ends for 120 mm. The application was completed by covering the textile strips with the second layer of matrix (5 mm thick, wet on wet application). The total thickness h_s of the coupons was 8 mm.

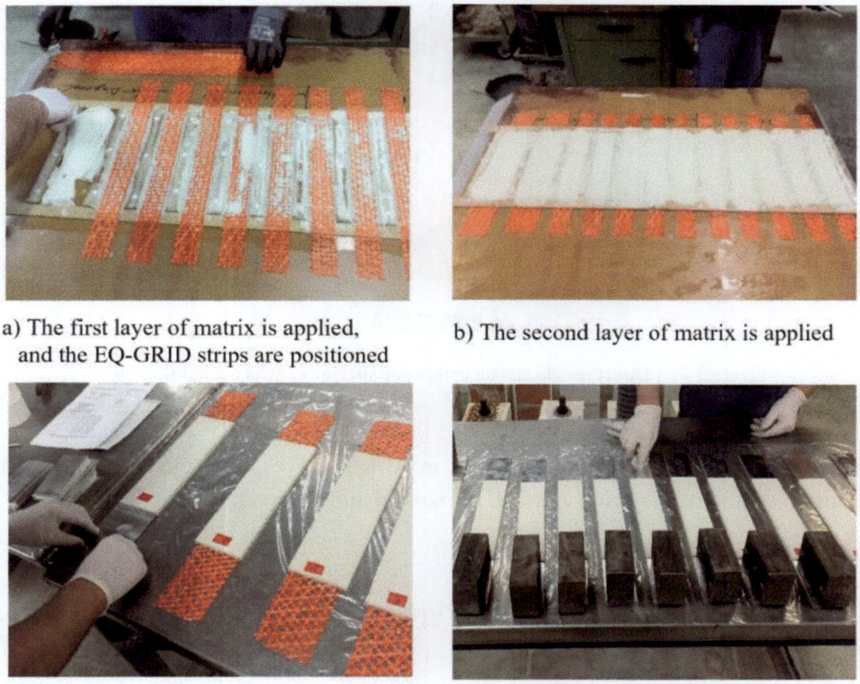

a) The first layer of matrix is applied, and the EQ-GRID strips are positioned

b) The second layer of matrix is applied

c) The aluminium tabs are glued

d) The composite coupons with the aluminium tabs are ready

Figure 3.6: Preparation of the composite specimens for the tensile tests with clamping grips

Finally, the samples were conditioned for 28 days in the laboratory climate room for the maturation of the inorganic matrix (25°C, 55% humidity). Moreover, sanded aluminium tabs were glued on the free ends of the samples to ensure homogeneous stress distribution near the gripping area of the testing machine. It is worth specifying that only the grid was clamped during the tests, as the ultimate tensile stress and strain of the composite specimens may vary, depending on the pressure applied in the gripping areas [50].

Figure 3.7: Direct tensile test on composite specimens with clamping grips

The test results are depicted in Figure 3.8, in which the response curves are shown for each main direction of the grid (vertical, horizontal and diagonal).

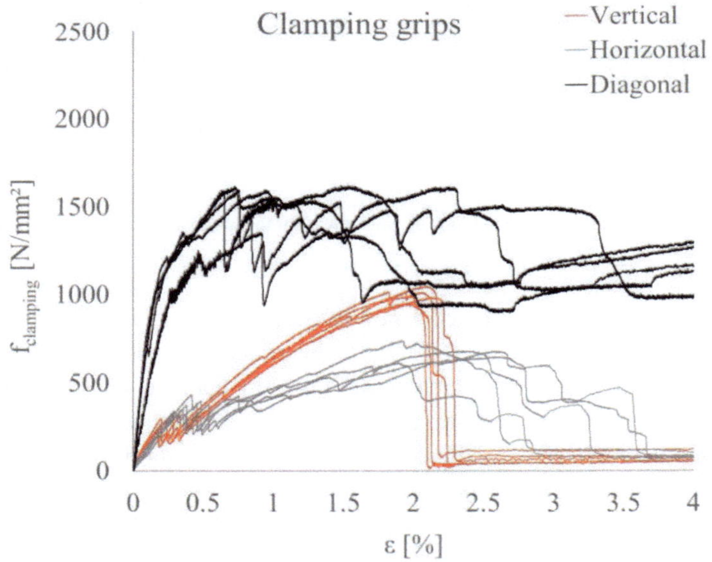

Figure 3.8: Experimental results of the direct tensile tests on EQ-GRID composite specimens with clamping-grips

The stresses are referred to the homogenized cross-section area of the bare textile A_{eq}. In this way, the results are not affected by any possible variations of the mortar cross-section. The strain is evaluated based on the global displacement measured between the wedge grips of the testing machine since no sliding occurred in the gripping areas.

As shown in Figure 3.7, the coupons were initially uncracked until the tensile strength of the mortar was reached (stage I). Therefore, the first cracking stress of the matrix $f_{cr,M}$ can be calculated:

$$f_{cr,M} = \frac{F_{cr,M}}{w_s \cdot h_s} \tag{3.10}$$

Where $F_{cr,m}$ is the first cracking load, w_s and h_s are respectively the samples' width and the thickness. Table 4 reports the mean matrix cracking stresses $f_{cr,M}$ for the three directions of the grid.

Table 4: First cracking stress of the matrix with clamping-grips

	Vertical	Horizontal	Diagonal
$f_{cr,M}$	1.00 N/mm²	1.34 N/mm²	0.96 N/mm²
Co. V.	15.8%	11.2%	5.7%

During the second test phase, cracks developed (stage II) and became wider. After that, the load was carried almost exclusively by the fibre wires up to failure (stage III). Therefore, the whole tensile response curve of the EQ-GRID system up to the fibres' breaking (Figure 3.9) could be obtained through the clamping-grip method.

Table 5: Mechanical properties of the composite specimens tested with clamping grips

Direction	$f_{clamping,cr}$	$\varepsilon_{clamping,cr}$	$f_{clamping,u}$	$\varepsilon_{clamping,u}$
Vertical	275 N/mm²	0.22%	996 N/mm²	2.0%
Co. V.	15.8%	22.1%	3.6%	4.5%
Horizontal	373 N/mm²	0.41%	673 N/mm²	2.3%
Co. V.	11.2%	12.5%	6.9%	15.4%
Diagonal	1261 N/mm²	0.37%	1591 N/mm²	2.1%
Co. V.	5.7%	24.0%	2.5%	27.6%

It is worth noting that, since the grid was embedded in the mortar, it was not free to strain as the bare textile. For this reason, the average ultimate stress and strain values, $f_{clamping,u}$

and $\varepsilon_{clamping,u}$, of the vertical and horizontal direction are lower than the ones of the bare textile (Table 2). On the contrary, the diagonal direction reached a tensile strength slightly higher than bare textile because of the horizontal wires' contribution. In this context, the horizontal fibres present in the diagonal composite specimens were also stressed for the following reasons:

- presence of the matrix, which transferred the tensile load to the fibre bundles after cracking;
- greater width of all the diagonal samples due to the greater aperture size of the grid in this direction (Figure 3.9);
- low inclination angle of the horizontal wires to the test direction (30°).

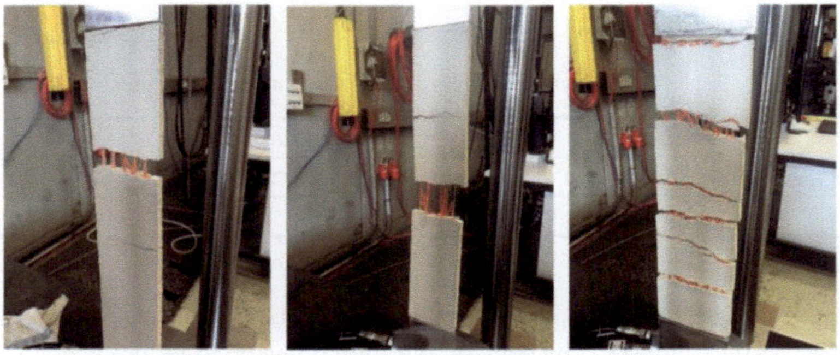

Vertical Horizontal Diagonal

Figure 3.9: Direct tensile tests on EQ-GRID composite specimens with clamping-grids

Finally, the average number of cracks and distance between these were equal to 2 and 10 cm for the samples in the vertical direction. The specimens in the horizontal direction developed 3 cracks with an average distance of 7.5 cm, and those in diagonal direction showed 2.4 cracks with a mean distance of 8.8 cm.

3.3.2 Direct tensile tests with clevis-type grips

Fifteen tensile tests were carried out on composite specimens with clevis-type grips (Figure 3.11) according to Annex A of the American standard AC434. This test method aims to simulate an FRCM application in which the grid is not anchored at its ends and whose response mainly depends on the bond between textile and matrix. Therefore, both tensile fibres' breaking or their slippage within the matrix are possible failures.

The preparation stages of the samples were the same as the first fifteen coupons, but in this case, the strips were completely embedded in the matrix for a length of 460 mm (no free ends of bare textile, Figure 3.10).

Therefore, the tensile load, the vertical displacement of the machine and LDTVs have been recorded. The results of the performed tensile tests are presented in terms of stress-strain curves in each main direction of the textile component (Figure 3.12).

Figure 3.10: Preparation of the EQ-grid composite specimens for the direct tensile tests with clevis-type grips

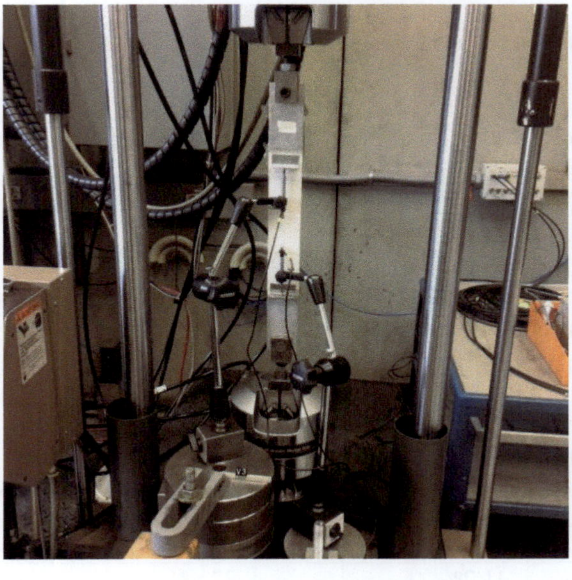

Figure 3.11: Direct tensile tests on EQ-GRID composite specimens with clevis-type grips

The load was transferred from the testing machine to the specimens by adhesion through the clevis-type gripping mechanism. Thus, the obtained tensile response curves are only influenced by the textile-matrix bond.

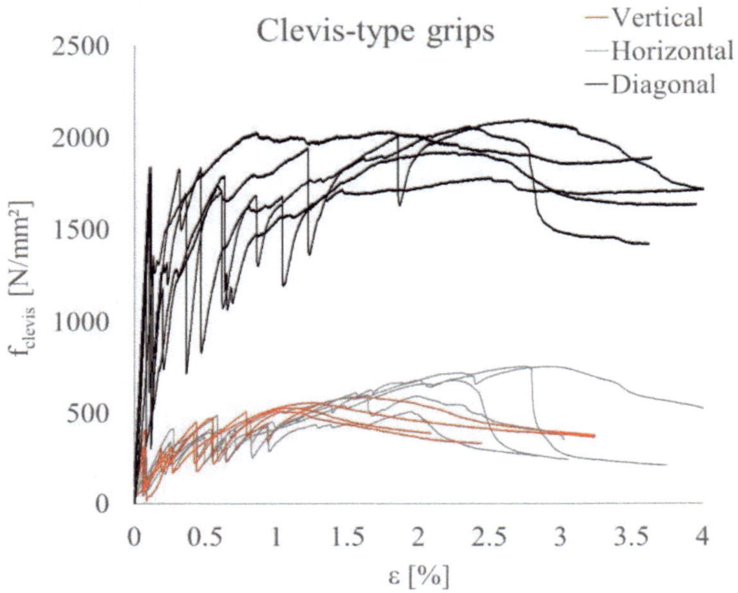

Figure 3.12: Results of the direct tensile tests on EQ-GRID composite specimens with clevis-type grips

As shown in Figure 3.12, the first phase of the system's response is linear elastic (stage I). The tensile load was mainly carried by the mortar, which remained uncracked until its tensile strength was reached.

The first cracking stress $f_{cr,M}$ of the mortar is calculated with Eq. (3.10), and it is reported in Table 6 for each direction of the grid. It is worth pointing out that the values of $f_{cr,M}$ in Table 6 are similar to each other and those obtained from direct tensile tests with clamping-grips (Table 4).

Table 6: First cracking stress of the mortar with clevis-type grips

	Vertical	Horizontal	Diagonal
$f_{cr,M}$	1.04 N/mm^2	0.99 N/mm^2	1.12 N/mm^2
Co.V.	41.2%	32.6%	17.7%

After stage I, cracks developed, and the load was gradually transferred from the matrix to the grid. Compared to the tensile tests with clamping grips, the cracking phase (stage II) was distributed over a wider strain range and the cracked phase (stage II) over a smaller one.

The specimens in the vertical direction presented an average number of cracks equal to 2.75. The distance between these was about 12.8 cm. Moreover, the samples in the horizontal direction showed 4 cracks with an average distance of about 8.2 cm and those in diagonal direction developed 3 cracks with a mean distance of 13.6 cm.

These results show that the horizontal direction has a better textile-matrix bond than the vertical one because of the higher number of cracks with lower distances. In this context, the horizontal AR glass wires are less smooth since the fibres cannot be uniformly bundled during the production process. Thus, the bond is improved, and the breaking of some threads could occur in two samples. As depicted in Figure 3.12, two curves of the horizontal direction are characterized by a strength drop after the peak stress.

Therefore, Table 7 reports the average first cracking stress $f_{clevis,cr}$, maximum strength $f_{clevis,u}$, and corresponding strains $\varepsilon_{clevis,cr}$ and $\varepsilon_{clevis,u}$.

Table 7: Mechanical properties of the EQ-GRID system tested with the clevis-type grips

Direction	$f_{clevis,cr}$	$\varepsilon_{clevis,cr}$	$f_{clevis,u}$	$\varepsilon_{clevis,u}$
Vertical	287.3 N/mm²	0.07%	551.3 N/mm²	1.2%
Co.V.	41.2%	12.8%	6.3%	20.1%
Horizontal	276.6 N/mm²	0.06%	665.9 N/mm²	2.4%
Co.V.	32.6%	26.9%	16.3%	18.5%
Diagonal	1482.7 N/mm²	0.11%	1843.9 N/mm²	2.5%
Co.V.	17.7%	18.4%	7.4%	9.3%

Also, in this case, the average diagonal tensile strength is higher than the bare textile since the horizontal wires present in the composite specimens were also stressed. The diagonal coupons' width was greater than the other ones (95 mm versus 60 mm), and the horizontal wires' inclination was only 30° to the test direction. Moreover, the grid was not mechanically anchored, and the tensile load was transferred from the matrix to the textile component only by adhesion. Therefore, the horizontal glass wires could give a contribution even greater than that observed with the clamping grips. In fact, $f_{clevis,u}$ and $f_{clamping,u}$ are respectively equal to 1843.9 N/mm² and 1591 N/mm².

3.4 Double-lap shear-bond tests

Twelve double-lap shear tests were carried out to investigate the EQ-GRID system's bonding behaviour when applied to a masonry unit. The adopted support was the hollow clay brick "WZI Poroton 2 DF 0,9/12" (240x115x113 mm), which was cut into two equal parts. The strengthening system was applied on both sides for a width of 113 mm. As depicted in Figure 3.13, three bonding lengths were investigated (4 cm, 8 cm and 12 cm). The first matrix layer 3 mm thick was applied to the dry masonry unit. Then, the grid strips were embedded in the still fresh mortar. Finally, the second matrix layer, 5 mm thick, was applied (wet on wet application). In this way, the grid was completely covered by the mortar, and the total system's thickness was 8 mm.

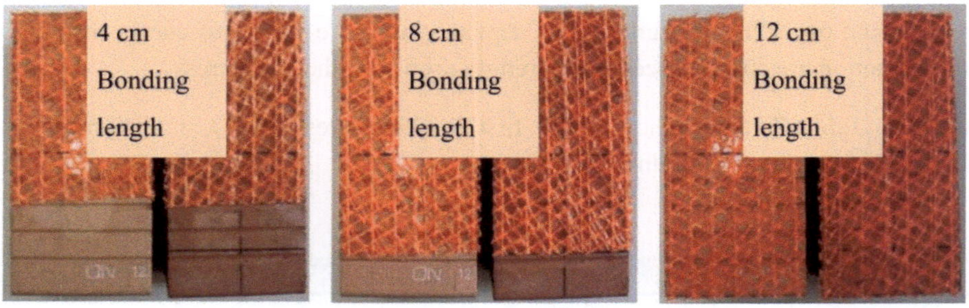

Figure 3.13: Double-lap shear-bond tests: bonding lengths

The experimental program included three tests for each main direction of the grid (Table 8). Three further tests were performed only for the specimens in the vertical direction, in which the diagonal fibre yarns were cut. Thus, their possible influence on the failure mode was also investigated.

The tests were carried out at ambient laboratory conditions with an MTS electro-hydraulic universal testing machine equipped with a load cell of 100 kN. They were performed under displacement control with a constant rate of 10 mm/min, and the tensile load was applied through steel plates bonded on the perforated surfaces of the masonry units with epoxy resin. The test setup is depicted in Figure 3.14. The tensile load and the vertical displacement of the machine were recorded.

The experimental results are expressed in terms of stress-displacement diagrams for each direction (Figure 3.15). The stresses are always referred to the homogenized equivalent cross-section of the bare textile. In this way, any possible variation of the matrix could not influence the results.

Table 8: Double-lap shear-bond tests: tested specimens

ID specimen	Masonry unit	Bonding length	Direction of the grid
V1	WZI Poroton 2 DF 0,9/12	4 cm	Vertical
V2		8 cm	
V3		12 cm	
V4		4 cm	
V5		8 cm	
V6		12 cm	
H1	WZI Poroton 2 DF 0,9/12	4 cm	Horizontal
H2		8 cm	
H3		12 cm	
D1	WZI Poroton 2 DF 0,9/12	4 cm	Diagonal
D2		8 cm	
D3		12 cm	

Figure 3.14: Double-lap shear-bond tests

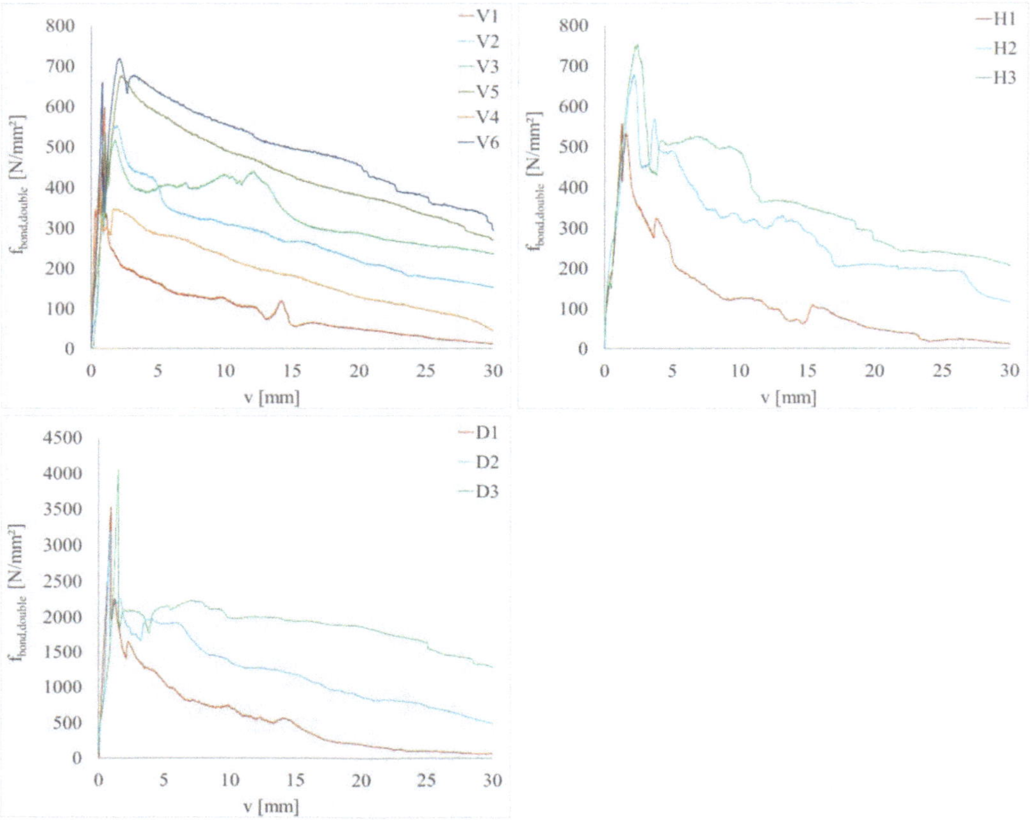

Figure 3.15: Double-lap shear-bond tests: stress - vertical displacement curves

Table 9 reports the maximum shear-bond stress $f_{bond,double,max}$ and the corresponding vertical displacement v_{max} achieved by each specimen.

It is worth noting that the strength attained by the V1, V2, V3 specimens is almost constant, and the post-peak values improve with increasing bonding length (Figure 3.15). Furthermore, the diagonal wires of the V4, V5, and V6 samples were cut, and their possible contribution was excluded. However, only the V4 specimen reached a lower shear-bond strength. The V5 and V6 samples attained higher values. The diagonal fibres could not substantially contribute because of the small samples' size and their inclination to the vertical (60°). All six specimens were characterized by the slippage of the fibres within the mortar, confirming the strong influence of the bond at the grid-matrix interface on the system's strength.

Table 9: Results of the double-lap shear-bond tests

ID	$f_{bond,double,max}$ [N/mm²]	v_{max} [mm]	Bonding length
V1	598.9	1.00	4 cm
V2	554.4	1.90	8 cm
V3	518.2	1.78	12 cm
V4	429.9	0.58	4 cm
V5	675.6	2.29	8 cm
V6	719.3	2.06	12 cm
H1	557.2	1.27	4 cm
H2	678.5	2.14	8 cm
H3	753.2	2.42	12 cm
D1	3529.9	0.93	4 cm
D2	3207.3	0.82	8 cm
D3	4060.0	1.50	12 cm

Furthermore, the specimens in the horizontal direction, H1, H2, H3, reached strength values similar to those obtained from the tensile tests with clevis-type grips. In fact, the mean shear bond strength is equal to 663 N/mm². The breaking of some glass wires characterized the failure mode. Therefore, compared to the vertical direction, the grid-matrix bond is higher in the horizontal one.

Finally, the samples D1, D2, D3 attained strength values greater than those obtained from the direct tensile tests with clevis-type grips. The load was also transferred from the matrix to the grid by adhesion, but the specimens' width was greater than the composite coupons (113 mm versus 95 mm). Therefore, the strength contribution of the horizontal yarns was even higher than in the direct tensile tests. Furthermore, the stress-displacement curves of the diagonal direction show a sudden strength drop after the peak (Figure 3.15), which corresponds to the breaking of some glass wires. Thus, the high bond of the grid with the matrix could also be confirmed for the diagonal direction.

It is important to stress that no debonding at the system-brick interface was observed in all the double-lap shear-bond tests. Since the mortar can penetrate the openings of the fibre mesh very well and has a very high adhesive resistance, only two failure modes were observed: slippage or partial wires' breaking.

3.5 Single-lap shear-bond tests

Five single-lap shear-bond tests were carried out for the vertical direction of the EQ-GRID system. As depicted in Figure 3.16, the textile strips (50x500 mm) were applied on a clay brick with dimensions of 240x115x71 mm, and the bonding length was 200 mm.

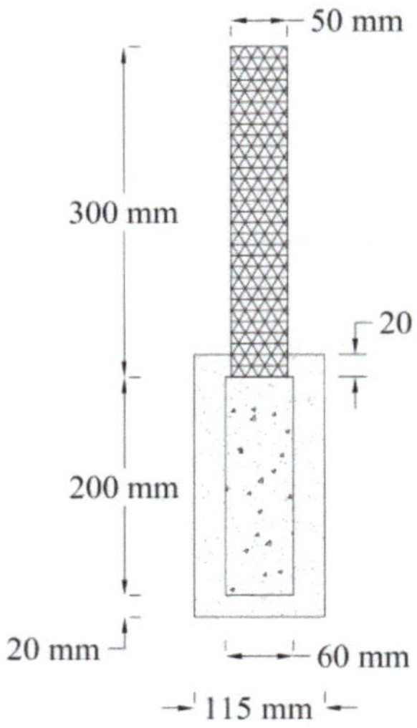

Figure 3.16: Dimensions of the specimens prepared for the single-lap shear-bond tests

The application's phases are illustrated in Figure 3.17. The first matrix layer was applied to the masonry unit. Then, the grid strip was embedded with the aid of a steel trowel in this still fresh layer. Finally, the application was completed with the second matrix layer.

Figure 3.17: Preparation's phases of the samples for the single-lap shear-bond tests

The specimens were left for 28 days at ambient laboratory conditions. Since the free part of the grid strip was clamped between wedge grips of the testing machine, aluminium-sanded tabs glued with epoxy resin were also adopted. Therefore, the samples were tested under displacement control with a constant rate of 0.5 mm/min using an MTS electro-hydraulic universal testing machine equipped with a load cell of 100 kN. In the adopted test setup (Figure 3.18), the masonry unit was restrained between two steel plates connected through four steel bars and the bottom plate was gripped to the testing machine. The displacement and the applied load were recorded continuously during the test. One LVDT reacted against a thin aluminium plate glued to the bare textile immediately outside the bonded length.

Figure 3.18: Single-lap shear bond tests

The results are expressed in terms of strength-displacement diagrams (Figure 3.19). The stresses are calculated considering the homogenized equivalent cross-sectional area of the bare textile. Four samples failed by fibre's breaking. Only the "Shear V2" sample showed the slippage of the grid into the matrix. After the peak, the curve of this specimen decreases progressively and doesn't have any sudden strength drop. The diagram depicted in Figure 3.19 confirms this behaviour.

Moreover, no debonding at the system-brick interface or detachment of the grid from the mortar layers was observed. The average maximum shear-bond strength $f_{bond,single,m}$ is equal to 971.2 N/mm² (Co.V. = 4.1%) at a mean vertical displacement value of 1.86 mm (Co.V. = 18.8%). It is worth noting that the average tensile strength obtained from the direct tensile tests on composite specimens with clamping grips is equal to 996 N/mm² and the width of the composite part was the same (60 mm). This result confirms that the fibre's breaking was the failure mode that influenced the shear-bond strength in the adopted test setup.

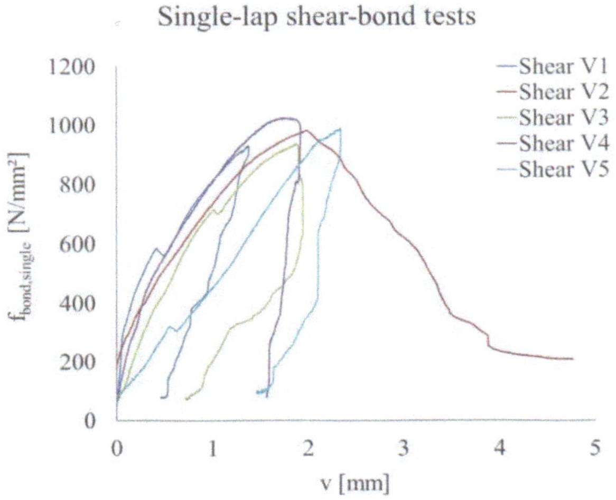

Figure 3.19: Stress-displacement diagrams obtained from the single-lap shear-bond tests

3.6 Cyclic lateral shear tests

The cyclic lateral shear tests allow evaluating the basic parameters of the seismic response of masonry panels, such as in-plane strength, displacement capacity and ductility. No technical standard exists for this type of test. Only a guideline of the RILEM technical committee gives indications [78].

For the foreseen experimental program, twelve masonry specimens (125 x 125 x 24 cm) with ten courses of bricks (Figure 3.20) were prepared. Four samples were in the original condition, and eight were reinforced with the application of the EQ-GRID system. Table 10 summarizes the specimens' characteristics.

Two types of masonry units were adopted: calcium-silicate and hollow clay bricks. The first was the "KS Heidelberger 4DF 20/2,0" (24 x 24 x 11.3 cm) with a 1.4% hole pattern. The second was the "WZI Poroton 3DF 0,9/12" (24 x 17,5 x 11.3 cm) with 34% rectangular perforation.

The mortar mix used to produce the specimens had a weight ratio of hydraulic lime CL 90, sand, cement and water equal to 1.1 kg : 1.2 kg : 14.9 kg : 3 kg. The mechanical properties were obtained through bending and uniaxial compression tests, according to

DIN EN 196-1. Therefore, six mortar prisms 40 x 40 x 160 mm were prepared and subjected to bending tests after 28 days of maturation. Then, the two halves of each specimen obtained from the bending tests were tested under uniaxial compression, as suggested by DIN EN 196-1. This method allowed performing compression tests on twelve mortar cubes 40 x 40 x 40 mm. Thus, the resulted mean values of the mortar flexural tensile strength $f_{M,t}$ and compressive strength f_M are shown in Table 10.

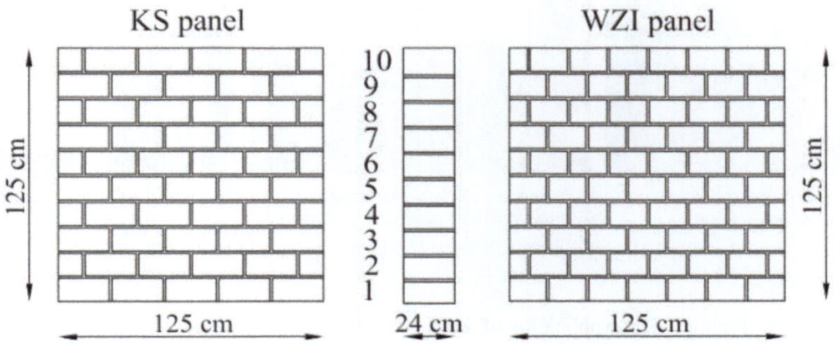

Figure 3.20: Geometry of the tested masonry walls

Table 10: Tested masonry specimens

ID	Masonry Unit	Mortar	EQ-GRID	Vertical load
KS_0_1	KS Heidelberger 4DF 20/2.0 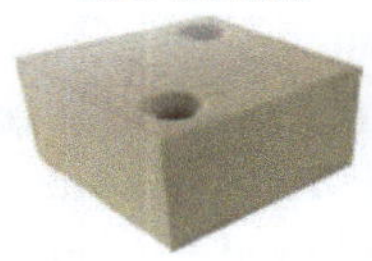	t_M = 12 mm f_M = 1.8 N/mm² $f_{M,t}$ = 0.58 N/mm²	-	240 kN
KS_0_2			-	120 kN
KS_1_1			On one side	240 kN
KS_1_2			On one side	120 kN
KS_2_1			On both sides	120 kN
KS_2_2			On both sides	120 kN
WZI_0_1	WZI Poroton 3DF 0.9/12 	t_M = 12 mm f_M = 1.8 N/mm² $f_{M,t}$ = 0.58 N/mm²	-	240 kN
WZI_0_2			-	120 kN
WZI_1_1			On one side	240 kN
WZI_1_2			On one side	120 kN
WZI_2_1			On both sides	120 kN
WZI_2_2			On both sides	120 kN

Finally, the preparation phases of the strengthened panels are illustrated in Figure 3.21. After applying the first matrix layer, the grid was embedded in the fresh mortar with the aid of a steel trowel. Then, a second layer of the same mixture was applied, covering the grid completely (wet on wet application). The total thickness of the finished reinforcement was about 8 mm.

a) Application of the first layer of matrix on the masonry panel

b) Positioning of the textile component of the EQ-GRID system and rolling out of the second layer of matrix

c) Applied system

Figure 3.21: Application of the EQ-GRID system on a masonry wall

3.6.1 Test setup of the cyclic lateral shear tests

The specimens were tested at the Karlsruhe Institute of Technology within the European research project "MULTITEXCO - High Performance Smart Multifunctional Technical Textiles for the Construction Sector". The adopted test setup is shown in Figure 3.22.

The first step of the testing procedure was to fix the panels to the existing structural floor. Then, a vertical hydraulic actuator of 500 kN with a relative load cell of 500 kN applied a constant force equal to 120kN or 240kN (Table 10) that was kept constant during the test. These values corresponded to applying a uniform compressive stress of 0.4 N/mm^2 (120 kN) and 0.8 N/mm^2 (240kN) at the top of the panels. In this respect, a load-distributing plate positioned on the top of the upper steel beam allowed the introduction of uniform normal stress in the specimens. Then, the cyclic displacements were applied through a horizontal hydraulic actuator of 1000 kN with a relative load cell of 1000 kN. The vertical and horizontal beam movements were decoupled thanks to a PTFE plate.

The experimental program provided for the application of alternated lateral displacements with increasing amplitude (1mm, 2mm, 4mm, 6mm, 8mm, 10mm, 12mm, 14mm, 16mm, 18mm, etc.) imposed quasi-statically up to the specimen collapse. The duration of each sinusoidal cycle was 120 seconds.

Figure 3.22: Cyclic lateral shear test: test setup

The displacements were introduced directly in the middle of the upper beam through a steel bolt. In this way, a symmetrical behaviour in compression as in the tensile was allowed. Furthermore, no mechanical fixing to the head beam was needed [79], and the deformation was transferred in the masonry specimen only by friction. Moreover, the panels were instrumented to get the most important experimental data. Forces and displacements of the vertical and horizontal cylinders were recorded. Two inductive displacement transducers (± 100 mm) positioned at the top of the specimens also measured the horizontal displacements. The free rotation of the head beam was prevented with four vertical threaded bars (two on the left and two on the right of the wall). For this reason, two inductive displacement transducers (± 100 mm) were placed vertically on the upper steel beam to record an eventual rotation during the tests. Each bar was pre-tensioned with a normal force of 9 kN, and the internal normal force was recorded with two load cells.

3.6.2 Interpretation of the experimental results of the cyclic shear tests

A common interpretation of the cyclic lateral shear tests' experimental data is to idealize the envelope of the hysteresis loops with a bilinear force-displacement curve, as shown in Figure 3.23.

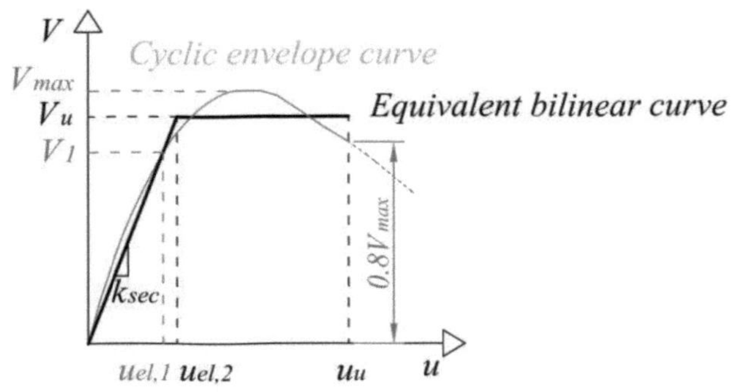

Figure 3.23: Cyclic envelope and equivalent bilinear curves

In this case, the definition of force-displacement relationships is relatively simple since the post-peak area is not taken into account completely. The first part of the equivalent response is linear elastic, and the secant stiffness k_{sec} can be calculated between zero and 70% of the maximum shear force V_{max} (V_1 in Figure 3.23). In this range of values, the panels are generally not cracked. Therefore, the behaviour can be approximated linear elastic [18]. Then, the ultimate displacement u_u is found on the cyclic envelope curve at a residual lateral strength equal to 80% of the maximum force ($0.8V_{max}$ in Figure 3.23). The ultimate shear force V_u is obtained through the equivalence of the energies dissipated from the experimental and the equivalent curve. It is worth pointing out that this interpretation is too conservative in the case of walls reinforced with the EQ-GRID system.

During the experimental campaign, it was observed that most of the strengthened specimens presented only slight damage at the displacement u_u or even no structural damage. The bilinear idealisation cannot completely consider the better crack pattern, the higher displacement capacity and residual strength after the peak of the reinforced panels. In this respect, the European Macroseismic Scale EMS-98 [80] defines five damage levels for masonry buildings under earthquake load, ranging from no structural damage to destruction. Thus, the cyclic envelope curves may be idealized through multilinear force-displacement relationships consistent with this phenomenological approach.

As shown in Table 11, five damage levels are identified for the tested masonry panels, and progressive strength drops are assigned by achieving the damage degrees DL3, DL4 and DL5. In this way, the post-peak structural behaviour is completely considered, and a more detailed equivalent response can be determined.

Therefore, starting from the linear elastic range of the multilinear curve (Figure 3.24), the secant stiffness k_{sec} is calculated between zero and V_1 ($V_1 = 0.7V_{max}$). Then, the horizontal displacements at DL3, DL4 and DL5, i.e. u_3, u_4, and u_5, can be identified at a defined percentage of residual lateral strength.

Table 11: Classification of damage for the tested masonry panels

DL1	No damage
DL2	Slight damage
DL3	Moderate damage
DL4	Heavy damage
DL5	Near collapse

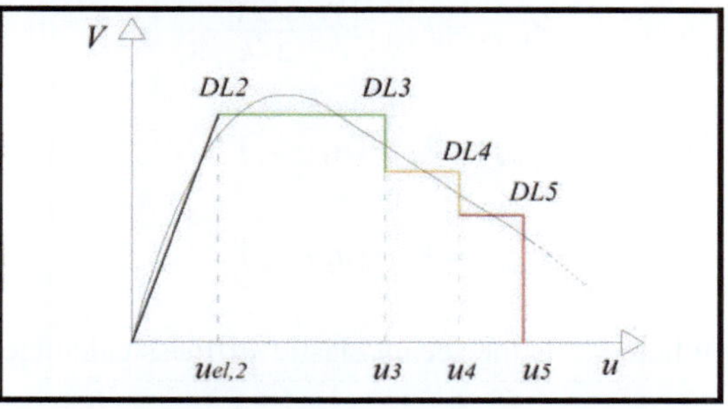

Table 12 shows the values chosen for the tested "KS" and "WZI" specimens based on the observed damage pattern.

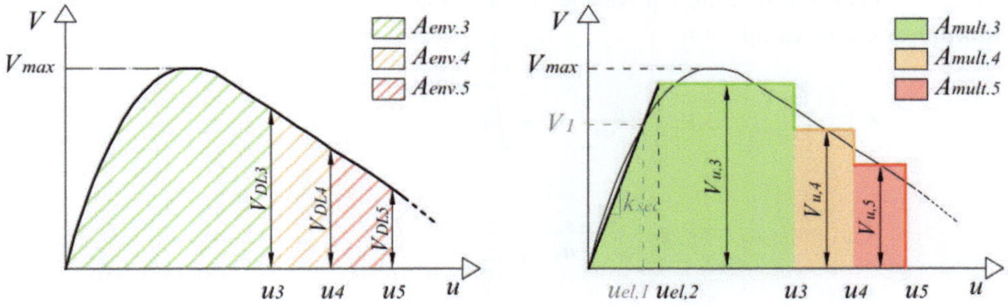

Figure 3.24: Cyclic envelope and equivalent multilinear curves

Table 12: Defined percentages of residual lateral strength V_{DL} at the DLs 3, 4 and 5

	EQ-GRID	V_{DL3}	V_{DL4}	V_{DL5}
"KS" specimens	-	$0.80V_{max}$	$0.60V_{max}$	$0.40V_{max}$
	on one side	$0.70V_{max}$	$0.50V_{max}$	$0.30V_{max}$
	on both sides	$0.70V_{max}$	$0.50V_{max}$	$0.25V_{max}$
"WZI" specimens	-	$0.80V_{max}$	$0.60V_{max}$	$0.40V_{max}$
	on one side	$0.80V_{max}$	$0.60V_{max}$	$0.40V_{max}$
	on both sides	$0.70V_{max}$	$0.50V_{max}$	$0.30V_{max}$

Furthermore, the ultimate shear forces $V_{u,3}$, $V_{u,4}$ and $V_{u,5}$ are calculated, ensuring the energy equivalence between the experimental and equivalent curves. To this aim, the area below the multilinear curve is divided into three parts, $A_{mult.3}$, $A_{mult.4}$ and $A_{mult.5}$, (Figure 3.24):

$$A_{mult.3} = V_{u,3} \cdot u_3 - \frac{V_{u,3}^2}{2 \cdot k_{sec}} \tag{3.11}$$

$$A_{mult.4} = V_{u,4} \cdot (u_4 - u_3) \tag{3.12}$$

$$A_{mult.5} = V_{u,5} \cdot (u_5 - u_4) \tag{3.13}$$

Where k_{sec} is the secant elastic stiffness calculated through Eq. (3.14):

$$k_{sec} = \frac{V_1}{u_{el,1}} \tag{3.14}$$

V_1 is the elastic shear force equal to 70% of V_{max} and $u_{el,1}$ is the corresponding experimental displacement value (Figure 3.24). Thus, the ultimate shear forces $V_{u,3}$, $V_{u,4}$ and $V_{u,5}$ can be obtained through the equivalence of the areas below the cyclic envelope and the multilinear curve, as follows:

$$A_{env.3} = A_{mult.3} \Rightarrow V_{u,3} = k_{sec} \cdot \left(u_3 - \sqrt{u_3^2 - \frac{2 \cdot A_{env.3}}{k_{sec}}} \right) \tag{3.15}$$

$$A_{env.4} = A_{mult.4} \Rightarrow V_{u,4} = \frac{A_{env.4}}{u_4 - u_3} \tag{3.16}$$

$$A_{env.5} = A_{mult.5} \Rightarrow V_{u,5} = \frac{A_{env.5}}{u_5 - u_4} \tag{3.17}$$

It is important to stress that each applied horizontal displacement was cyclically imposed three times during the tests. For this reason, six backbones and six equivalent curves are calculated for each specimen. Since this interpretation aims to identify just one equivalent curve, the average value of the ultimate shear forces $V_{u,3}$, $V_{u,4}$, $V_{u,5}$ is considered. On the contrary, the minimum one is chosen for the ultimate displacements u_3, u_4, and u_5. Therefore, the ductility μ reached at each DL can be determined with Eq. (3.18):

$$\mu_{DL3} = \frac{u_3}{u_{el,2}}; \quad \mu_{DL4} = \frac{u_4}{u_{el,2}}; \quad \mu_{DL5} = \frac{u_5}{u_{el,2}} \tag{3.18}$$

Where the elastic limit displacement $u_{el,2}$ is calculated through Eq. (3.19):

$$u_{el,2} = \frac{V_{u,3}}{k_{\sec}} \tag{3.19}$$

Finally, the deformation capacity of the tested panels is expressed in the following in terms of drift. Since the rotation of the upper steel beam was negligible, the drift d is defined as the ratio between the horizontal displacement u and the height of the specimen h:

$$d_2 = \frac{u_{el,2}}{h}; \quad d_3 = \frac{u_3}{h}; \quad d_4 = \frac{u_4}{h}; \quad d_5 = \frac{u_5}{h} \tag{3.20}$$

3.6.3 Experimental results of the "KS" masonry panels

3.6.3.1 Unstrengthened "KS" specimens

The unstrengthened panels in calcium-silicate bricks were the "KS_0_1" and "KS_0_2". The samples were characterized by the same geometrical and mechanical properties. The first one was tested under a constant vertical force of 240 kN and the second one under a constant force of 120 kN. The hysteresis and the three envelope curves of the first specimen are depicted in Figure 3.25.

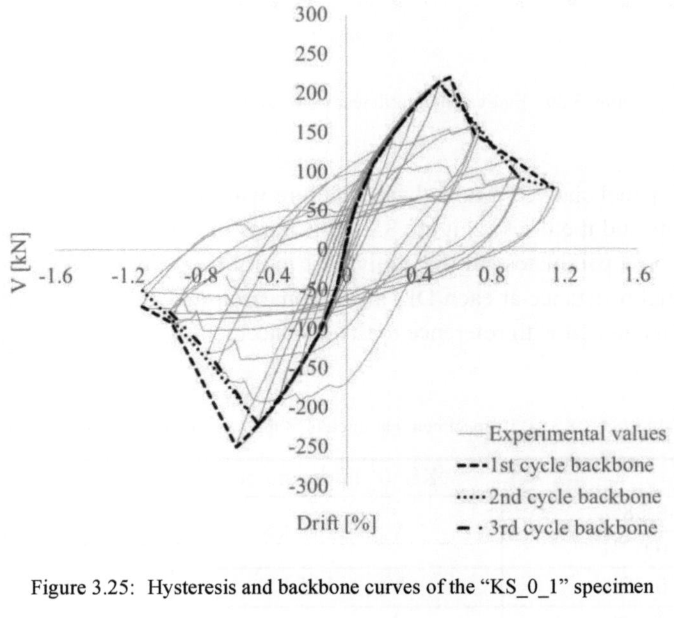

Figure 3.25: Hysteresis and backbone curves of the "KS_0_1" specimen

Nine displacement steps (from 1 mm to 16 mm) were applied. During the last one, at the end of the second cycle, the panel collapsed. The maximum horizontal force V_{max} recorded during the test was 250.6 kN at a horizontal displacement of 7.57 mm. The equivalent multilinear curve is shown in Figure 3.26. The displacement values u_3, u_4, u_5 are identified on the backbone curves at a residual lateral strength equal to 80%, 60% and 40% of V_{max}.

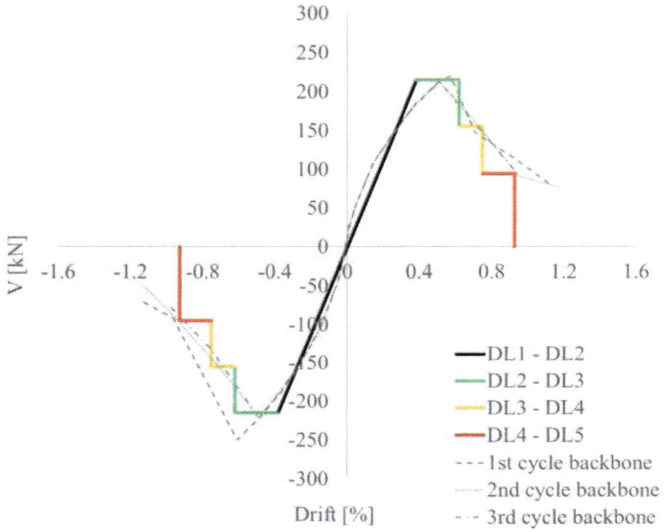

Figure 3.26: Equivalent multilinear curve of the "KS_0_1" specimen

The "KS_0_1" panel showed a mixed shear failure with diagonal cracks passing through the mortar joints and the bricks. Figure 3.27 illustrates the DLs reached by the specimen. Furthermore, some parameters of the equivalent multilinear curve, such as the values of drift and residual resistance at each DL, are summarized in Table 13, where the ductility μ_{DLi} is always calculated with reference to the displacement $u_{el,2}$.

Table 13: "KS_0_1" specimen: parameters of the equivalent multilinear curve

"KS_0_1" specimen		
d_2 = 0.38%	$V_{u,3}$ = 215 kN	μ_{DL3} = 1.6
d_3 = 0.62%		
d_4 = 0.75%	$V_{u,4}$ = 155 kN	μ_{DL4} = 2.0
d_5 = 0.93%	$V_{u,5}$ = 95 kN	μ_{DL5} = 2.4

DL3: side A

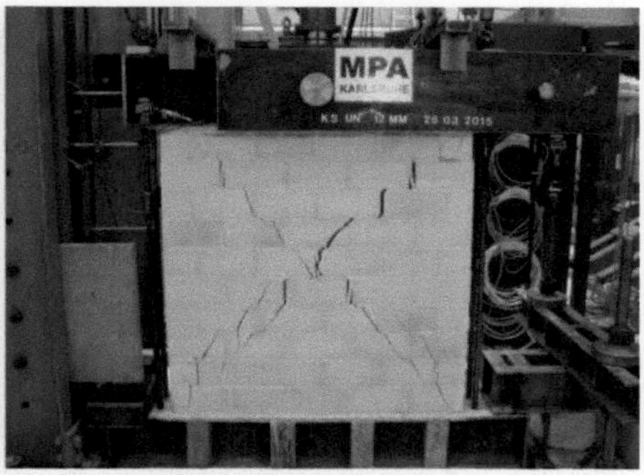

DL4: side A

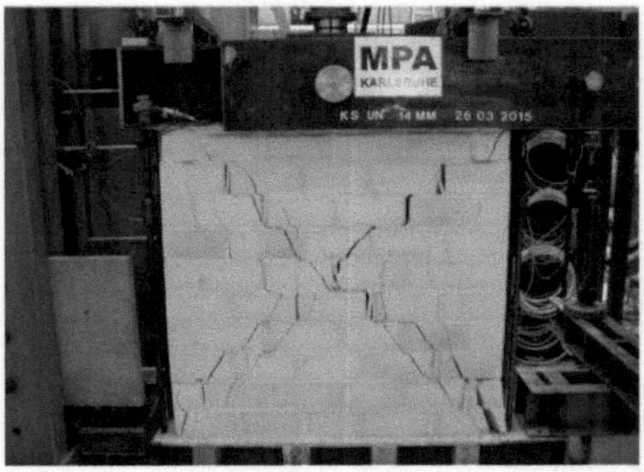

DL5: side A

Figure 3.27: Damage levels reached by the "KS_0_1" specimen

The "KS_0_2" specimen also failed by shear with diagonal cracks from corner to corner, but compared to the previous sample, they developed mainly through the mortar joints. Some inclined cracks occurred in the masonry units at the centre and compressed toes of the panel (Figure 3.30), areas in which a higher concentration of stress was provided.

The hysteresis and envelope curves are shown in Figure 3.28. Six displacement steps (from 1 mm to 10 mm) were applied. The maximum horizontal force V_{max} was equal to 158.58 kN, and it was recorded at a horizontal displacement of 7.44 mm. It is worth noting that, compared to the "KS_0_1" specimen, the value of V_{max} is about 100 kN lower since the second specimen showed "stepped-stairs" cracks due to the lower applied normal stress. On the contrary, the failure mechanism of the "KS_0_1" specimen was influenced by the achievement of the brick tensile strength. After attaining the maximum shear force, a resistance degradation occurred, and the test was stopped. Therefore, there was insufficient experimental data to calculate the equivalent multilinear curve, and only the bilinear one is determined (Figure 3.29).

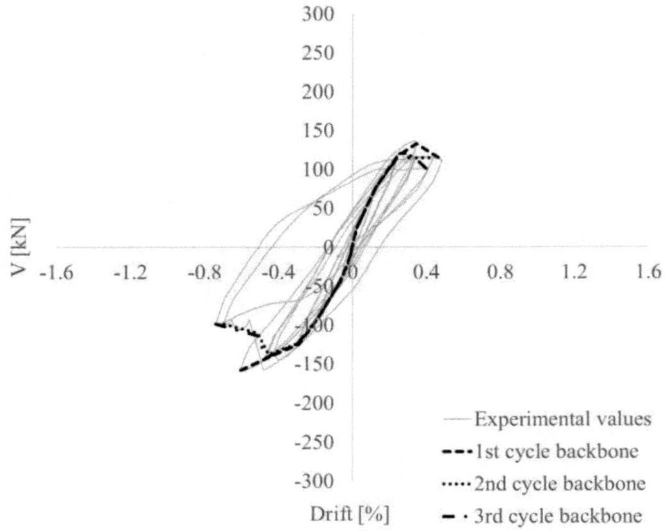

Figure 3.28: Hysteresis and envelope curves of the "KS_0_2" specimen

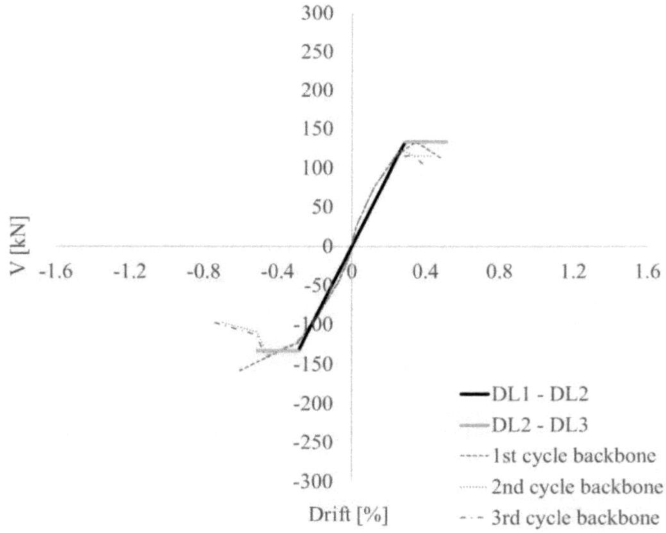

Figure 3.29: Equivalent curve of the "KS_0_2" specimen

Table 14: "KS_0_2" specimen: parameters of the equivalent bilinear curve

"KS_0_2"		
$d_2 = 0.29\%$	$V_{u,3} = 133.30$ kN	$\mu_{DL3} = 1.8$
$d_3 = 0.51\%$		

The final cracking pattern of the "KS_0_2" sample is shown in Figure 3.30, in which the "stepped-stairs" cracks are evident. Therefore, the parameters of the equivalent bilinear curve are indicated in Table 14. The displacement u_3 is identified on the backbone curves at a residual lateral strength equal to 80% of V_{max}.

Figure 3.30: "KS_0_2" specimen: cracking pattern at the end of the test

3.6.3.2 "KS" specimens strengthened on one side

The "KS" specimens strengthened on one side were the "KS_1_1" and "KS_1_2". They had the same geometrical and mechanical properties as the URM panels. The first one was tested under a constant vertical force of 240 kN and the second one under a constant value of 120 kN. The hysteresis loops with the three envelope curves of the "KS_1_1" sample are depicted in Figure 3.31.

Compared to the URM specimen, the "KS_1_1" panel showed higher lateral strength and displacement capacity. Ten displacement steps from 1 mm until 18 mm were applied, and the maximum attained shear force was 271.4 kN. This value was recorded at a horizontal displacement equal to 9.85 mm. Moreover, the residual strength measured during the third cycle of the last step was about 50 kN.

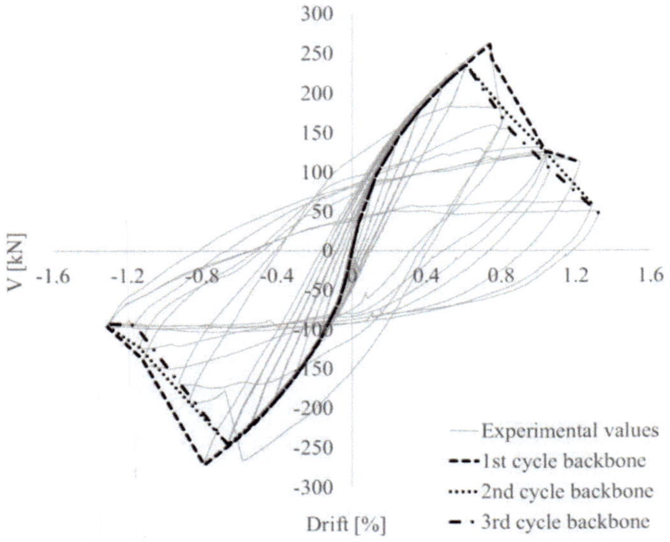

Figure 3.31: "KS_1_1" specimen: hysteresis and backbone curves

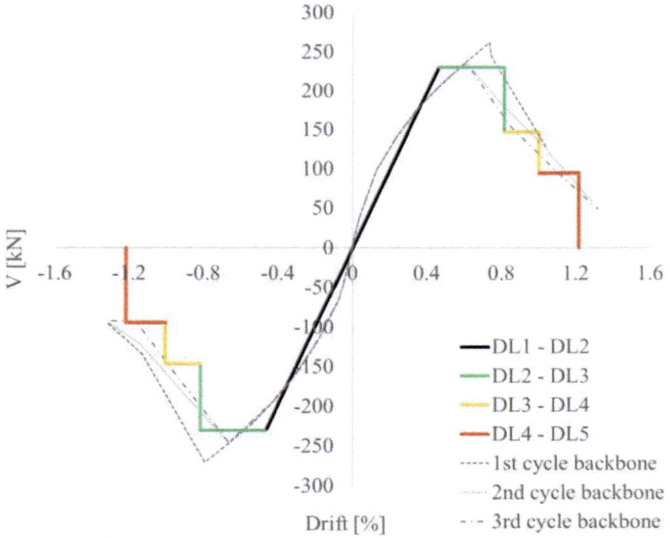

Figure 3.32: "KS_1_1" specimen: equivalent multilinear curve

The panel showed a mixed shear failure with diagonal cracks that developed along the mortar joints and the bricks, and no debonding of the EQ-GRID system was observed. The different states of damage are depicted in Figure 3.33

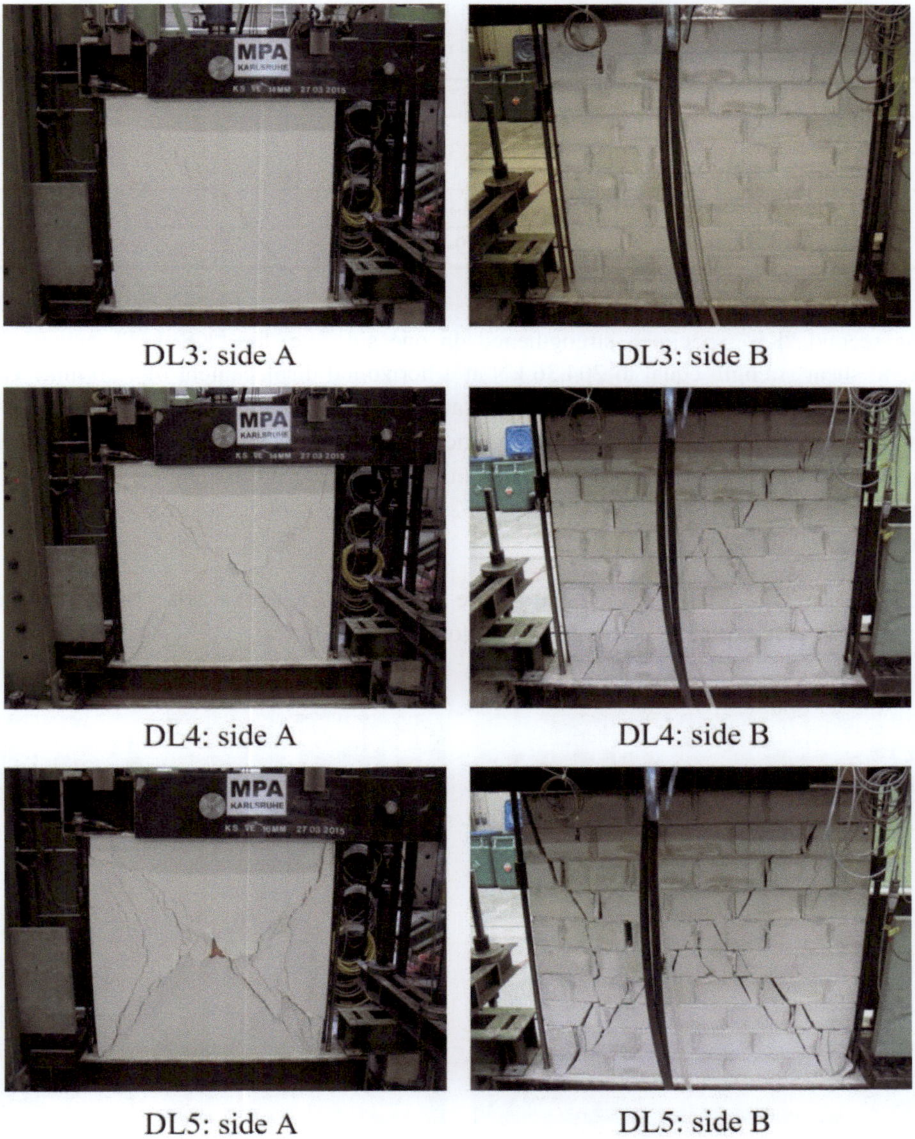

Figure 3.33: "KS_1_1" specimen: damage levels

Concerning the calculation of the equivalent multilinear curve, the horizontal displacements u_3, u_4, u_5 are identified at a residual lateral strength equal to 70%, 50% and 30% of V_{max}, since the EQ-GRID system improves the cracking pattern of the panel. Compared to the URM sample, the achievement of the DLs 3, 4 and 5 occurred at greater drift values.

Table 15: "KS_1_1" specimen: Values of drift and residual resistance for the different states of damage

"KS_1_1" specimen		
$d_2 = 0.46\%$	$V_{u,3} = 230.69$ kN	$\mu_{DL3} = 1.8$
$d_3 = 0.82\%$		
$d_4 = 1.00\%$	$V_{u,4} = 146.35$ kN	$\mu_{DL4} = 2.2$
$d_5 = 1.22\%$	$V_{u,5} = 94.41$ kN	$\mu_{DL5} = 2.6$

The second "KS" specimen strengthened on one side, i.e. the "KS_1_2", achieved a lateral shear strength equal to 204.36 kN at a horizontal displacement of 7.50 mm. This resistance value is lower than the previous sample since the panel was subjected to lower normal stress. In this case, the constant vertical load was 120kN. Thus, the displacement capacity of the "KS_1_2" specimen was a little bit higher than the one of the "KS_1_1" panel. Eleven displacement steps, from 1 to 20 mm, were applied, and the damage mechanism was a little less brittle.

The final damage pattern is shown in Figure 3.34. It was characterized by a mixed shear failure with cracks mainly along the mortar joints.

Figure 3.34: Final cracking pattern of the "KS_1_2" specimen

Moreover, some masonry units were cracked in the higher stress concentration areas, i.e. the panel's centre and the compressed toes.

Since the test data in the post-peak phase were not completely reliable, only the first part of the hysteresis curve until the maximum shear force achievement is depicted in Figure 3.35.

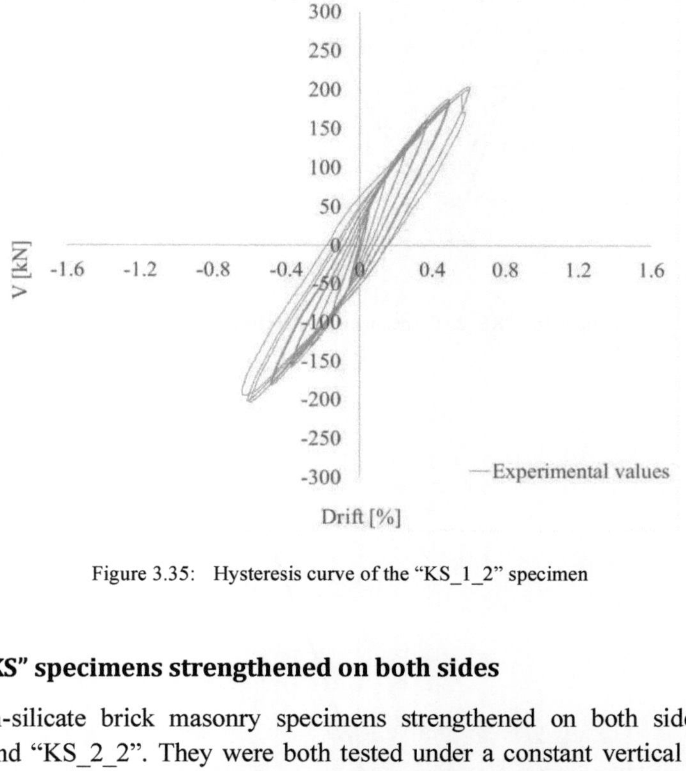

Figure 3.35: Hysteresis curve of the "KS_1_2" specimen

3.6.3.3 "KS" specimens strengthened on both sides

The calcium-silicate brick masonry specimens strengthened on both sides were the "KS_2_1" and "KS_2_2". They were both tested under a constant vertical load of 120 kN. The hysteresis and the backbone curves of the first sample are depicted in Figure 3.36. The maximum shear force reached from this specimen is equal to 271.5 kN, and it was recorded at a horizontal displacement of 12.78 mm. It is worth noting that, thanks to the EQ-GRID system, fourteen displacement steps (from 1 mm to 32 mm) were applied, and the resistance increased by 71% compared to the unreinforced sample "KS_0_2". Moreover, the maximum achieved drift is equal to 2.4%, and it was recorded during the first cycle of the last displacement step. The equivalent multilinear curve is shown in Figure 3.37. The horizontal displacements u_3, u_4, u_5 are defined on the backbones at a residual shear strength equal to 70%, 50% and 25% of V_{max}.

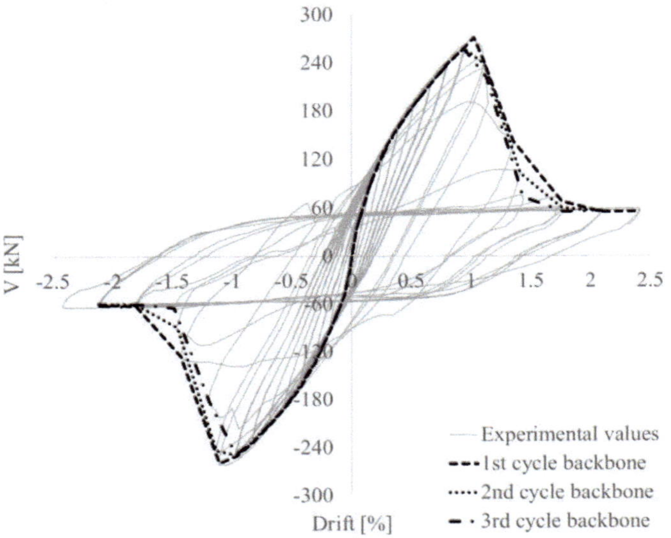

Figure 3.36: "KS_2_1" specimen: hysteresis and backbone curves

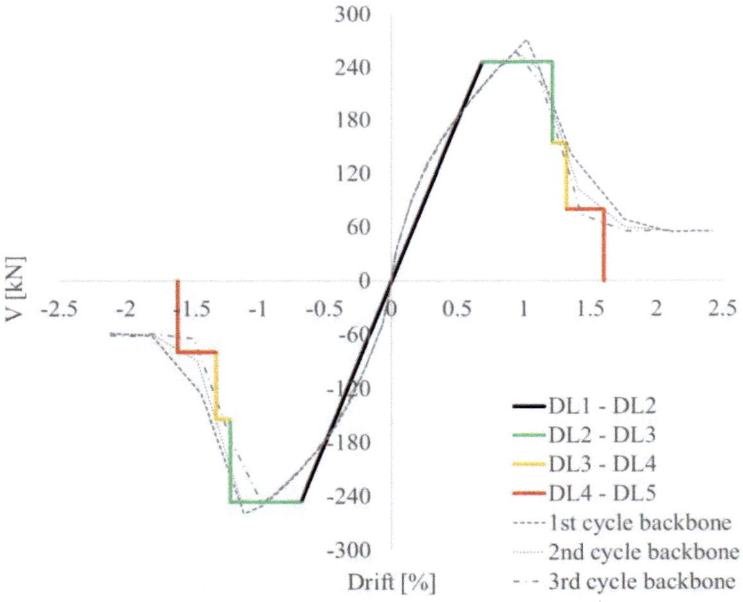

Figure 3.37: "KS_2_1" specimen: the equivalent multilinear curve

The "KS_2_1" specimen failed with diagonal cracks from corner to corner, and no debonding of the strengthening system was observed. Thanks to the high bond properties of EQ-GRID with the masonry support, the panel structure was held together by the system. Thus, larger displacements could be applied without losing the stability of the panel. Moreover, horizontal cracks occurred in the lower bed joint after the drift value of 1.8% (Figure 3.38), and large displacements with a relative constant strength of about 60 kN were measured. The panel began to show sliding behaviour. Therefore, the residual lateral strength was relatively constant in the latest steps.

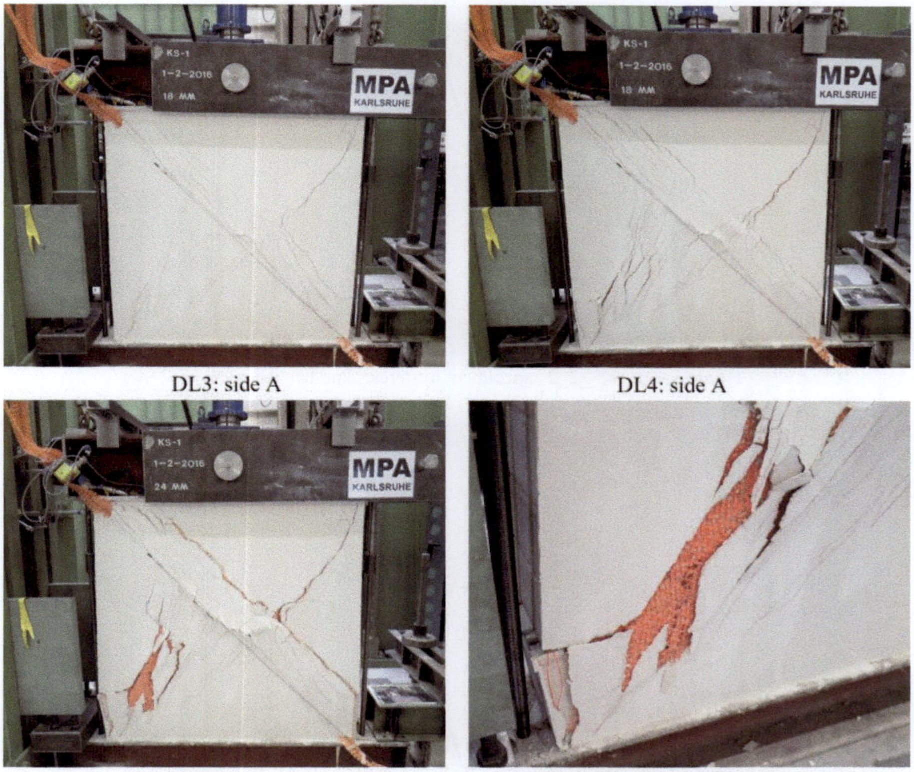

DL3: side A | DL4: side A

DL5: side A | DL5: bottom left corner of the side A

Figure 3.38: "KS_2_1" specimen: damage levels

The "KS_2_2" panel also failed with two diagonal cracks from corner to corner. The hysteresis and the envelope curves are depicted in Figure 3.39. The maximum shear force recorded during the test is equal to 252.5 kN, and it was measured at a horizontal displacement of 9.69 mm.

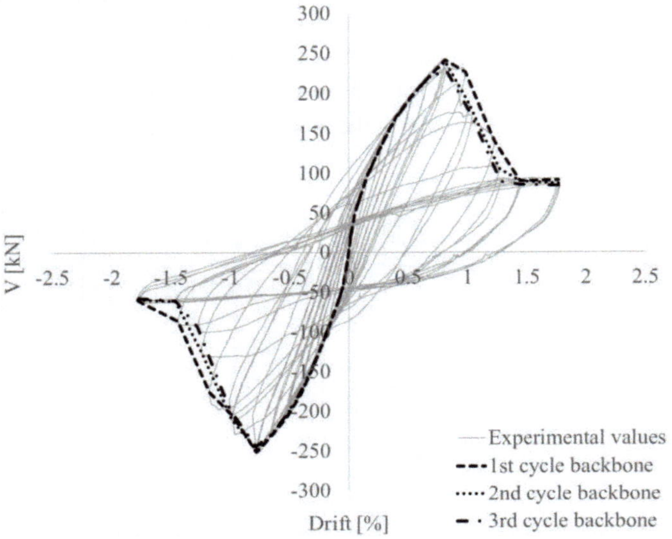

Figure 3.39: "KS_2_2" specimen: hysteresis and backbone curves

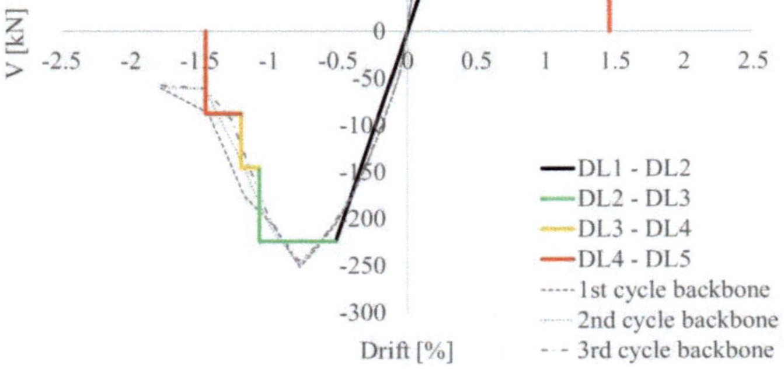

Figure 3.40: "KS_2_2" specimen: the equivalent multilinear curve

Compared to the URM specimen "KS_0_2", the resistance increased by about 60%. It is worth noting that the hysteresis loops appear to be non-symmetrical in the last phase of the test, with positive strength values greater than the negative ones. The reason is to be found in the non-symmetrical damage that the specimen showed. As illustrated in Figure 3.41, three horizontal cracks developed in the left bottom corner during the latest displacement steps, and only one horizontal crack formed in the right bottom corner. Therefore, the panel showed a non-symmetrical sliding behaviour.

The equivalent multilinear curve is depicted in Figure 3.40. The horizontal displacements u_3, u_4, u_5 are defined on the backbones at a residual strength equal to 70%, 50%, 25% of V_{max}.

DL3: side A

DL4: side A

DL5: side A

Figure 3.41: "KS_2_2" specimen: damage levels

Finally, Table 16 reports the parameters of the multilinear curves for the "KS_2_1" and "KS_2_2" specimens.

Table 16: Values of drift and residual resistance for the "KS" specimens strengthened on both sides

"KS_2_1"			"KS_2_2"		
$d_2 = 0.70\%$	$V_{u,3} = 252$ kN	$\mu_{DL3} = 1.6$	$d_2 = 0.53\%$	$V_{u,3} = 231$ kN	$\mu_{DL3} = 1.9$
$d_3 = 1.14\%$			$d_3 = 0.99\%$		
$d_4 = 1.27\%$	$V_{u,4} = 179$ kN	$\mu_{DL4} = 1.8$	$d_4 = 1.14\%$	$V_{u,4} = 171$ kN	$\mu_{DL4} = 2.1$
$d_5 = 1.37\%$	$V_{u,5} = 127$ kN	$\mu_{DL5} = 2.0$	$d_5 = 1.26\%$	$V_{u,5} = 120$ kN	$\mu_{DL5} = 2.4$

It is worth pointing out that both panels reached the DL3 at drift values higher than the "KS_0_1" specimen at the DL5 ($d_5 = 0.93\%$). Compared to the URM sample, the average values of d_3, d_4 and d_5 increased by 72%, 61%, 41%.

3.6.4 Experimental results of the "WZI" masonry panels

3.6.4.1 "WZI" unstrengthened specimens

The "WZI_0_1" and "WZI_0_2" specimens were unstrengthened panels in hollow clay brick characterized by the same geometrical and mechanical properties. They were tested under a constant vertical force of 240 kN and 120 kN, respectively. Compared to the unreinforced "KS" panels, both "WZI" samples showed more brittle behaviour. In this regard, seven displacement steps, from 1 mm to 12 mm, were applied to the "WZI_0_1" specimen, which presented a very brittle behaviour with a sudden diagonal cracking up to the collapse. The specimen reached the DLs 3, 4 and 5, respectively, during the first, second and third cycle of the last displacement step. As shown in Figure 3.44, the panel collapsed completely, losing its ability to resist any further horizontal action.

Figure 3.42 illustrates the hysteresis and the backbone curves of this specimen. The maximum attained shear force is equal to 218.3 kN, and it was recorded when the first diagonal cracks appeared from the centre to the corners, mainly through the bricks. The horizontal displacement value at the peak shear force was 8.0 mm.

The equivalent multilinear curve is shown in Figure 3.43. Table 17 reports the drift and residual resistance values at the different DLs.

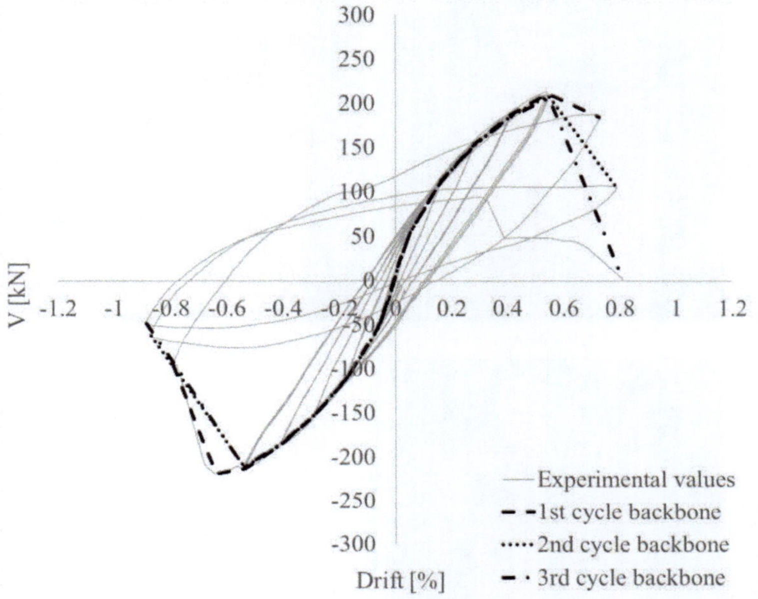

Figure 3.42: Hysteresis and backbone curves of the "WZI_0_1" specimen

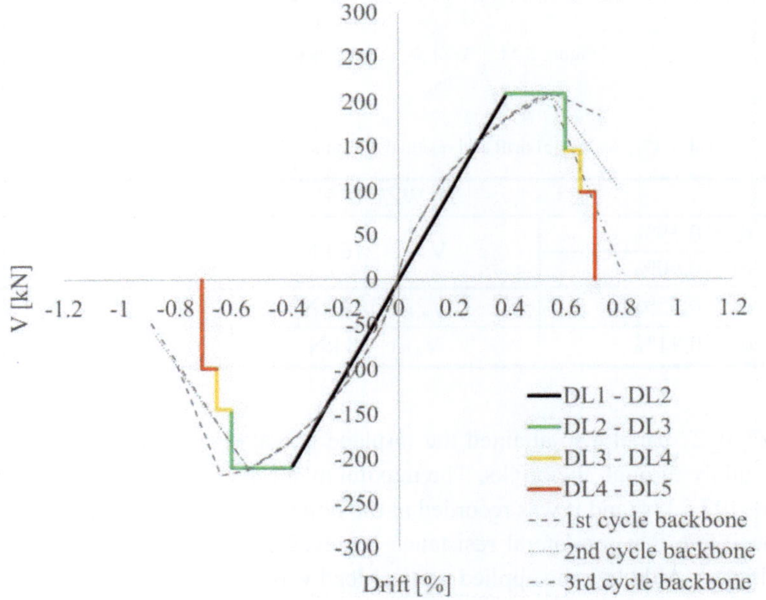

Figure 3.43: Equivalent multilinear curve of the "WZI_0_1" specimen

DL3: side A

DL4: side A

DL5: side A

Figure 3.44: "WZI_0_1" specimen: damage levels

Table 17: Values of drift and residual resistance for the "WZI_0_1" specimen

"WZI_0_1"		
$d_2 = 0.39\%$	$V_{u,3} = 216$ kN	$\mu_{DL3} = 1.5$
$d_3 = 0.60\%$		
$d_4 = 0.65\%$	$V_{u,4} = 144$ kN	$\mu_{DL4} = 1.7$
$d_5 = 0.71\%$	$V_{u,5} = 88$ kN	$\mu_{DL5} = 1.8$

The "WZI_0_2" panel also attained the in-plane lateral resistance with diagonal cracks forming mainly through the bricks. The maximum shear force reached by this specimen is equal to 197.5 kN, and it was recorded at the horizontal displacement of 7.37 mm. It is worth specifying that the lateral resistance achieved from the previous panel was only 20.8 kN higher. Although the applied vertical load was very different, both "WZI" panels presented diagonal cracking with tensile failure of the masonry units. For this reason, the strength values of these two URM specimens are not so different from each other.

However, the “WZI_0_2” panel showed a less brittle behaviour because of the lower constant vertical force. Eight displacement steps, from 1 mm to 14 mm, were applied. The sample collapsed during the first cycle of the last step. The final cracking pattern was characterized by the partial loss of the panel’s structure (Figure 3.45), as for the “WZI_0_1” specimen (Figure 3.44) but slightly less severe.

Figure 3.45: Final cracking pattern of the “WZI_0_2” specimen

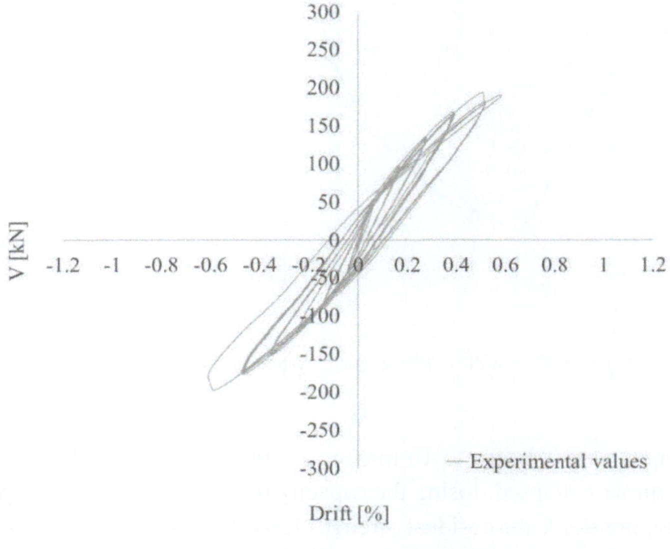

Figure 3.46: Hysteresis curve of the “WZI_0_2” specimen

Figure 3.46 shows the first part of the hysteresis curve since the experimental data in the post-peak phase were not completely reliable. Therefore, the equivalent multilinear curve is not determined in this case.

3.6.4.2 "WZI" specimens strengthened on one side

The "WZI" specimens strengthened on one side were the "WZI_1_1" and "WZI_1_2" panels. They had the same geometrical and mechanical properties. The first one was tested under a constant vertical load of 240 kN and the second one under a constant load of 120 kN. The in-plane lateral resistance achieved by the "WZI_1_1" specimen was equal to 251.45 kN, and it was recorded at a horizontal displacement value of 9.77 mm. The hysteresis and the three envelope curves of the "WZI_1_1" panel are depicted in Figure 3.47.

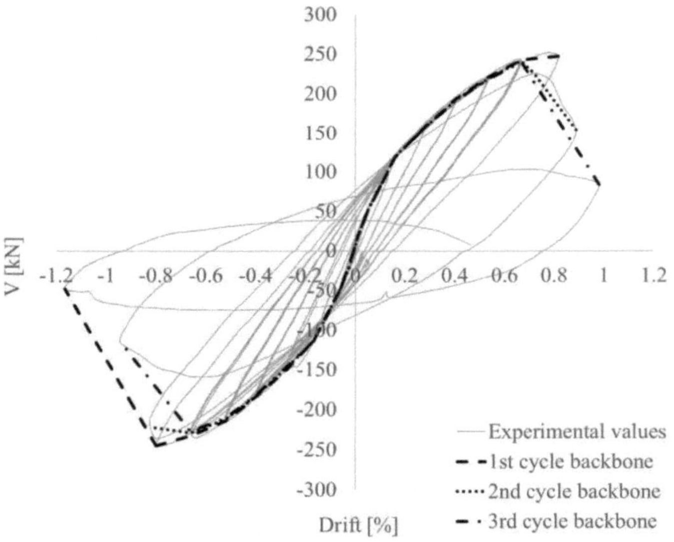

Figure 3.47: "WZI_1_1" specimen: hysteresis and backbone curves

Nine displacement steps (1 mm to 16 mm) were imposed. During the first cycle of the last step, the sample collapsed, losing the capacity to resist further horizontal actions. As illustrated in Figure 3.47, the residual strength tends to zero in the last hysteresis loop. However, the curve shows a greater energy dissipation than the unreinforced specimen "WZI_0_1" (Figure 3.42). The presence of the system on one side allowed reaching higher drift values and a better crack pattern at the same time. In this regard, the average maximum drift achieved by the "WZI_0_1" and "WZI_1_1" specimens were 0.81% and

0.96%. Moreover, the in-plane strength was also improved by the EQ-GRID system. The "WZI_0_1" sample recorded a value of 218.3 kN, while the reinforced specimen "WZI_1_1" attained a lateral resistance of 251.45 kN.

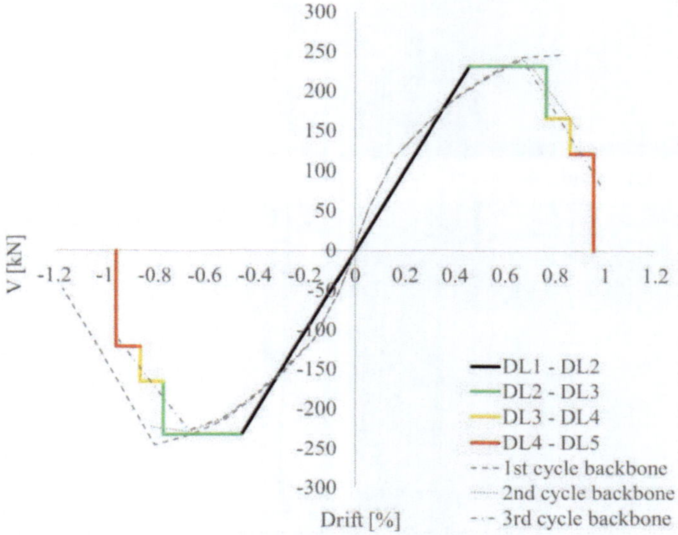

Figure 3.48: "WZI_1_1" specimen: the equivalent multilinear curve

Figure 3.48 illustrates the equivalent multilinear curve. The drifts and residual resistances are indicated in Table 18.

Table 18: Values of drift and residual resistance for the "WZI_1_1" specimen

"WZI_1_1"		
$d_2 = 0.46\%$	$V_{u,3} = 232$ kN	$\mu_{DL3} = 1.7$
$d_3 = 0.76\%$		
$d_4 = 0.86\%$	$V_{u,4} = 166$ kN	$\mu_{DL4} = 1.9$
$d_5 = 0.95\%$	$V_{u,5} = 121$ kN	$\mu_{DL5} = 2.1$

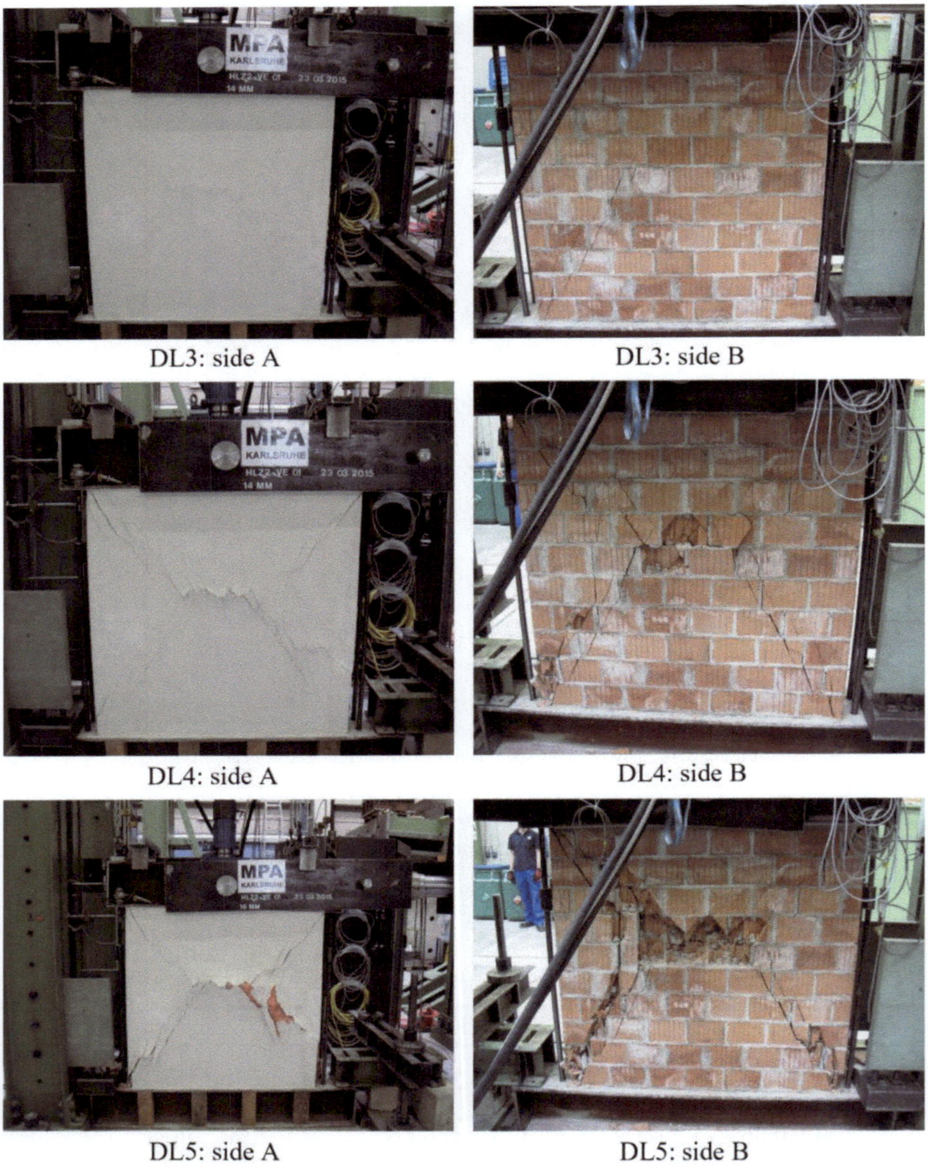

Figure 3.49: "WZI_1_1" specimen: damage levels

The DLs reached by the "WZI_1_1" sample are shown in Figure 3.49. Like the URM specimens, the failure was characterized by diagonal cracks passing mainly through the bricks. The breaking of some masonry units at the centre and the bottom corners of the panel was also evident. Therefore, the in-plane lateral resistance was influenced by the achievement of the brick tensile strength.

The failure of the second specimen strengthened on one side, i.e. the "WZI_1_2", was less brittle than the "WZI_1_1" sample since the applied vertical load was lower (120 kN). Compared to the unreinforced panel "WZI_0_2", the "WZI_1_2" specimen showed similar behaviour in terms of maximum shear force. The lateral resistance is equal to 195.7 kN, but the post-peak strength and displacement capacity are higher than the "WZI_0_2" panel. No debonding failure of the EQ-GRID system was observed. The adhesion of the mortar matrix to the masonry substrate was very strong. The strengthening system kept the panel's structure together, allowing dissipating more energy. Moreover, it improved the final cracking pattern since the damage state was less severe (Figure 3.50).

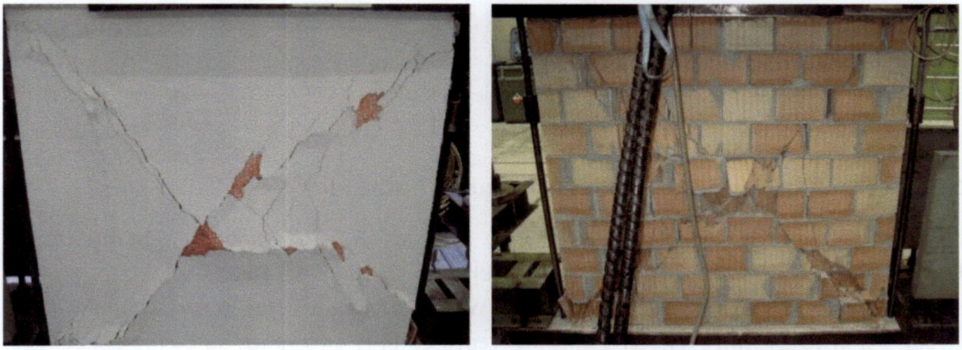

Figure 3.50: "WZI_1_2" specimen: cracking pattern at the end of the test

Compared to the "WZI_0_2" specimen, one more displacement step was applied for a total of eight steps (1 mm to 14 mm). The residual lateral strength was about 30 kN in the last cycle, like the URM sample, but it was recorded at a higher displacement value.

Finally, the experimental data were not completely reliable in the post-peak phase. Therefore, only the first part of the hysteresis loops is shown in Figure 3.51, and the equivalent multilinear curve is not determined in this case.

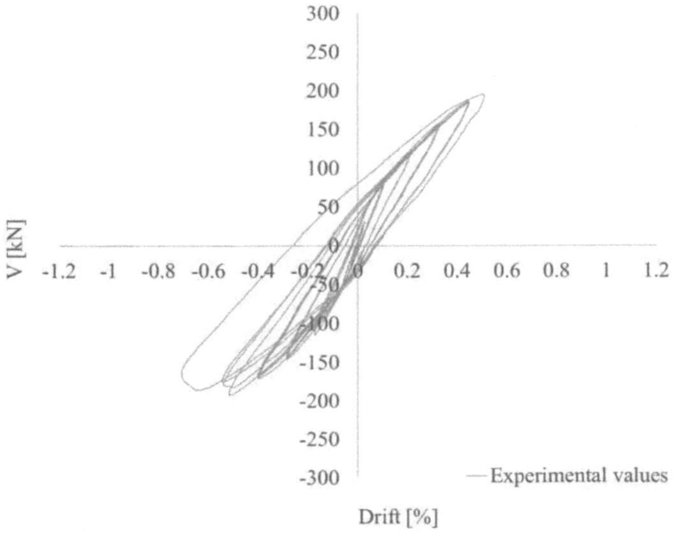

Figure 3.51: Hysteresis and backbone curves of the "WZI_1_2" specimen

3.6.4.3 "WZI" specimens strengthened on both sides

The "WZI" specimens strengthened on two sides were tested under a constant vertical load of 120 kN. Compared to the URM panels, the "WZI_2_1" and "WZI_2_2" samples showed higher dissipative behaviour. They were subjected to eleven displacement steps (from 1 mm to 24 mm). The last one was repeated only two times since the panels collapsed with a residual lateral strength equal to 36 kN and 32.5 kN, respectively. Moreover, the specimens showed good performance also in terms of load capacity. They attained a lateral strength of 280.47 kN and 255.33 kN at a horizontal displacement equal to 15.1 mm and 10.2 mm, respectively. Thus, compared to the "WZI_0_2" sample, the average improvement of load-bearing capacity is about 40%.

The hysteresis and the three envelope curves of the "WZI_2_1" panel are depicted in Figure 3.52. The maximum drift value achieved during the last hysteresis loop is 1.85%. The equivalent multilinear curve is shown in Figure 3.53. The horizontal displacements u_3, u_4, u_5 are defined on the backbones at a residual shear strength equal to 70%, 50% and 30% of V_{max}.

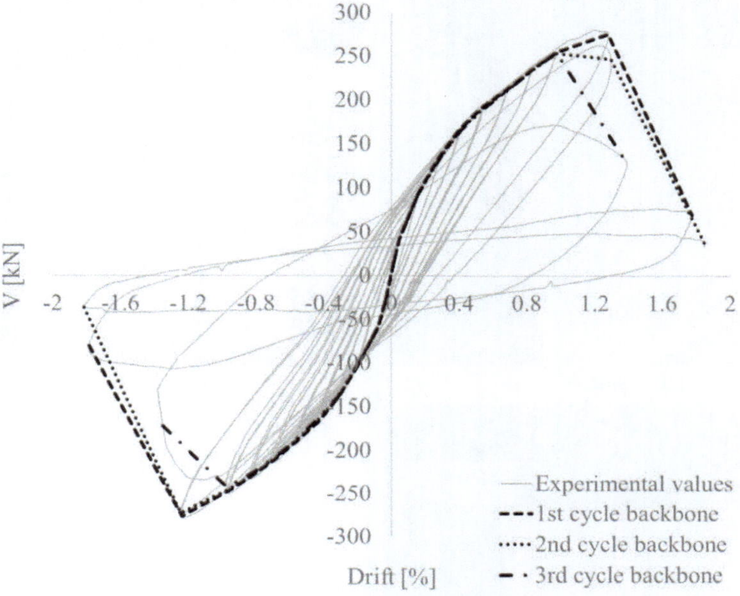

Figure 3.52: Hysteresis curve of the "WZI_2_1" specimen

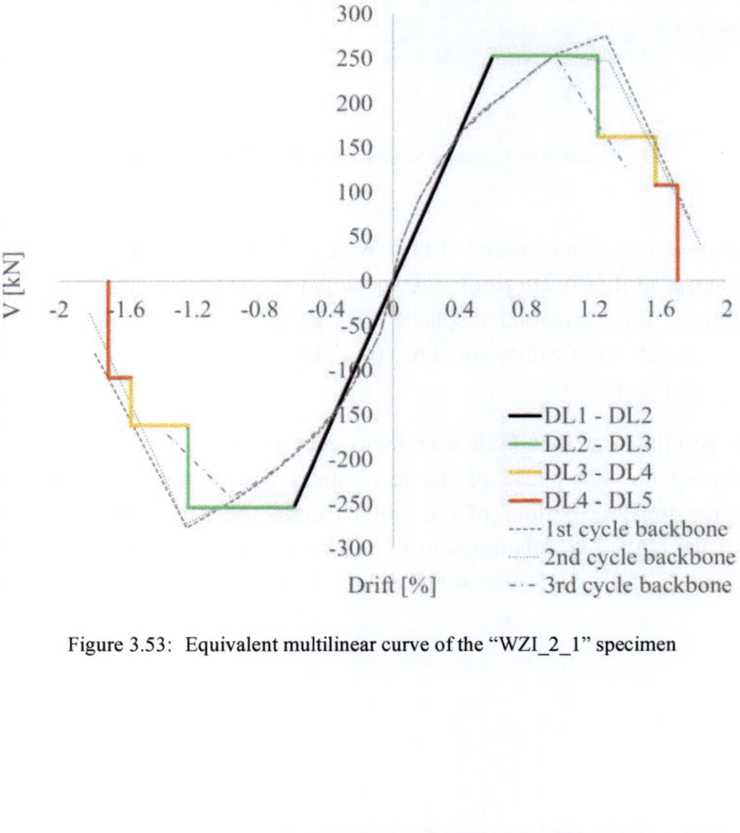

Figure 3.53: Equivalent multilinear curve of the "WZI_2_1" specimen

DL3: side A

DL4: side A

DL5: side A

Figure 3.54: Damage levels of the "WZI_2_1" specimen

The hysteresis and envelope curves of the "WZI_2_2" panel are depicted in Figure 3.55. The equivalent multilinear curve of this specimen is shown in Figure 3.56. As for the previous sample, the horizontal displacements u_3, u_4, u_5 are defined on the envelope curves at residual strength values equal to 70%, 50% and 30% of V_{max}. The achieved DLs are illustrated in Figure 3.57.

It is worth pointing out that both specimens presented sub-vertical cracks across the thickness during the last phase of the test. Stress concentrations due to the bending moment at the top and bottom of the panel caused the achievement of the masonry compressive strength. No delamination of the reinforcing system was observed. Only local detachments of the outer mortar layer at the centre of the panels occurred.

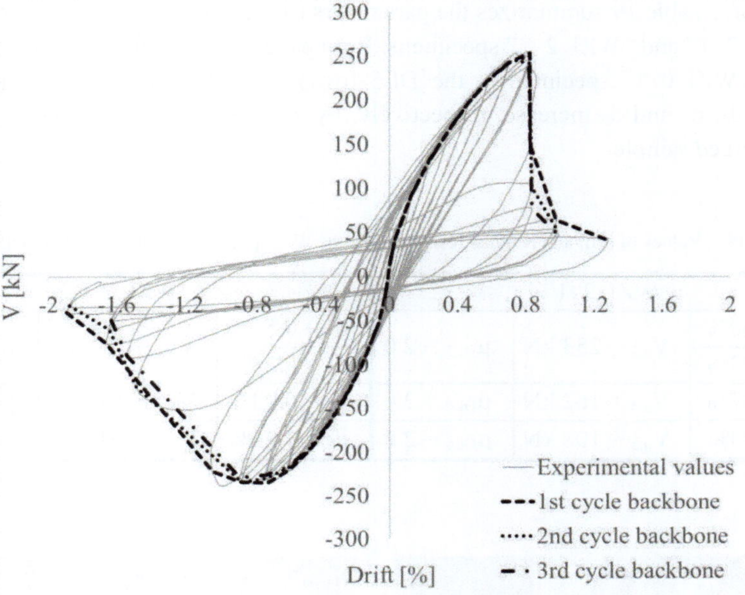

Figure 3.55: Hysteresis curve of the "WZI_2_2" specimen

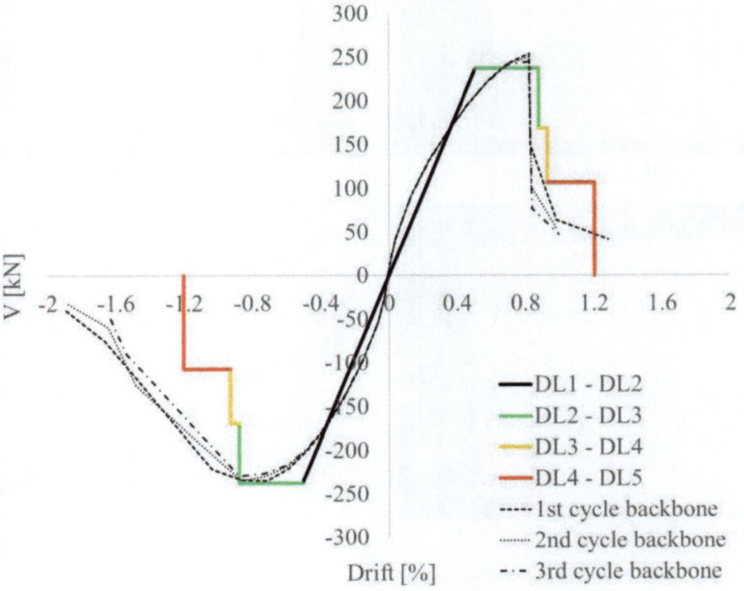

Figure 3.56: Equivalent multilinear curve of the "WZI_2_2" specimen

Furthermore, Table 19 summarizes the parameters of the equivalent multilinear curves of the "WZI_2_1" and "WZI_2_2" specimens. Both panels attained the DL3 at drifts higher than the "WZI_0_1" specimen at the DL5 (d_{5,WZI_0_1} = 0.73%). Finally, the average values of d_3, d_4 and d_5 increase, respectively, by 76%, 92% and 99% compared to the non-reinforced sample.

Table 19: Values of drift and residual resistance for the WZI specimens strengthened on both sides

"WZI_2_1"			"WZI_2_2"		
d_2 = 0.60%	$V_{u,3}$ = 253 kN	μ_{DL3} = 2.0	d_2 = 0.51%	$V_{u,3}$ = 237 kN	μ_{DL3} = 1.7
d_3 = 1.23%			d_3 = 0.88%		
d_4 = 1.57%	$V_{u,4}$ = 162 kN	μ_{DL4} = 2.6	d_4 = 0.93%	$V_{u,4}$ = 169 kN	μ_{DL4} = 1.8
d_5 = 1.70%	$V_{u,5}$ = 108 kN	μ_{DL5} = 2.8	d_5 = 1.20%	$V_{u,5}$ = 107 kN	μ_{DL5} = 2.4

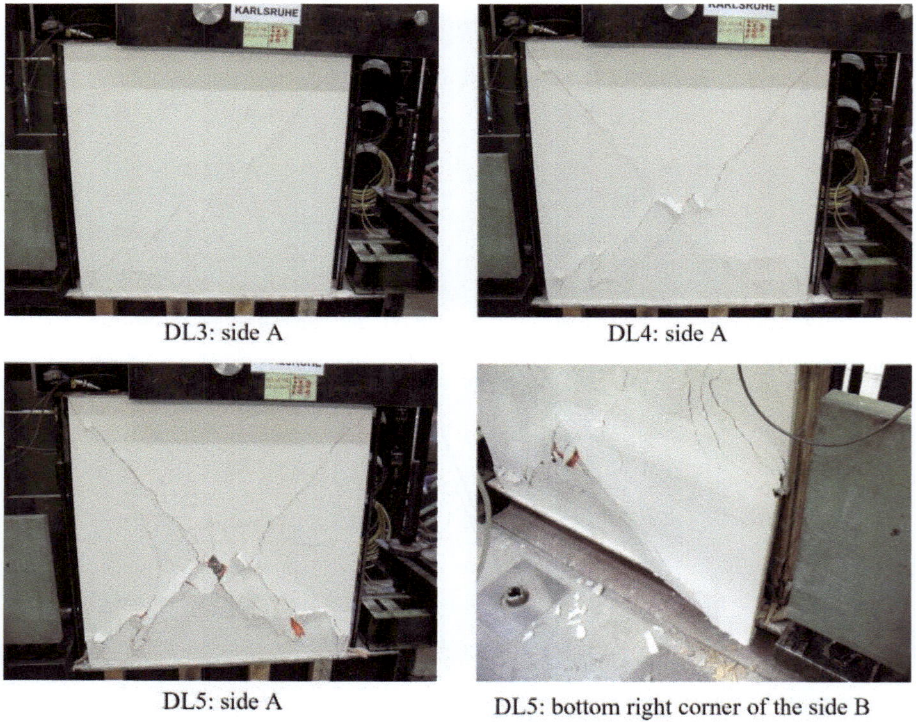

Figure 3.57: Damage levels of the "WZI_2_2" specimen

Chapter 4

In-plane strength of the tested "KS" and "WZI" masonry panels

The in-plane load-bearing capacity of the tested "KS" and "WZI" masonry piers is analytically determined in this chapter. The shear and flexural strength criteria described in § 1.3 are adopted for the URM panels. Furthermore, two different calculation approaches are used to determine the shear resistance of the specimens reinforced with EQ-GRID. The cross-sectional analysis is also performed to evaluate their flexural bearing capacity. Therefore, the in-plane failure domains are obtained from the envelope of the adopted shear and bending strength criteria in the N -V plane. Finally, the analytical results are compared to the experimental ones for validation.

4.1 In-plane failure domains of the unstrengthened "KS" and "WZI" masonry panels

The Mann and Müller theory [23] illustrated in § 1.3 is adopted to calculate the in-plane shear resistance of the unstrengthened masonry panels. The following formulations are used:

$$V_{Rd} = \min\left(V_t; V_{t,\lim}\right) \tag{4.1}$$

$$V_t = \frac{l \cdot t}{b} \cdot \left(\frac{f_{vm0}}{1+\mu \cdot \phi} + \frac{\mu}{1+\mu \cdot \phi} \cdot \sigma_v \right) \tag{4.2}$$

$$V_{t,\lim} = \frac{l \cdot t}{b} \cdot \frac{f_{bt}}{2.3} \cdot \sqrt{1+\frac{\sigma_v}{f_{bt}}} \tag{4.3}$$

where:

- l and t are the length and thickness of the panels, i.e. 1250 mm and 240 mm;

- b is equal to 1 since the height of the tested panels is 1250 mm;

Furthermore, the flexural resistance is calculated with Eq. (4.4), obtained neglecting the masonry tensile strength and considering a stress-block distribution of the compressive stresses at one end section, as described in § 1.3:

$$M_{Rd} = \frac{l^2 \cdot t}{2} \cdot \sigma_v \cdot \left(1 - \frac{\sigma_v}{0.85 \cdot f_{m,u}}\right) \tag{4.4}$$

Thus, the ultimate shear force under bending failure is calculated as follows:

$$V_{Rd}^F = \frac{M_{Rd}}{h'} = \frac{M_{Rd}}{0.5h} \tag{4.5}$$

Where h is the height of the panels, i.e. 1250 mm.

Finally, the average material properties adopted for the URM "KS" and "WZI" specimens are defined in § 4.1.1.

4.1.1 Experimental validation

The masonry compressive strength is calculated after Eurocode 6 [81] and the German National Annex [28] since no experimental data were available. Therefore, the characteristic values $f_{m,k,KS}$ and $f_{m,k,WZI}$ are determined through Eqs. (4.6) and (4.7):

$$f_{m,k,KS} = 0.95 \cdot f_{b,KS}^{0.585} \cdot f_M^{0.162} = 0.95 \cdot 25^{0.585} \cdot 1.8^{0.162} = 6.87 N/mm^2 \tag{4.6}$$

$$f_{m,k,WZI} = 0.69 \cdot f_{b,WZI}^{0.585} \cdot f_M^{0.162} = 0.69 \cdot 15^{0.585} \cdot 1.8^{0.162} = 3.70 N/mm^2 \tag{4.7}$$

Where $f_{b,KS}$, $f_{b,WZI}$ and f_M are the units and mortar compressive strength.

The average values $f_{m,u,KS}$ and $f_{m,u,WZI}$ are calculated with Eqs. (4.8) and (4.9):

$$f_{m,u,KS} = \frac{f_{m,k,KS}}{0.7} = \frac{6.87}{0.7} = 9.81 N/mm^2 \tag{4.8}$$

$$f_{m,u,WZI} = \frac{f_{m,k,WZI}}{0.7} = \frac{3.70}{0.7} = 5.28 N/mm^2 \tag{4.9}$$

Moreover, the Young's modulus E perpendicular to the bed joints and the shear modulus G are obtained from the characteristic masonry compressive strength through the formulations provided by Eurocode 6 ([81], [28]):

$$E_{KS} = 950 \cdot f_{m,k,KS} = 950 \cdot 6.87 = 6522 N / mm^2 \tag{4.10}$$

$$G_{KS} = 0.4 \cdot E_{KS} = 0.4 \cdot 6522 = 2608.8 N / mm^2 \tag{4.11}$$

$$E_{WZI} = 1100 \cdot f_{m,k,WZI} = 1100 \cdot 3.70 = 4068 N / mm^2 \tag{4.12}$$

$$G_{WZI} = 0.4 \cdot E_{WZI} = 0.4 \cdot 4068 = 1627.2 N / mm^2 \tag{4.13}$$

Concerning the initial shear strength f_{vm0}, it is assumed 0.17 N/mm² for the "KS" specimens and 0.30 N/mm² for the "WZI" panels, as suggested from the German masonry manual "Mauerwerk Kalender" for these types of masonry [82]. The interlocking parameters ϕ_{KS} and ϕ_{WZI} are respectively equal to 0.94 and 1.29, and they correspond to a "stepped-stair" cracks inclination angle of about 45° and 53° (Figure 4.1).

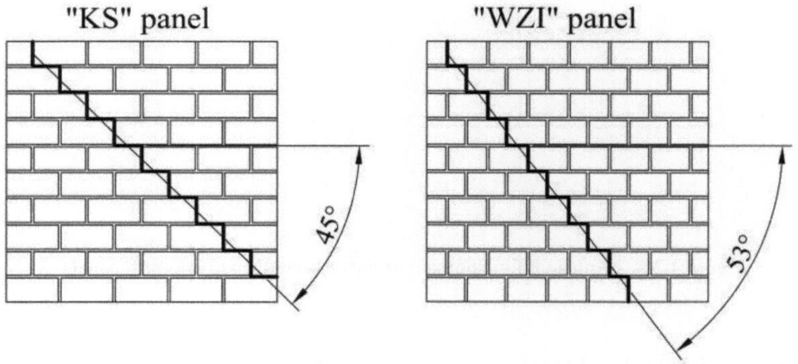

Figure 4.1: "Stepped-stair" cracks inclination

Moreover, since the masonry has a good texture, both local coefficients of friction, μ_{KS} and μ_{WZI}, are assumed to be 0.8, obtaining in this way an equivalent coefficient of friction μ_{eq} of about 0.4. Regarding the tensile strength of the units, $f_{bt,KS}$ and $f_{bt,WZI}$, the first one is assumed to be $0.063 f_{b,KS}$, as suggested from the German masonry manual "Mauerwerk Kalender" [82] for the calcium-silicate unit, where $f_{b,KS}$ is the compressive strength of the brick. The second one is assumed $0.06 f_{b,WZI}$. Since the specimen "WZI_0_1" attained the shear resistance with the formation of diagonal cracks passing only through the bricks, it has been possible to calibrate this value so that the sample's dominant failure mode was predicted by Eq.(4.3). Therefore, Table 20 summarizes the average values of the mechanical properties necessary to determine the in-plane strength domains of the tested panels shown in Figure 4.2.

Table 20: Average values of the mechanical properties of the unstrengthened masonry panels

"KS" unstrengthened panels						
$f_{m,u,KS}$ [N/mm²]	$f_{vm0,KS}$ [N/mm²]	μ_{KS} [-]	ϕ_{KS} [-]	$f_{vm0,red,KS}$ [N/mm²]	$\mu_{red,KS}$ [-]	$f_{bt,KS}$ [N/mm²]
9.81	0.17	0.8	0.94	0.10	0.46	1.58
"WZI" unstrengthened panels						
$f_{m,u,WZI}$ [N/mm²]	$f_{vm0,WZI}$ [N/mm²]	μ_{WZI} [-]	ϕ_{WZI} [-]	$f_{vm0,red,WZI}$ [N/mm²]	$\mu_{red,WZI}$ [-]	$f_{bt,WZI}$ [N/mm²]
5.28	0.30	0.8	1.29	0.15	0.39	0.90

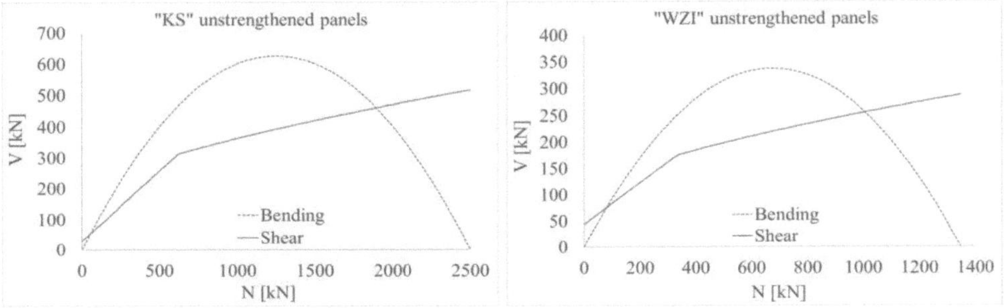

Figure 4.2: In-plane failure domain of the unstrengthened masonry panels

Finally, the experimental values V_{Ed} are compared to the analytical ones for each URM specimen in Table 21, and they are also depicted in Figure 4.3.

As illustrated in § 3.6.3.1, the "KS_0_1" panel presented a mixed shear failure with diagonal cracks passing through the mortar joints and the bricks. The analytical shear resistance V_t is equal to 208.47 kN, and it is lower than the experimental one, V_{Ed}, since it corresponds to the diagonal cracking only through the mortar joints.

Table 21: Experimental values and analytical estimation of the shear strength of the unstrengthened masonry panels

ID specimen	V_{Ed} [kN]	V_t [kN]	$V_{t,lim}$ [kN]	V_{Rd} [kN]
"KS_0_1"	250.61	208.47	278.06	208.47
"KS_0_2"	158.58	143.98	254.35	143.98
"WZI_0_1"	218.29	190.21	180.86	180.86
"WZI_0_2"	197.49	158.47	169.10	158.47

It is worth noting that the limit shear resistance $V_{t,lim}$ (278.06 kN) is higher than V_{Ed} (250.61 kN); thus, the mixed failure is validated.

Furthermore, there is a very good match between the analytical and experimental shear resistances for the “KS_0_2” specimen. In fact, the maximum horizontal force recorded during the test was 158.58 kN, while the expected value V_{Rd} is 143.98 kN. This result confirms the formation of “stepped-stair” cracks.

Regarding the “WZI” specimens, it is worth pointing out that they showed a more brittle behaviour than observed in the “KS” panels. The experimental lateral resistance of the “WZI_0_1” sample was attained by achieving the brick tensile strength, as it failed with diagonal cracks only through the masonry units. This result is analytically validated since V_{Rd} is equal to the limit value $V_{t,lim}$ (180.86 kN). Finally, the “WZI_0_2” panel presented a mixed shear failure with diagonal cracks passing through the mortar joints and the bricks. The shear resistance V_{Rd} is equal to 158.47 kN, and it is lower than the experimental value V_{Ed} since it corresponds to the diagonal cracking only through the mortar joints. The mixed failure is also, in this case, validated because the limit shear resistance $V_{t,lim}$ (169.10 kN) is higher than V_{Ed} (197.49 kN).

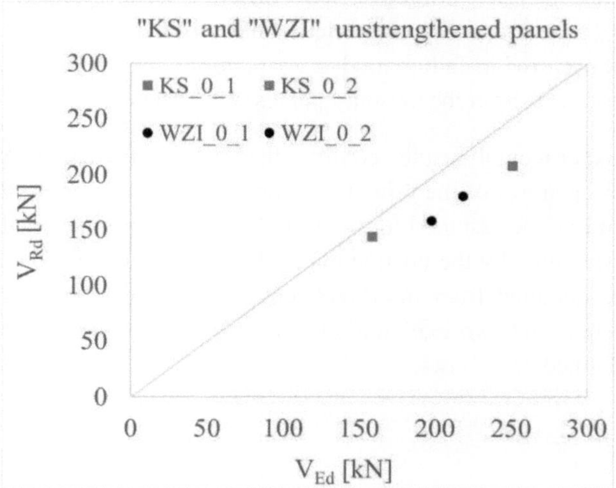

Figure 4.3: Experimental and analytical in-plane shear resistance of the “KS” and “WZI” unstrengthened panels

4.2 In-plane failure domains of the "KS" and "WZI" masonry panels strengthened with EQ-GRID

As observed in § 3.6, the EQ-GRID system improves the in-plane load-bearing capacity of masonry panels. Therefore, the definition of its tensile strength is the first necessary step for determining the in-plane failure domain of the tested panels. Two different approaches to calculate the shear resistance are described in the following paragraphs. Finally, a cross-section analysis is performed to consider the increase of in-plane bending resistance due to the system.

4.2.1 Determination of the EQ-GRID system's tensile strength

The average tensile strength of the EQ-GRID system is defined based on the results of the performed experimental campaign. The double- and single-lap shear-bond tests have shown that the mortar has a very high adhesive strength since no debonding was observed at the system-brick interface. Moreover, two different masonry units were adopted as standard support (hollow and full clay brick), and only the fibres breaking, or slippage within the matrix occurred as failure modes. Also, during the cyclic lateral shear tests, no debonding of the system from the masonry panels was observed.

Therefore, the experimental results confirm that the mortar can penetrate very well through the mesh openings of the grid. The latter never detached from the matrix layers. For this reason, it may be assumed for the tested masonry that the tensile strength of the system is influenced only by the bond at the grid-matrix interface. To this aim, the stress and strain values obtained from the direct tensile tests with clevis-type and clamping grips are considered, and two exploitation ratios of the bare textile strength, η_{clevis} and $\eta_{clamping}$, can be defined, as follows:

$$\eta_{clevis} = \frac{f_{clevis,u}}{\sigma_{u,m}} \tag{4.14}$$

$$\eta_{clamping} = \frac{f_{clamping,u}}{\sigma_{u,m}} \tag{4.15}$$

Where $f_{clevis,u}$ and $f_{clamping,u}$ are the mean tensile strength of the composite specimens, according to the adopted test method, and $\sigma_{u,m}$ is the mean tensile strength of the bare textile. These values are directly compared in Table 22.

Table 22: Mechanical properties of the EQ-GRID system from the performed tensile tests.

Test method	Composite				Bare textile		Exploitation ratios	
	Clevis-type grips		Clamping grips		Clamping grips			
Direction	$f_{clevis,u}$ [N/mm²]	$\varepsilon_{clevis,u}$ [%]	$f_{clamping,u}$ [N/mm²]	$\varepsilon_{clamping,u}$ [%]	$\sigma_{u,m}$ [N/mm²]	$\varepsilon_{u,m}$ [%]	η_{clevis} [%]	$\eta_{clamping}$ [%]
V	551	1.2	996	2.0	1387	2.8	40	72
H	666	2.4	673	2.3	1046	3.2	64	64
D	1844	2.5	1591	2.1	1458	3.7	100	100

It is worth noting that the average tensile strengths of the EQ-GRID system in the horizontal direction, $f_{clevis,u}$ and $f_{clamping,u}$, are both equal to 64% of the bare textile's one. The corresponding failure strains are 2.4% and 2.3%, respectively.

On the contrary, the exploitation ratios for the vertical direction, η_{clevis} and $\eta_{clamping}$, depend on the adopted test method since the failure modes of the composite specimens were different. In the test setup with clevis-type grips, the tensile load was transferred from the matrix to the textile strip by adhesion. Therefore, the wires could slip after reaching the maximum strength. The diagrams of the vertical direction depicted in Figure 4.4 (clevis-type grips) confirm this behaviour. The curves have a softening branch after the peak and not the sudden strength drop typical by the fibres' tensile failure (Figure 4.4, clamping grips). For this reason, the exploitation ratios, η_{clevis} and $\eta_{clamping}$, in the vertical direction are equal to 40% and 72%. The corresponding ultimate strains also are quite different, 1.2% and 2.0% for the first and second test method.

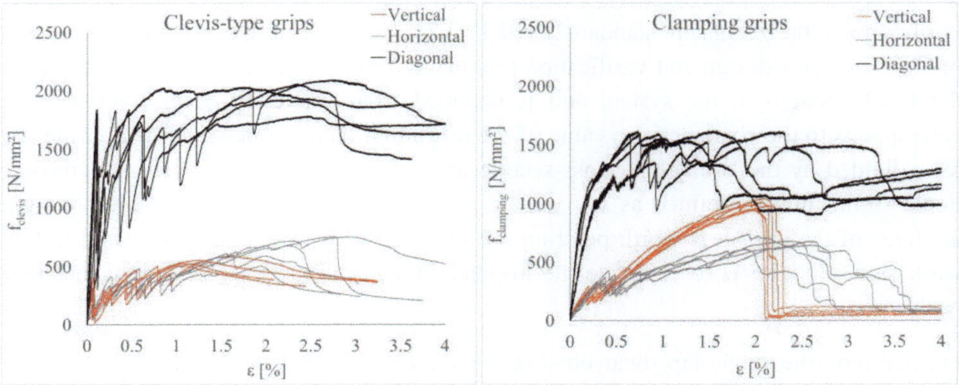

Figure 4.4: Stress-strain curves of the EQ-GRID system obtained from tensile tests carried out with clevis-type and clamping grips

Furthermore, it is possible to assume that both exploitation ratios are equal to 100% in the diagonal direction since the tensile strengths, $f_{clevis,u}$ and $f_{clamping,u}$, are higher than the bare textile's one, $\sigma_{f,u}$ (Table 22). It is worth specifying that the composite samples' width was greater (95 mm) in this direction than the other composite specimens (60 mm) because of the different wires' arrangement. Since the axial load was transferred only by adhesion in the test setup with the clevis-type grips, the horizontal fibres present in the composite specimens were also stressed, as their inclination angle was 30° to the test direction. Moreover, Figure 4.4 (clevis-type grips) shows that the cracking phase of the diagonal composite samples is distributed over a wider range of deformations than the other two directions. The maximum strength values are achieved at the end of the cracking phase; after that, the tensile stresses decrease.

On the contrary, the diagonal specimens' tensile strength, $f_{clamping,u}$, was attained during the cracking phase. It is important to highlight that, in this case, the load was directly transmitted from the clamping grips to the grid, which in turn transferred it to the matrix. Therefore, the uncracked mortar layers allowed the activation of the horizontal fibres present in the samples. In this way, they could contribute to the strength of the composite specimens during the cracking phase. It is worth noting that this contribution decreases with the formation of cracks and disappears in the samples' cracked state. Therefore, the strain value, $\varepsilon_{clamping,u}$, is lower than the bare textile since it corresponds to the strain at the average maximum stress attained during the cracking phase and not at the breaking of the glass fibre. The latter happened just at the end of the formation of the cracks; after that, only the PP fibres carried the tensile load. The complete diagrams of the samples tested with clamping grips in the diagonal direction are depicted in Figure 4.5. The tensile strength achieved in the cracked state is not considered since the strain values are very high in this phase (until 26%).

Furthermore, the technical standard CNR-DT 215/2018 defines a conventional stress limit, $\sigma_{lim,conv}$, for design and verification problems. As illustrated in § 2.2.1, it represents the bond strength of the system and is obtained from single-lap shear-bond tests. It corresponds to the characteristic value of the maximum tensile force recorded during the tests divided by the equivalent cross-section area of the bare textile. Then, the conventional strain limit is defined as $\varepsilon_{lim,conv} = \sigma_{lim,conv} / E_f$, where E_f is the average Young's modulus of the grid. It is worth pointing out that the mean values are considered in this work since the aim is to calculate the in-plane lateral resistance of the tested masonry panels.

Furthermore, the single-lap shear-bond tests were carried out only for the vertical direction of the EQ-GRID system. The average maximum shear-bond strength $f_{bond,single,m}$ is equal to 971.2 N/mm², and it is very close to the tensile strength $f_{clamping,u}$ obtained for this direction (996 N/mm²). It is worth noting that in both test setups, the free ends of the textile strip were clamped between the wedge grips of the testing machine, and the fibre's breaking was the failure mode. Therefore, these results may be adopted when the grid is applied with mechanical connectors or extended up to a significant distance (e.g. at least 20 cm) from the cross-section with maximum tensile stress or turned over the corners of the walls.

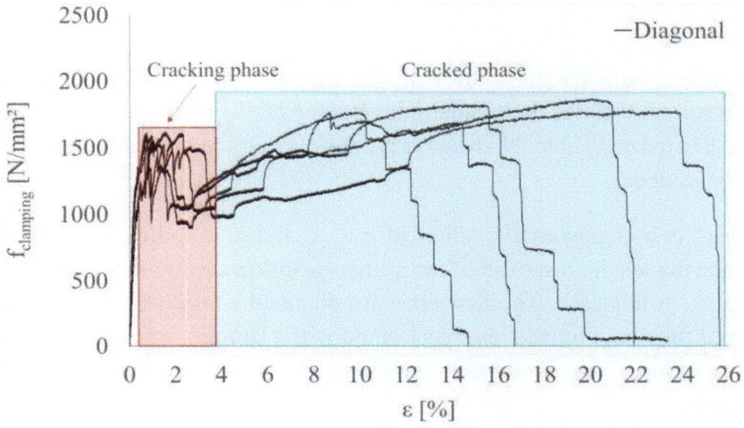

Figure 4.5: Stress-strain curves of the composite specimens in the diagonal direction from the tensile tests with clamping grips

Moreover, since the specimens were tested with different bonding lengths (4 cm, 8 cm and 12 cm), it is impossible to obtain an average strength from the double-lap shear-bond tests. However, the results have shown that only the bond between textile and matrix influences the resistance. Thus, it is reasonable to adopt the mean tensile strengths, $f_{clevis,u}$, as conventional stress limit for calculating the in-plane failure domains of the tested "KS" and "WZI" masonry panels. It is worth noting that this assumption is coherent since the EQ-GRID system was applied without any mechanical connectors, and it was not anchored to the upper or bottom steel beam or turned over the thickness of the specimen (Figure 3.21).

Therefore, Table 23 shows the average values of $\sigma_{lim,conv}$, $\varepsilon_{lim,conv}$ used to determine the in-plane strength domains of the tested "KS" and "WZI" masonry panels. It is worth pointing out that the grid's tensile properties are adopted for the diagonal direction since the exploitation ratio η_{clevis} is 100% and the average tensile modulus E_f corresponds to the secant one reported in Table 2.

Table 23: Average values of $\sigma_{lim,conv}$, $\varepsilon_{lim,conv}$ and E_f of the EQ-GRID system adopted for the calculation of the in-plane failure domains of the tested "KS" and "WZI" masonry panels

Direction	$\sigma_{lim,conv}$ (avg.)	$\varepsilon_{lim,conv}$ (avg.)	E_f
V	551 N/mm^2	1.27 %	43269 N/mm^2
H	666 N/mm^2	1.78 %	37516 N/mm^2
D	1458 N/mm^2	3.64 %	40067 N/mm^2

Furthermore, the characteristic values of $\sigma_{lim,conv}$ and $\varepsilon_{lim,conv}$ can be calculated as the average ones minus k_n times the standard deviation *STD*, as recommended by Annex D of Eurocode 0. Therefore, Table 24 shows a proposal since the available experimental data are limited in number.

In this respect, two characteristic values of $\sigma_{lim,conv}$ and $\varepsilon_{lim,conv}$ are given for the vertical direction since the tensile response of the system is influenced by the mechanical anchorage of the grid, in this case. The first ones are obtained considering the single-lap shear bond tests and the direct tensile tests performed with clamping grips, as the experimental results are very close. The second ones are obtained from the tensile tests performed with clevis-type grips.

Regarding the horizontal direction, the characteristic conventional stress and strain limits are calculated, considering the direct tensile tests with clevis and clamping type grips. As shown in Figure 4.4, the obtained experimental response is very similar; thus, both results are considered.

Finally, the characteristic values of $\sigma_{lim,conv}$ and $\varepsilon_{lim,conv}$ for the diagonal direction are obtained from the tensile tests performed on the bare textile specimens since the grid's tensile strength exploitation ratio is 100%.

Table 24: Proposal of characteristic values of $\sigma_{lim,conv}$, $\varepsilon_{lim,conv}$ of the EQ-GRID system

Direction	$\sigma_{lim,conv}$ (char.)	$\varepsilon_{lim,conv}$ (char.)	$n_{specimens}$	k_n
V	470 N/mm^2	1.09 %	5 (clevis)	2.33
V	869 N/mm^2	2.01 %	10 (clamping + single-lap shear tests)	1.92
H	517 N/mm^2	1.38 %	10 (clevis + clamping)	1.92
D	1311 N/mm^2	3.27 %	8 (grid)	2.00

4.2.2 Proposal of modification of the masonry shear strength with the EQ-GRID system

The cyclic lateral shear tests described in § 3.6 have shown that the EQ-GRID system can improve the in-plane load-bearing capacity of masonry panels. Therefore, a possible way to consider the presence of the textile reinforcement is to increase the masonry material properties, as also suggested by Urban in [83]. In this respect, Urban proposed to modify the masonry initial shear strength f_{vmo} through the EQ-GRID system's diagonal tensile resistance. Furthermore, he suggested improving the brick tensile strength f_{bt} using the system's resistance in the three main directions (vertical, horizontal and diagonal).

The approach presented in this work originates from the formulations proposed in [83], but it is based on the experimental observations illustrated in Chapter 3. During the cyclic lateral shear tests, it was evident that the system was mainly stressed in the direction parallel to the shear force when diagonal cracks opened. Therefore, the proposed approach is based on modifying the initial shear strength f_{vmo} and the brick tensile strength f_{bt} through the EQ-GRID system's resistance in the direction parallel to the acting shear force. Figure 4.6 schematically shows the deformation of the horizontal and diagonal fibre yarns when a shear crack opens. It is worth noting that, compared to a common biaxial textile mesh, EQ-GRID has the advantage that the directions activated during the shear failure are two and not only one. In this regard, the diagonal fibre yarns are tilted 30° to the horizontal to maximize the strengthening effect under shear failure [83].

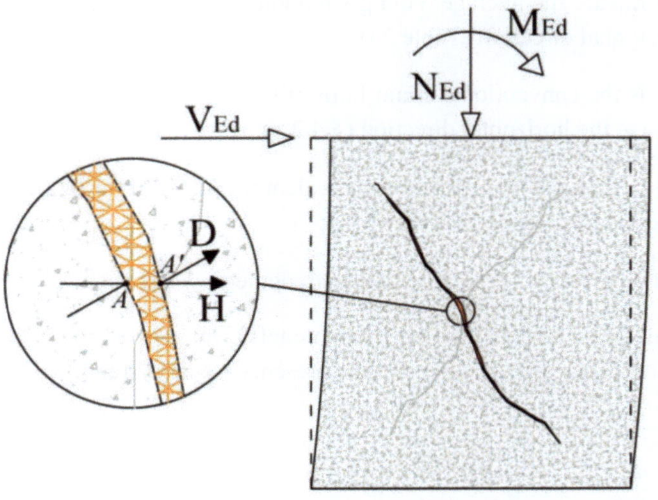

Figure 4.6: Diagonal cracking of a masonry pier strengthened with EQ-GRID

Furthermore, as shown in § 3.6, the EQ-GRID system can also increase the deformation capacity of the reinforced structural element. In fact, the polypropylene wires arranged in the diagonal direction have a very high ultimate tensile strain (> 20%) and allow the panel structure to be held together even in the advanced cracking stage, improving the element's stability.

Therefore, Eqs. (4.16) and (4.17) are proposed to modify the initial shear strength f_{vmo} and the brick tensile strength f_{bt} of a masonry pier through the EQ-GRID system's resistance in the direction parallel to the acting shear force:

$$f_{vm0,s} = f_{vm0} + n_f \cdot \left(\frac{E_{f,h} \cdot \varepsilon_{\lim,conv,h} \cdot t_{eq,h} + E_{f,d} \cdot \varepsilon_{\lim,conv,h} \cdot t_{eq,d}}{t} \right) \tag{4.16}$$

$$f_{bt,s} = f_{bt} + n_f \cdot \left(\frac{E_{f,h} \cdot \varepsilon_{\lim,conv,h} \cdot t_{eq,h} + E_{f,d} \cdot \varepsilon_{\lim,conv,h} \cdot t_{eq,d}}{t} \right) \tag{4.17}$$

Where:

- f_{vmo} and f_{bt} are the average initial shear strength and brick tensile strength of the unstrengthened masonry;
- n_f is the total number of textile layers. It is equal to 1 when the system is applied only on one side and 2 for application on both sides;
- $E_{f,h}$ and $E_{f,d}$ are the average Young's modulus of the bare textile in the horizontal and diagonal direction (Table 23).
- $\varepsilon_{lim,conv,h}$ is the conventional strain limit of the fibres arranged parallel to the shear force. i.e. the horizontal direction (§ 4.2.1);
- $t_{eq,h}$ and $t_{eq,d}$ are the homogenized equivalent thickness of the textile in the horizontal and diagonal direction (Eq. (3.6));
- t is the thickness of the panel without strengthening system.

Finally, it is important to note that the friction coefficient μ is not modified since it is an intrinsic property of the mortar joint, and the presence of the system does not influence it.

4.2.3 In-plane strength of masonry panels reinforced with the EQ-GRID system according to CNR-DT 215/2018

As shown in § 2.2.2, the technical standard CNR-DT 215/2018 proposes to calculate the in-plane shear resistance of a strengthened masonry panel, $V_{Rd,s}$, as the sum of the shear resistance of the URM panel under diagonal cracking, V_{URM}, and the increase of strength due to the reinforcing system, $V_{t,f}$:

$$V_{Rd,s} = V_{URM} + V_{t,f} \tag{4.18}$$

Since the shear failure with diagonal cracking stresses the EQ-GRID system mainly parallel to the shear force, the formulation of $V_{t,f}$ (Eq. (2.7)) can be expressed considering the tensile strength of the horizontal and diagonal fibres, as follows:

$$V_{t,f} = \frac{1}{\gamma_{Rd}} \cdot n_f \cdot l_f \cdot \alpha_t \cdot \left(E_{f,h} \cdot \varepsilon_{\lim,conv,h} \cdot t_{eq,h} + E_{f,d} \cdot \varepsilon_{\lim,conv,h} \cdot t_{eq,d} \right) \tag{4.19}$$

Where:

- $E_{f,h}$ and $E_{f,d}$ are the average Young’s modulus of the bare textile in the horizontal and diagonal direction (Table 23);
- $\varepsilon_{lim,conv,h}$ is the conventional strain limit of the fibres arranged parallel to the shear force, i.e. the horizontal direction (§ 4.2.1);
- $t_{eq,h}$ and $t_{eq,d}$ are the homogenized equivalent thickness of the textile in the horizontal and diagonal direction (Eq. (3.6));

Moreover, CNR-DT 215/2018 suggests reducing the shear contribution $V_{t,f}$ by at least 30% for application of the strengthening system only on one side of the wall, and connectors shall be applied to fix the grid to the wall. As shown by the experimental results, the matrix can penetrate very well in the mesh openings of EQ-GRID, and it has a very high pull-off strength. Furthermore, no debonding of the system from the masonry substrate or detachment of the grid from the matrix layers was observed during the cyclic lateral shear tests. Therefore, it is generally applied without transversal connectors, and the reduction of the shear contribution $V_{t,f}$ is not considered in this case.

Regarding the in-plane flexural strength domain of a masonry pier strengthened with EQ-GRID, it can be determined through a cross-sectional analysis performed under the following assumptions:

- plane sections remain plane after loading;
- strain compatibility between masonry and strengthening system;

- contribution of the strengthening system in compression is neglected.

It is worth noting that the in-plane bending failure of a masonry pier is generally characterized by sub-vertical cracks at the compressed toe and sub-horizontal cracks at the tensioned part (Figure 4.7). Therefore, the reinforcing system is mainly stressed in the direction parallel to the axis of the structural element, i.e. the vertical direction. As depicted in Figure 4.7, when the cracks appear at the tensioned part, the vertical and diagonal fibres can have the same deformation. Furthermore, the stress-strain diagram of EQ-GRID is assumed linear elastic (Figure 4.8), and the ultimate tensile strain is the conventional strain limit $\varepsilon_{lim,conv,v}$ defined in § 4.2.1. Thus, the corresponding conventional strength limit $f_{t,u,}$ takes into account both strength contributions and is defined as follows:

$$f_{t,u} = E_{f,v} \cdot \varepsilon_{\lim,conv,v} \cdot t_{eq,v} + E_{f,d} \cdot \varepsilon_{\lim,conv,v} \cdot t_{eq,d} \quad (4.20)$$

Where:

- $E_{f,v}$ and $E_{f,d}$ are the average elastic modulus of the bare textile in the vertical and diagonal direction (Table 23);
- $\varepsilon_{lim,conv,v}$ is the mean conventional strain limit for the vertical direction (§ 4.2.1);
- $t_{eq,h}$ and $t_{eq,d}$ are the homogenized equivalent thickness of the textile in the horizontal and diagonal direction (Eq. (3.6)).

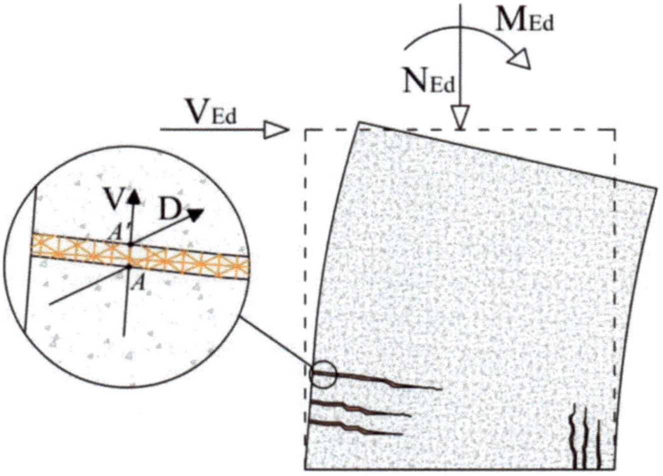

Figure 4.7: Bending failure of a masonry pier strengthened with EQ-GRID

Moreover, the adopted masonry compressive stress diagram is rectangular with a uniform stress equal to $\alpha f_{m,u}$ (Figure 4.8). This latter is distributed over an equivalent compressive zone equal to βy_n, where y_n is the distance from the outermost compressed fibre to the neutral axis. The coefficients α and β are respectively 0.85 and 0.8. The masonry ultimate strain, $\varepsilon_{m,u}$, can be assumed 3.5 ‰ unless experimental data are available and its tensile strength is neglected.

Therefore, the in-plane resistance domain of a masonry pier reinforced with the EQ-GRID system can be determined in two steps:

- Identification of the failure mode;
- Calculation of the normal force $N_{Rd,s}$ and bending moment $M_{Rd,s}$ with two equations of equilibrium of the internal forces.

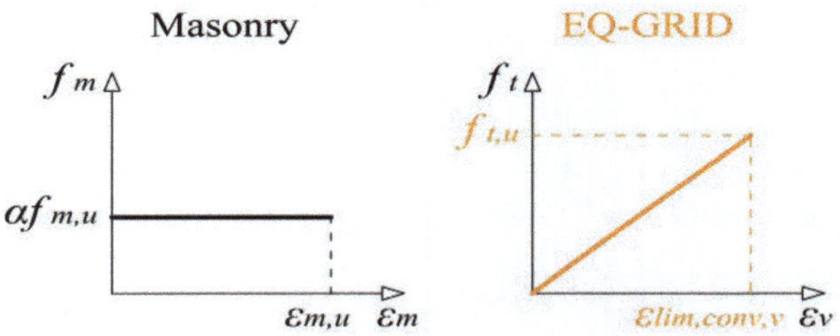

Figure 4.8: Adopted stress-strain diagrams of the masonry and the EQ-GRID system for the bending failure

As shown in Figure 4.9, the system is applied along the entire length l of the panel. Therefore, the distance d_f between the outermost reinforcement fibres that attain the conventional strain limit is equal to l.

Five failure modes can be identified by the following conditions, in which ε_1 and ε_2 are respectively the tensile and compressive strain attained by the outermost fibres of the panel's cross-section:

- *Failure mode 0*: the cross-section of the panel is uniformly tensioned, $\varepsilon_1 = \varepsilon_2 = \varepsilon_{lim,conv,v}$;
- *Failure mode 1*: the cross-section of the panel fully tensioned, $\varepsilon_1 = \varepsilon_{lim,conv,v}$ and $\varepsilon_2 = 0$;
- *Failure mode 2*: the masonry and the EQ-GRID system achieve their ultimate strains, $\varepsilon_1 = \varepsilon_{lim,conv,v}$ and $\varepsilon_2 = \varepsilon_{m,u}$;

- *Failure mode 3*: the cross-section of the panel is fully compressed, $\varepsilon_1 = 0$ and $\varepsilon_2 = \varepsilon_{m,u}$;
- *Failure mode 4:* the cross-section of the panel is uniformly compressed, $\varepsilon_1 = \varepsilon_2 = \varepsilon_{m,u}$.

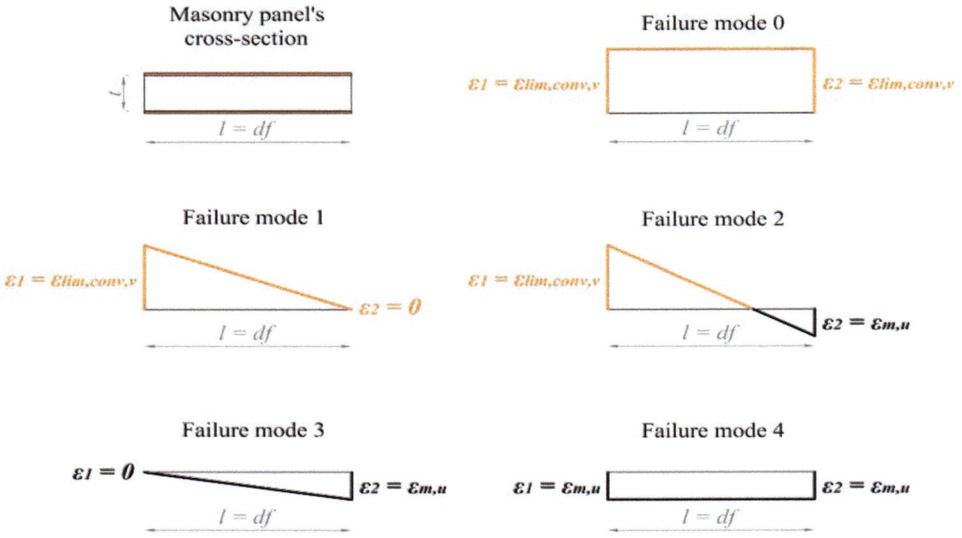

Figure 4.9: Possible failure strain distributions of a masonry cross-section reinforced with the EQ-GRID system

Failure mode 0 is the ideal situation of uniform tension, both stress and strain distributions are constant, and the neutral axis y_{n0} lies outside the section at infinity (Figure 4.10).

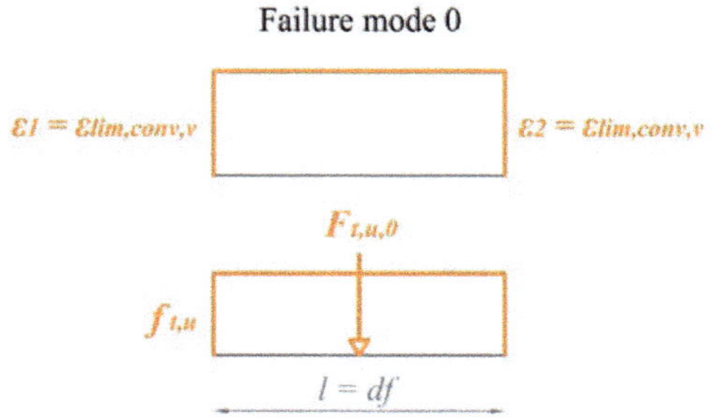

Figure 4.10: Stress-strain distributions of *failure mode 0*

Therefore, the resultant of the compressive stresses, $F_{m,u,0}$, is zero, and that of the tensile stresses, $F_{t,u,0}$, is calculated as follows:

$$F_{t,u,0} = \left(E_{f,v} \cdot \varepsilon_{\lim conv,v} \cdot n_f \cdot t_{eq,v} + E_{f,d} \cdot \varepsilon_{\lim conv,v} \cdot n_f \cdot t_{eq,d}\right) \cdot d_f \tag{4.21}$$

Where n_f is the total number of textile layers (for one side application is 1, while for both sides application is 2).

To ensure the force and moment equilibrium, the ultimate normal force $N_{Rd,0}$ is equal to $F_{t,u,0}$ and the bending moment $M_{Rd,s,0}$ is 0 since it is calculated at the cross-section's centre of gravity, $l/2$.

$$N_{Rd,s,0} = F_{t,u,0} \tag{4.22}$$

$$M_{Rd,s,0} = 0 \tag{4.23}$$

With *failure mode 1*, the cross-section is still completely tensioned, but the strain and stress distributions are linear (Figure 4.11). The system attains its tensile strain limit, $\varepsilon_{lim,conv,v}$, and ε_2 is 0. Thus, the neutral axis depth y_{n1} is equal to d_f.

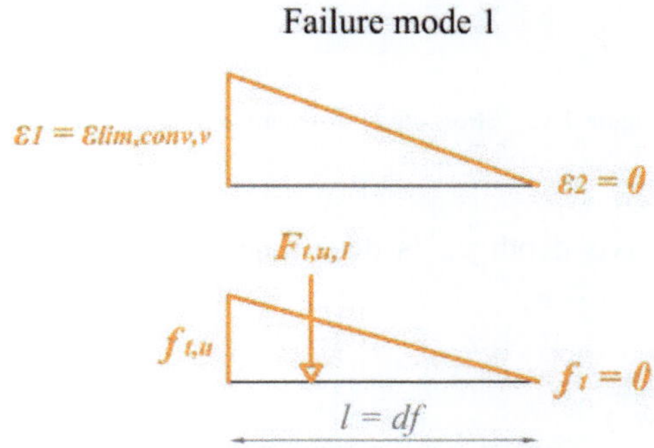

Figure 4.11: Stress-strain distributions of *failure mode 1*

The resultant of the compressive stresses, $F_{m,u,1}$ is zero, while $F_{t,u,1}$ can be calculated as follows:

$$F_{t,u,1} = \frac{\left(E_{f,v} \cdot \varepsilon_{\lim conv,v} \cdot n_f \cdot t_{eq,v} + E_{f,d} \cdot \varepsilon_{\lim conv,v} \cdot n_f \cdot t_{eq,d}\right) \cdot d_f}{2} \tag{4.24}$$

The ultimate normal force $N_{Rd,1}$ and bending moment $M_{Rd,s,1}$ can be easily obtained; the first one is equal to $F_{t,u,1}$ and the second one is calculated at $l/2$:

$$N_{Rd,s,1} = F_{t,u,1} \tag{4.25}$$

$$M_{Rd,s,1} = \frac{F_{t,u,1} \cdot d_f}{6} \tag{4.26}$$

Failure mode 2 is defined under the assumption that both masonry and system attain their strain limits ($\varepsilon_1 = \varepsilon_{lim,conv,v}$ and $\varepsilon_2 = \varepsilon_{m,u}$).

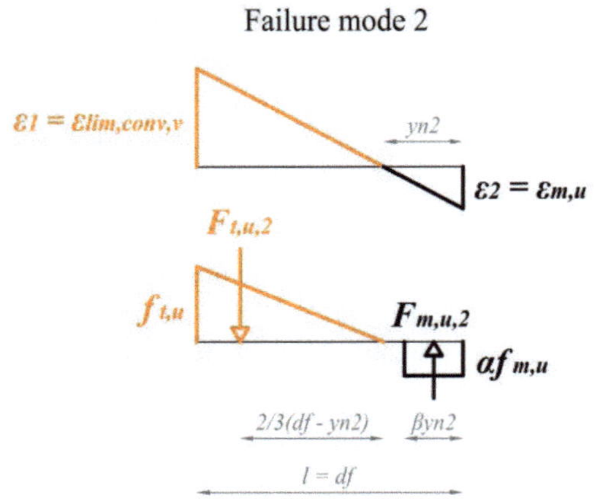

Figure 4.12: Stress-strain distribution of *failure mode 2*

Therefore, the neutral axis depth y_{n2} is determined according to the strain distribution shown in Figure 4.12:

$$y_{n2} = \frac{d_f \cdot \varepsilon_2}{(\varepsilon_1 + \varepsilon_2)} \tag{4.27}$$

The resultant of the compressive and tensile stresses, $F_{m,u,2}$ and $F_{t,u,2}$, are calculated with the following equations:

$$F_{m,u,2} = \left(\alpha f_{m,u} \cdot \beta \cdot t\right) \cdot y_{n2} \tag{4.28}$$

$$F_{t,u,2} = \frac{\left(E_{f,v} \cdot \varepsilon_{\lim conv,v} \cdot n_f \cdot t_{eq,v} + E_{f,d} \cdot \varepsilon_{\lim conv,v} \cdot n_f \cdot t_{eq,d}\right) \cdot \left(d_f - y_{n2}\right)}{2} \tag{4.29}$$

The ultimate normal force $N_{Rd,s,2}$ and bending moment $M_{Rd,s,2}$ are obtained from the force and moment equilibrium equations, as follows:

$$N_{Rd,s,2} = F_{m,u,2} - F_{t,u,2} \tag{4.30}$$

$$M_{Rd,s,2} = F_{m,u,2} \cdot \left(\frac{d_f}{2} - \frac{\beta y_{n2}}{2} \right) + F_{t,u,2} \cdot \left(y_{n2} + \frac{2}{3}(d_f - y_{n2}) - \frac{d_f}{2} \right) \tag{4.31}$$

Failure mode 3 is characterized only by masonry crushing. As depicted in Figure 4.13, the strain ε_2 attains its ultimate value $\varepsilon_{m,u}$ and the system is not subjected to tension ($\varepsilon_1 = 0$). Thus, the neutral axis depth y_{n3} is equal to d_f.

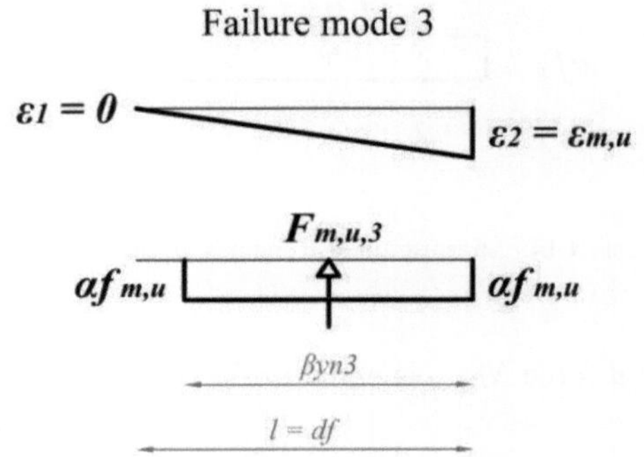

Figure 4.13: Stress-strain distribution of *failure mode 3*

The resultant of the tensile stresses $F_{t,u,3}$ is zero, while $F_{m,u,3}$ can be calculated through Eq. (4.32):

$$F_{m,u,3} = \alpha f_{m,u} \cdot t \cdot \beta d_f \tag{4.32}$$

Therefore, the ultimate normal force $N_{Rd,s,3}$ is equal to $F_{m,u,3}$ and the bending moment $M_{Rd,s,3}$ can be determined with Eq. (4.34):

$$N_{Rd,s,3} = F_{m,u,3} \tag{4.33}$$

$$M_{Rd,s,3} = F_{m,u,3} \cdot \left(\frac{d_f}{2} - \frac{\beta d_f}{2} \right) \tag{4.34}$$

Finally, the cross-section is uniformly compressed with *failure mode 4*. As a result, both stress and strain diagrams are constant (Figure 4.14), and the neutral axis y_{n4} lies outside the section at infinity.

The resultant of the tensile stresses, $F_{t,u,4}$ is zero, and that of the compressive stresses, $F_{m,u,4}$ is calculated as follows:

$$F_{m,u,4} = \alpha f_{m,u} \cdot t \cdot d_f \tag{4.35}$$

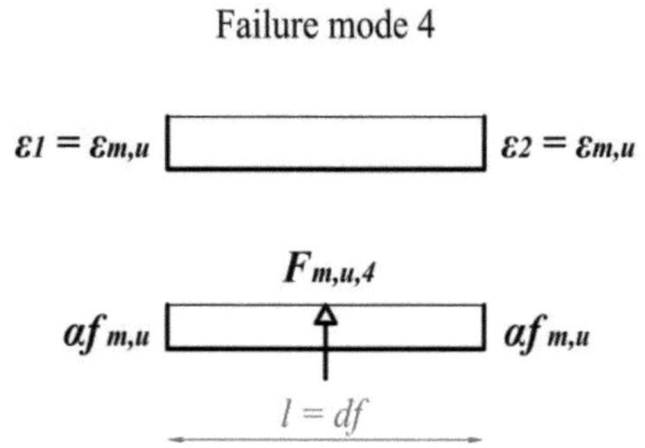

Figure 4.14: Stress-strain distributions of *failure mode 4*

Thus, the ultimate normal force $N_{Rd,s,4}$ is equal to $F_{m,u,4}$ and the bending moment $M_{Rd,s,4}$ is zero:

$$N_{Rd,s,4} = F_{m,u,4} \tag{4.36}$$

$$M_{Rd,s,4} = 0 \tag{4.37}$$

Therefore, to obtain the in-plane flexural failure domain, it is necessary to calculate the ultimate shear force with Eq. (4.38):

$$V^F_{Rd,s} = \frac{M_{Rd,s}}{h'} \tag{4.38}$$

Where $M_{Rd,s}$ is the ultimate bending moment and *h'* is the shear length, defined as the distance between the end section of the panel and that with zero moment (i.e., *h/2* for elements with double-fixed boundary conditions and *h* for cantilevers).

Figure 4.15 shows two in-plane flexural strength domains of a masonry pier. The first one is with EQ-GRID and the second one is without EQ-GRID. It is worth noting that, thanks to the strengthening system, the structural element can also be subjected to normal tensile forces, i.e. $N_{Rd,s,0}$ and $N_{Rd,s,1}$. In fact, the first part of the domain is shifted towards the negative values of *N*. Moreover, the contribution of the strengthening system to the in-plane flexural resistance of the panel is maximum between *failure mode 1* and *2* since the system attains its limit strain. After that, the resistance improvement decreases since the masonry achieves its ultimate strain. In this regard, the strengthening effect of the system becomes smaller values at higher values of N, as the compressed part of the cross-section increases. Finally, from *failure mode 3*, the masonry is completely compressed, and the system cannot give any strength contribution.

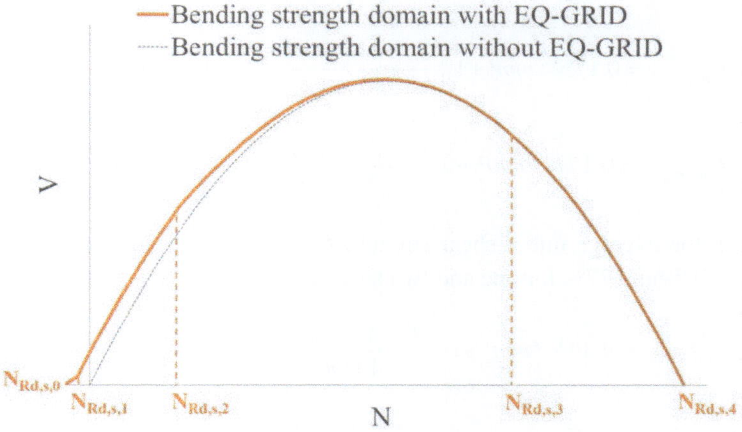

Figure 4.15: In-plane flexural strength domains of a masonry pier with and without the EQ-GRID system

4.2.4 Experimental validation

The in-plane shear strength of the tested "KS" and "WZI" masonry panels is compared to the experimental values for validation. Both calculation approaches discussed in the previous paragraphs, i.e. modification of the masonry shear strength and the formulations after the CNR-DT 215/2018, are adopted. To this aim, the masonry's average material properties defined in § 4.1 are assumed. The mean conventional strain limit $\varepsilon_{lim,conv,h}$ and the Young's secant modulus, $E_{f,h}$ and $E_{f,d}$, of the bare textile reported in Table 23 are used for the EQ-GRID system.

Therefore, the mechanical properties of the masonry are improved with the following formulations:

$$f_{vm0,s} = f_{vm0} + n_f \cdot \left(\frac{E_{f,h} \cdot \varepsilon_{\lim,conv,h} \cdot t_{eq,h} + E_{f,d} \cdot \varepsilon_{\lim,conv,h} \cdot t_{eq,d}}{t} \right) \tag{4.39}$$

$$f_{bt,s} = f_{bt} + n_f \cdot \left(\frac{E_{f,h} \cdot \varepsilon_{\lim,conv,h} \cdot t_{eq,h} + E_{f,d} \cdot \varepsilon_{\lim,conv,h} \cdot t_{eq,d}}{t} \right) \tag{4.40}$$

It is worth noting that the product $\left(\varepsilon_{\lim,conv,h} \cdot E_{f,h} \cdot t_{eq,h} + \varepsilon_{\lim,conv,h} \cdot E_{f,d} \cdot t_{eq,d}\right)$ is equal to 27.74 N/mm, which corresponds to the system's average strength limit under diagonal cracking. Thus, the mean initial shear strength f_{vm0} of the unreinforced "KS" panels improves by 68% and 136% when the EQ-GRID is applied on one or two sides of the panel:

$$f_{vm0,KS,s1} = 0.17N/mm^2 + 1 \cdot \left(\frac{27.74N/mm}{240mm}\right) = 0.29N/mm^2 \qquad (4.41)$$

$$f_{vm0,KS,s2} = 0.17N/mm^2 + 2 \cdot \left(\frac{27.74N/mm}{240mm}\right) = 0.40N/mm^2 \qquad (4.42)$$

Furthermore, the average initial shear strength f_{vm0} of the unstrengthened "WZI" panels increases by 39% and 77% for one and two sides applications:

$$f_{vm0,WZI,s1} = 0.30N/mm^2 + 1 \cdot \left(\frac{27.74N/mm}{240mm}\right) = 0.42N/mm^2 \qquad (4.43)$$

$$f_{vm0,WZI,s2} = 0.30N/mm^2 + 2 \cdot \left(\frac{27.74N/mm}{240mm}\right) = 0.53N/mm^2 \qquad (4.44)$$

Concerning the mean brick tensile strength of the unreinforced "KS" panels, it has an improvement of 7% and 15% when the system is applied on one or two sides:

$$f_{bt,KS,s1} = 1.575N/mm^2 + 1 \cdot \left(\frac{27.74N/mm}{240mm}\right) = 1.69N/mm^2 \qquad (4.45)$$

$$f_{bt,KS,s2} = 1.575N/mm^2 + 2 \cdot \left(\frac{27.74N/mm}{240mm}\right) = 1.81N/mm^2 \qquad (4.46)$$

Finally, the average brick tensile strength of the unstrengthened "WZI" panels increases by 13% and 26% in the case of one and both sides application, respectively:

$$f_{bt,WZI,s1} = 0.90N/mm^2 + 1 \cdot \left(\frac{27.74N/mm}{240mm}\right) = 1.02N/mm^2 \qquad (4.47)$$

$$f_{bt,WZI,s2} = 0.90N/mm^2 + 2 \cdot \left(\frac{27.74N/mm}{240mm}\right) = 1.13N/mm^2 \qquad (4.48)$$

Therefore, Table 25 shows the modified mechanical properties, $f_{vm0,s}$ and $f_{bt,s}$, obtained with the proposed formulations, the friction coefficient μ, the interlocking ϕ, the length l and thickness t of the tested specimens.

Table 25: Mechanical properties of the panels strengthened with EQ-GRID

"KS" strengthened panels						
n_f [-]	$f_{vm0,KS,s}$ [N/mm²]	μ [-]	ϕ [-]	$f_{bt,KS,s}$ [N/mm²]	l [mm]	t [mm]
1	0.29	0.8	0.94	1.69	1250	240
2	0.40	0.8	0.94	1.81	1250	240
"WZI" strengthened panels						
n_f [-]	$f_{vm0,WZI,s}$ [N/mm²]	μ [-]	ϕ [-]	$f_{bt,WZI,s}$ [N/mm²]	l [mm]	t [mm]
1	0.42	0.8	1.29	1.02	1250	240
2	0.53	0.8	1.29	1.13	1250	240

Furthermore, to validate the calculated values, the maximum shear force attained during the laboratory tests, V_{Ed}, is directly compared with the analytical shear resistance, $V_{Rd,mod}$. To this aim, the Mann and Müller formulation is used with the modified material properties, $f_{vm0,s}$ and $f_{bt,s}$, as follows:

$$V_{Rd,\mathrm{mod}} = \min\left(V_{t,\mathrm{mod}}; V_{t,\lim,\mathrm{mod}}\right) \tag{4.49}$$

$$V_{t,\mathrm{mod}} = \frac{l \cdot t}{b} \cdot \left(\frac{f_{vm0,s}}{1+\mu \cdot \varphi} + \frac{\mu}{1+\mu \cdot \varphi} \cdot \sigma_v \right) \tag{4.50}$$

$$V_{t,\lim,\mathrm{mod}} = \frac{l \cdot t}{b} \cdot \frac{f_{bt,s}}{2.3} \cdot \sqrt{1+\frac{\sigma_v}{f_{bt,s}}} \tag{4.51}$$

Moreover, the in-plane shear resistance $V_{Rd,s}$ is calculated after CNR-DT 215/2018 with Eq. (4.52):

$$V_{Rd,s} = \min\left(V_t; V_{t,\lim}\right) + \frac{1}{\gamma_{Rd}} \cdot n_f \cdot l_f \cdot \left(\varepsilon_{\lim,conv,h} \cdot E_{f,h} \cdot t_{eq,h} + \varepsilon_{\lim,conv,h} \cdot E_{f,d} \cdot t_{eq,d} \right) \tag{4.52}$$

where:

- γ_{Rd} is a model safety factor equal to 2;
- n_f is the total number of applied textile layers; for one side applications is 1 while for both sides application is 2;

- l_f is the design length of the applied reinforcement. It is measured orthogonally to the shear force, and in any case, it cannot be assumed to be greater than the dimension of the wall H indicated in Figure 2.10. For all the tested panels, l_f is equal to the height h of the specimens (1.25 m);

- $\varepsilon_{lim,conv,h}$ is the conventional strain limit of the fibres arranged parallel to the shear force, i.e. the horizontal direction (Table 23);

- $E_{f,h}$ and $E_{f,d}$ are the average Young's modulus of the bare textile in the horizontal and diagonal direction (Table 23);

- $t_{eq,h}$ and $t_{eq,d}$ are the homogenized equivalent thickness of the system in the horizontal and diagonal directions.

As for the previous method, also, in this case, the product $\left(\varepsilon_{\lim,conv,h} \cdot E_{f,h} \cdot t_{eq,h} + \varepsilon_{\lim,conv,h} \cdot E_{f,d} \cdot t_{eq,d}\right)$ corresponds to the average strength limit of the system under diagonal cracking, and it is equal to 27.74 N/mm. Thus, the strength contribution $V_{t,f}$ is calculated in the case of one and both sides application, as follows:

$$V_{t,f} = \frac{1}{2} \cdot 1 \cdot 1.25m \cdot 27.74kN/m = 17.3kN \text{ (one side)} \tag{4.53}$$

$$V_{t,f} = \frac{1}{2} \cdot 2 \cdot 1.25m \cdot 27.74kN/m = 34.7kN \text{ (both sides)} \tag{4.54}$$

Finally, Table 26 directly compares the experimental and analytical in-plane shear strengths of the tested panels.

Table 26: Experimental and analytical shear strengths of the "KS" and "WZI" panels strengthened with EQ-GRID

ID specimen	V_{Ed} [kN]	$V_{t,mod}$ [kN]	$V_{t,lim,mod}$ [kN]	$V_{Rd,mod}$ [kN]	$V_{Rd,s}$ [kN]	EQ-GRID
"KS_1_1"	271.40	234.86	296.16	234.86	232.43	On one side
"KS_1_2"	204.36	193.73	281.18	193.73	191.29	
"KS_2_1"	271.49	234.46	304.51	234.46	229.58	On both sides
"KS_2_2"	252.50	235.97	305.06	235.97	231.09	
"WZI_1_1"	251.45	224.52	203.56	203.56	204.29	On one side
"WZI_1_2"	195.68	173.60	184.34	173.60	173.89	
"WZI_2_1"	280.47	224.56	213.54	213.54	215.63	On both sides
"WZI_2_2"	255.33	218.11	211.07	211.07	213.30	

Analyzing the results of the "KS" specimens, it is possible to conclude that the experimental values V_{Ed} are between the theoretical ones, $V_{t,mod}$ and $V_{t,lim,mod}$, confirming the mixed shear failure shown by the specimens. Only for the "KS_1_2" panel, V_{Ed} is slightly lower than the theoretical one $V_{t,mod}$. On the other hand, the shear strength of the "WZI" samples is reached analytically by the tensile failure of the masonry units ($V_{t,lim,mod}$). Only the theoretical resistance of the "WZI_1_2" panel corresponds to the diagonal cracking along the mortar joints ($V_{t,mod}$). Thus, the analytical results confirm the experimental observations. Moreover, a very good match is found between the resistance obtained with the proposed approach ($V_{Rd,mod}$) and after the CNR-DT 215/2018 ($V_{Rd,s}$). The maximum percentage difference between $V_{Rd,mod}$ and $V_{Rd,s}$ is +2.1% for the "KS" panels and -1.0% for the "WZI" specimens.

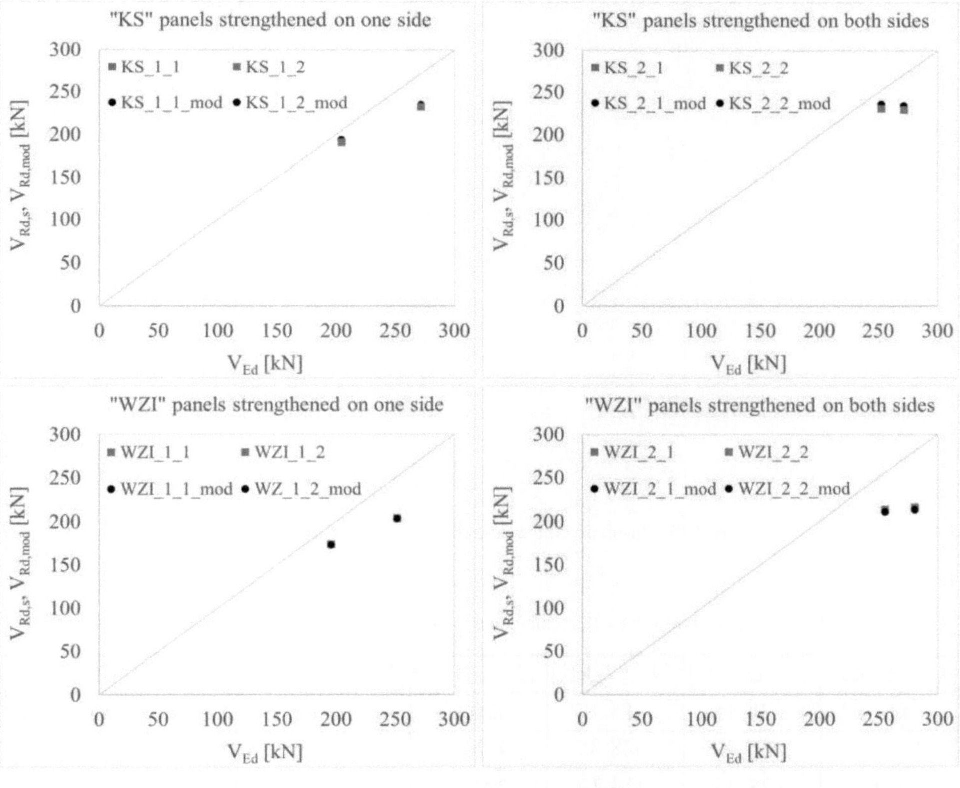

Figure 4.16: Experimental validation of the theoretical in-plane shear strength of the tested panels

In this regard, Figure 4.16 shows the experimental shear strength V_{Ed} versus the theoretical one for one or two sides applications of EQ-GRID, in which the values are properly marked to differentiate $V_{Rd,mod}$ (red points) from $V_{Rd,s}$ (light blue squares). This comparison confirms that both calculation methods provide a safe estimate of the experimental results. Furthermore, the cross-section analysis described in the previous paragraph is also performed. Figure 4.17 shows the in-plane flexural and shear strength domains of the "KS" and "WZI" panels reinforced with EQ-GRID. Since the shear failure domain is calculated with the proposed approach and after CNR-DT 215/2018, a very good agreement between the calculation methods is evident. Finally, the values of the normal forces $N_{Rd,s}$, calculated for each failure mode, are indicated in Table 27.

Figure 4.17: In-plane failure domains of the tested masonry panels

Table 27: Normal forces $N_{Rd,s}$ calculated at the flexural failure modes from 0 to 4

ID panel	$N_{Rd,s,0}$ [kN]	$N_{Rd,s,1}$ [kN]	$N_{Rd,s,2}$ [kN]	$N_{Rd,s,3}$ [kN]	$N_{Rd,s,4}$ [kN]	EQ-GRID
"KS_1"	-28.35	-14.17	420.07	2000.82	2501.03	One side
"KS_2"	-56.70	-28.35	408.95	2000.82	2501.03	Both sides
"WZI_1"	-28.35	-14.17	221.16	1077.84	1347.30	One side
"WZI_2"	-56.70	-28.35	210.04	1077.84	1347.30	Both sides

Chapter 5

Modelling of cyclic lateral shear tests with the Equivalent frame method

5.1 Description of the modelling

The numerical modelling results of the cyclic lateral shear tests illustrated in § 3.6 are presented in this chapter. The panels are modelled as masonry piers through the nonlinear ML-BEAM element implemented in the TREMURI software (§ 1.5.2, § 2.4). As shown in Figure 5.1, this element is characterized by two nodes, *i* and *j*, three kinematic variables (*u, v, φ*) and three generalized forces (*V, N, M*) at each node. The assumed element's height *h*, width *l* and thickness *t* are equal to the dimensions of the specimens, i.e. 1250 x 1250 x 240 mm.

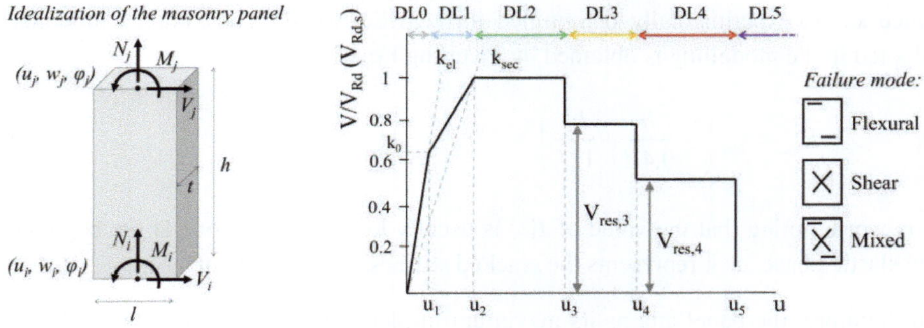

Figure 5.1: Idealization of the single panel and monotonic constitutive law of the ML-BEAM element implemented in TREMURI, from [14], [46]

The panels' elastic behaviour is modelled through the secant stiffness k_{sec} resulting from the experimental multilinear curves (Eq. (3.14)). In this regard, the horizontal elastic displacement $u_{j,e}$ of the ML-BEAM element is calculated after the beam theory through the shear and flexural stiffness. Thus, Eq. (5.1) is obtained when the element's base is fixed ($u_i = v_i = \varphi_i = 0$) and the upper rotation constrained ($u_j \neq 0;\ v_j \neq 0;\ \varphi_j = 0$):

$$u_{j,e} = u_{j,e,S} + u_{j,e,F} = \chi \frac{V_j h}{GA} + \frac{V_j h^3}{12 E_{\mathrm{sec}} I} \tag{5.1}$$

Where:

- χ is the shear coefficient equal to 1.2 for rectangular cross-sections,
- V_j is the shear force,
- h is the height of the panel,
- G and E_{sec} are the elastic shear and secant moduli,
- A and I are the area and moment of inertia of the panel's cross-section.

By assigning the unit value to V_j, the elastic secant stiffness k_{sec} of the ML-BEAM element can be determined as follows:

$$k_{\mathrm{sec}} = \frac{1}{u_{j,e}} = \frac{1}{u_{j,e,S} + u_{j,e,F}} = \frac{1}{\chi \frac{h}{GA} + \frac{h^3}{12 E_{\mathrm{sec}} I}} = \frac{1}{\chi \frac{h}{0.4 E_{\mathrm{sec}} A} + \frac{h^3}{12 E_{\mathrm{sec}} I}} \tag{5.2}$$

Since k_{sec} is experimentally determined through Eq. (3.14), the Young's modulus E_{sec} adopted in the modelling is obtained by inverting Eq. (5.2):

$$E_{\mathrm{sec}} = k_{\mathrm{sec}} \left(\chi \frac{h}{0.4A} + \frac{h^3}{12I} \right) \tag{5.3}$$

It is worth noting that the value of E_{sec} is usually lower than the one at the beginning of the elastic phase, as it represents the cracked stiffness of the element.

Furthermore, the panel attains its maximum in-plane resistance at the end of the elastic response (Figure 5.1). V_{Rd} (or $V_{Rd,s}$ for the reinforced element) is calculated by the program as the minimum between the chosen failure criteria and considering the current axial force acting at each load step. In this respect, the Mann and Müller theory [23] is adopted for the shear failure (Eqs. (1.18) - (1.19)), and the increase in strength is computed after CNR-DT 215/2018 (§ 4.2.3). Concerning the in-plane flexural failure, the resistance of the URM panels is calculated, neglecting the masonry tensile strength and assuming a stress block normal distribution at the compressed toe (Eqs. (1.25) - (1.26)). For the specimens reinforced with EQ-GRID, the cross-section analysis shown in § 4.2.3

is performed. It is important to note that the ML-BEAM element can only reproduce the same cyclic behaviour and type of failure as the tested panels if the improved in-plane flexural capacity is also taken into account. For this reason, the strengthening effect of the EQ-GRID system is computed after CNR-DT 215/2018 and not through the proposed modification of the masonry material properties. As shown in § 4.2.4, the adoption of the modified masonry material properties, $f_{vm0,s}$ and $f_{bt,s}$ (Eqs. (4.16), (4.17)), leads to maximum shear resistances very similar to those obtained after CNR-DT 215/2018 (Eqs. (4.18), (4.19)). However, some strengthened panels showed a hybrid behaviour during the post-peak phase that only with the increased flexural capacity can be reproduced. In this respect, the mixed failure mode is possible in the modelling through the parameters a_1 and a_2 illustrated in Figure 1.40. They are assumed 0.95 and 1.25, respectively, and define two areas close to the intersection points of the flexural and shear domains in which the hybrid behaviour can occur.

Furthermore, the nonlinear response of the tested specimens is modelled by assigning progressive strength degradations ($V_{res,3}$, $V_{res,4}$ in Figure 5.1) at defined drift values until very severe damage levels. Different values can be assumed through the ratio $V_{res,i}$ / V_{Rd} ($V_{Rd,s}$) to describe the bending and shear response of the panels. They are indicated for each specimen in the next paragraphs. Thus, when the hybrid failure mode occurs, the average values of $V_{res,i}$ / V_{Rd} ($V_{Rd,s}$) are calculated by the program starting from those assigned for the shear and bending failure.

Concerning the hysteretic response of the tested panels, the parameters c_1, c_2, c_3 for the shear behaviour (Table 28) are calibrated based on the experimental results, as all the samples showed diagonal shear cracks. Moreover, since no specimen failed by pure bending, it was impossible to calibrate the parameters c_1, c_2, c_3, c_4 for the flexural failure. Therefore, the values reported in Table 28 are consistent with those adopted in [46] for the non-linear modelling and time-history analyses of masonry Italian code-conforming buildings.

Table 28: Parameters adopted for the cyclic response of the ML-BEAM element

“KS” specimens						
Shear			Bending			
$c_1 = 0.4$	$c_2 = 0.6$	$c_3 = 0$	$c_1 = 0.9$	$c_2 = 0.8$	$c_3 = 0.6$	$c_4 = 0.5$
“WZI” specimens						
Shear			Bending			
$c_1 = 0.6$	$c_2 = 0.8$	$c_3 = 0$	$c_1 = 0.9$	$c_2 = 0.8$	$c_3 = 0.6$	$c_4 = 0.5$

Regarding the modelling phases, all the analyses are performed by first applying the panel's self-weight and the constant vertical load N_v, as imposed during the test. Then, the experimental values of the horizontal displacements recorded at the top of the walls, u, are applied to the node j of the element through incremental static analyses performed under displacement control. Concerning the boundary conditions, the base of the element is considered fixed ($u_i = v_i = \varphi_i = 0$) while the top has only one d.o.f. constrained ($u_j \neq 0$; $v_j \neq 0$; $\varphi_j = 0$). This latter assumption is reasonable since the experimental values of top rotation are so small that they may be neglected in the numerical model. As depicted in Figure 5.3, the rotation of the steel beam positioned on the top of the panel was limited by four threaded rods with a diameter of 32 mm (two bars on each side of the sample). For example, Figure 5.2 shows the top rotation - drift diagram of the unstrengthened specimen "WZI_0_1".

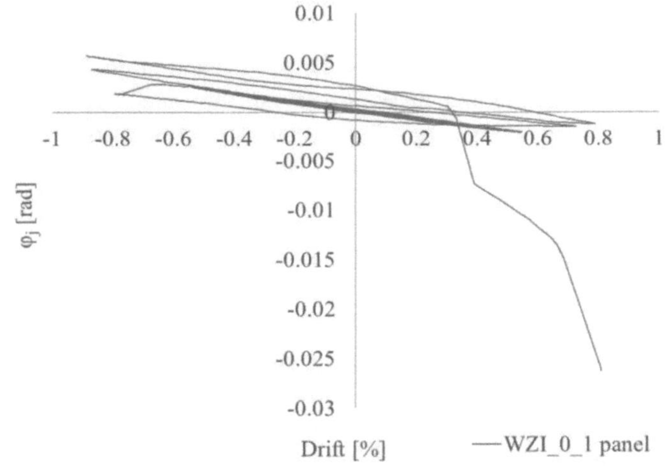

Figure 5.2: Specimen "WZI_0_1": rotation at the top of the panel

In this case, the rotation values are equal to about ±0.005 rad, and only at the end of the test, when the specimen collapsed, the maximum was -0.03 rad. Therefore, it is reasonable to assume that they may be neglected for the modelling.

Moreover, the axial force developed in the four threaded rods during the tests (Figure 5.3) was recorded by two load cells.

Thus, the total normal force acting on the node j of the element, N_j (Figure 5.3), can be calculated as follows:

$$N_j = -\left(\frac{G_w}{2} + N_v + 2N_{v1} + 2N_{v2}\right) \tag{5.4}$$

Where:

- G_w is the panel's self-weight,
- N_v is the normal force applied by the vertical actuator,
- $2N_{v1}$, $2N_{v2}$ are the forces developed in the threaded bars.

It is worth noting that N_v is a constant value, while N_{v1} and N_{v2} are variable since the tests were cyclically performed.

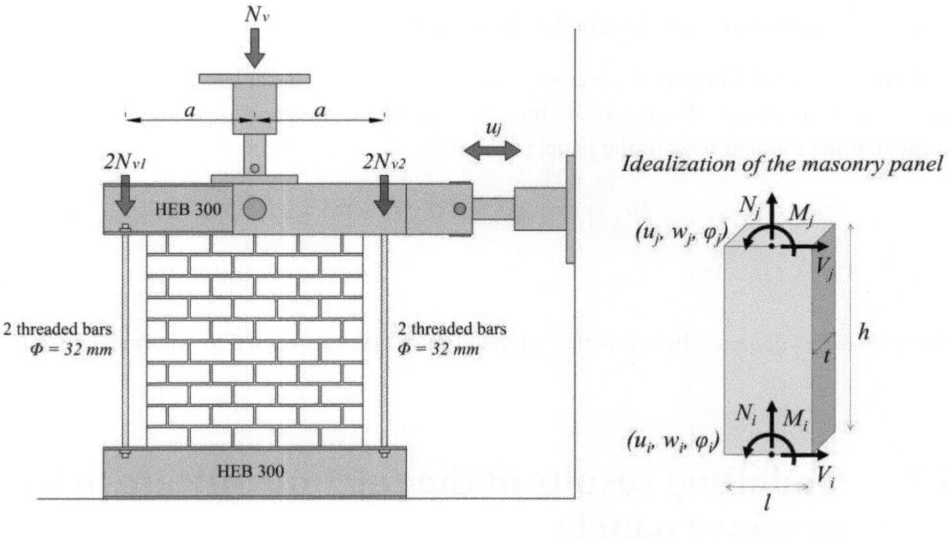

Figure 5.3: Scheme of the normal forces acting on the tested masonry panels

Therefore, as the vertical stress influences the load-bearing capacity of a masonry pier, it is important to consider this variation during the modelling. To this aim, the normal force N_j defined with Eq. (5.4) is also equal to Eq. (5.5). The latter is calculated through the axial stiffness of the ML-BEAM element:

$$N_j = -\frac{E_{sec}A}{h}v_i + \frac{E_{sec}A}{h}v_j \tag{5.5}$$

Since the experimental values of N_j are known and the node i is constrained ($v_i = 0$), the vertical displacement of the node j under the normal force N_j can be easily obtained by inverting Eq. (5.5):

$$v_j = N_j \frac{h}{E_{\text{sec}} A} \tag{5.6}$$

In this way, the variation of normal force can be defined in the input by assigning both displacements, u_j and v_j, simultaneously, through incremental static analyses performed under displacement control. It is worth noting that the TREMURI program calculates the normal force N_j with Eq. (5.5) at each load step and, if necessary, performs a nonlinear correction based on the comparison with the limit resistance $N_{Rd,4}$ of the element (Table 27). Since $N_{Rd,4}$ is never attained during the modelling, the elastic prediction of N_j through Eq. (5.5) leads to results that concur exactly with the experimental ones even when the element is in the non-linear phase under shear forces.

Therefore, the modelling results are presented in the next paragraphs in terms of internal forces' and moments' diagrams. In this regard, the experimental values of bending moment at the top and base of the panel, M_j and M_i, are so calculated:

$$M_j = 2\left(N_{v2} - N_{v1}\right) a \tag{5.7}$$

$$M_i = V_j h - M_j \tag{5.8}$$

Where a is the distance between the threaded bars and the centre of the panel (Figure 5.3).

5.2 Modelling results of the calcium-silicate brick masonry panels

5.2.1 "KS" unstrengthened specimens

The modelling results of the cyclic lateral shear tests performed on the "KS_0_1" and "KS_0_2" specimens are presented in this paragraph. The secant elastic stiffness of the ML-BEAM element, k_{sec}, is experimentally obtained from the equivalent multilinear curves of both samples. The masonry mechanical properties are the same as those adopted in § 4.1.1 to calculate the in-plane failure domain of the unreinforced "KS" panels. In this respect, Table 29 reports the average values of masonry compressive strength, $f_{m,u,KS}$, initial shear strength, $f_{vm0,KS}$, local coefficient of friction, μ_{KS}, interlocking parameter, ϕ_{KS} and tensile strength of the brick, $f_{bt,KS}$, adopted for the modelling.

Since the "KS_0_1" specimen failed with diagonal cracks, the drift and residual resistance values defined for the shear failure (Table 30) are obtained from the equivalent multilinear curve determined in § 3.6.3.1. Concerning flexural behaviour, the drifts are assumed to be twice those adopted for the shear, as no experimental data were available for this type of failure.

Table 29: Average values of the mechanical properties of the unstrengthened "KS" masonry panels

"KS" unstrengthened panels						
$f_{m,u,KS}$ [N/mm²]	$f_{vm0,KS}$ [N/mm²]	μ_{SK} [-]	ϕ_{KS} [-]	$f_{vm0,red,KS}$ [N/mm²]	$\mu_{red,KS}$ [-]	$f_{bt,KS}$ [N/mm²]
9.81	0.17	0.8	0.94	0.10	0.46	1.58

It is important to stress that this assumption is consistent with the indicative values proposed by the Italian technical standard CNR-DT 212/2013 "Guide for the Probabilistic Assessment of the Seismic Safety of Existing Buildings" [25]. The residual resistances for the flexural damage are also coherent with CNR-DT 212/2013.

Table 30: "KS_0_1" specimen: values of drift and residual resistance

	Drift			Residual resistance ($V_{res,i}$ / V_{Rd})	
Damage	DL3	DL4	DL5	DL3 → DL4	DL4 → DL5
Shear	$d_{3,S} = 0.62\%$	$d_{4,S} = 0.75\%$	$d_{5,S} = 0.93\%$	0.72	0.44
Bending	$d_{3,B} = 1.24\%$	$d_{4,B} = 1.5\%$	$d_{5,B} = 1.86\%$	1.0	0.85

The experimental results are compared to the numerical ones in terms of internal forces, V_j, N_j, and moments, M_i, M_j. In Figure 5.4 are shown the results of the "KS_0_1" specimen, which was tested under the constant vertical load of 240 kN.

It is worth noting that a very good match between the experimental and numerical diagrams is provided. The values of normal force N_j concur exactly with those recorded during the test. Moreover, the maximum shear force $V_{j,max}$ is 208.62 kN. It is very close to the one calculated in § 4.1.1 (V_{Rd} = 208.47), and the hysteretic response reliably reproduces the experimental curve. Concerning the bending moments, M_i and M_j, the first one is underestimated, while the second is overestimated. This approximation is due to the assumption that the rotation at the top of the panel, φ_j, is constrained. However, it does not influence the shear force V_j since the latter is always in equilibrium with the sum of the moments (Eq. (5.9)).

$$V_j = \frac{M_i + M_j}{h} \tag{5.9}$$

Where h is the height of the element.

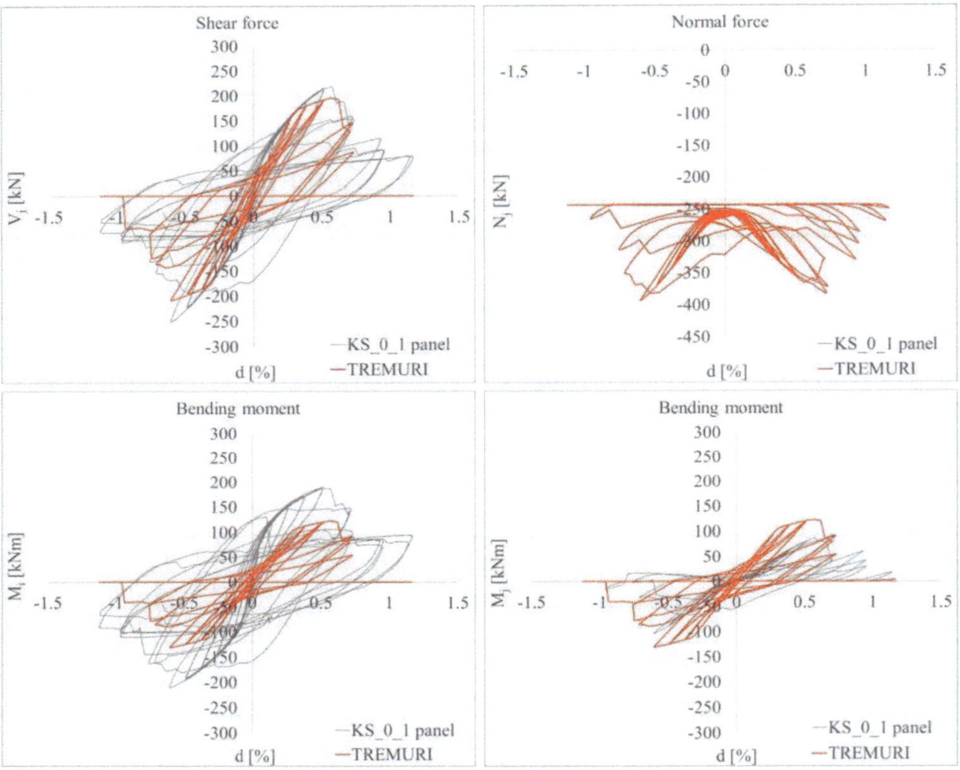

Figure 5.4: Specimen "KS_0_1": internal forces and moments

The "KS_0_1" panel attains the in-plane strength with shear failure from DL2 to DL5, reproducing correctly the cracking pattern observed during the test. The failure domain in Figure 5.5 also confirms this result since N_j varies between 243 kN and 393 kN. In this range, no mixed failure can occur. The latter is only possible at normal force values between 51 kN and 93 kN and between 1690 kN and 1930 kN.

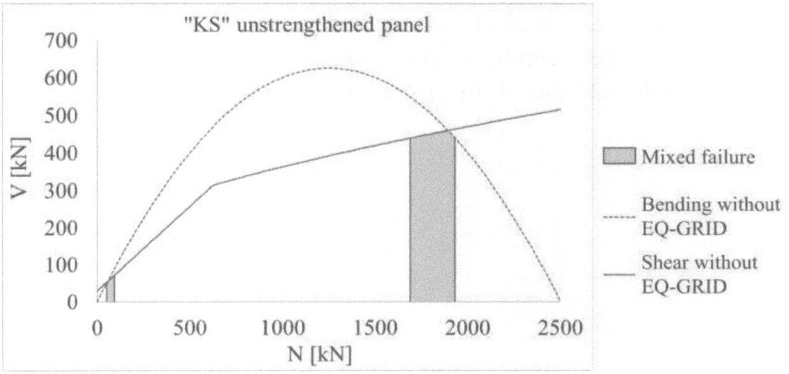

Figure 5.5: In-plane failure domain of the "KS" unstrengthened panels

Concerning the masonry properties of the second unstrengthened "KS" specimen, they are the same as those adopted for the previous sample. The drifts and residual resistances assumed in the modelling with TREMURI are shown in Table 31.

Table 31: "KS_0_2" specimen: values of drift and residual resistance

	Drift			Residual resistance ($V_{res,i}$ / V_{Rd})	
Damage	DL3	DL4	DL5	DL3 → DL4	DL4 → DL5
Shear	$d_{3,S}$ = 0.51%	$d_{4,S}$ = 0.75%	$d_{5,S}$ = 0.93%	0.72	0.44
Bending	$d_{3,B}$ = 1.0%	$d_{4,B}$ = 1.5%	$d_{5,B}$ = 1.86%	1.0	0.85

As described in § 3.6.3.1, the "KS_0_2" panel was tested under the constant vertical load of 120 kN, and the test was stopped shortly after achieving the maximum horizontal force. Therefore, only the equivalent bilinear curve could be determined, from which the drift $d_{3,S}$ was obtained since the specimen failed by diagonal cracking ($d_{3,F}$ is assumed to be twice). For this reason, the drifts and residual resistances defined for the DLs 4, 5 are the same as those assumed for the previous sample. The in-plane strength domain in Figure 5.5 confirms that no hybrid behaviour can occur since the normal force N_j varies between 123 kN and 273 kN.

Furthermore, as shown in Figure 5.6, the numerical results reproduce very well the experimental ones. The element achieves the in-plane strength by shear failure. The maximum shear force $V_{j,max}$ is equal to 144.09 kN, and it is very close to the expected one (V_{Rd} = 143.98). The cyclic behaviour is also very well reproduced from DL2 to DL3 (DL4 is not reached). As depicted in Figure 3.30, the panel showed a final experimental cracking pattern between DL3 and DL4. Thus, it is possible to conclude that the numerical results find an excellent correspondence with the experimental ones, although the imposed values of drift and residual strength at DLs 4, 5 are those of the previous sample.

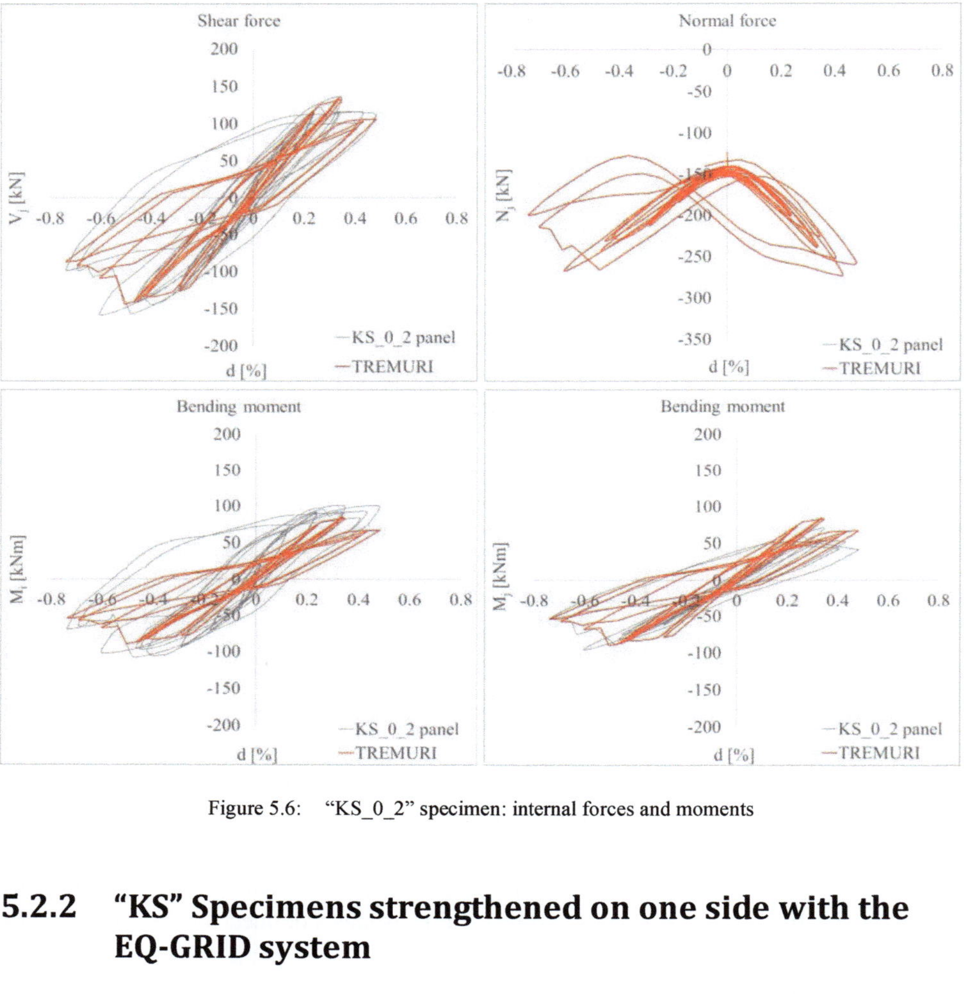

Figure 5.6: “KS_0_2” specimen: internal forces and moments

5.2.2 “KS” Specimens strengthened on one side with the EQ-GRID system

The first calcium-silicate brick masonry panel strengthened EQ-GRID and modelled in TREMURI is the “KS_1_1” specimen. The adopted masonry mechanical properties are reported in Table 30. The increase in shear strength is computed after CNR-DT 215/2018 through Eqs. (4.18) - (4.19), where the total number of applied textile layers n_f is one since the grid was applied only on one side. In this respect, the FRCM design length l_f is

1.25 m, and the average strength of the system under diagonal cracking is equal to 27.74 kN/m $\left(= \varepsilon_{\lim,conv,h} \cdot E_{f,h} \cdot t_{eq,h} + \varepsilon_{\lim,conv,h} \cdot E_{f,d} \cdot t_{eq,d}\right)$.

Moreover, the TREMURI program checks that the shear force calculated at each load step does not exceed the diagonal crushing resistance $V_{t,c}$, as suggested by CNR-DT 215/2018. For the "KS" masonry, $V_{t,c}$ is equal to:

$$V_{t,c} = 0.25 \cdot f_{m,u,KS} \cdot t \cdot d_f = 0.25 \cdot 9.81 \cdot 240 \cdot 1250 = 735.75kN \tag{5.10}$$

Where:

- $f_{m,u,KS}$ is the average masonry compressive strength (Table 30);
- t is the thickness of the specimen (240 mm);
- d_f is the distance between the compressed edge of the panel and the fibre of the reinforcement that attains the highest tensile strain. As the EQ-GRID system is applied on the entire panel's surface, d_f corresponds to the sample's length (1250 mm).

Furthermore, the adopted drifts and residual resistances are shown in Table 33. The values for the shear damage are obtained from the equivalent multilinear curve calculated in § 3.6.3.2 since the sample failed by diagonal cracking. Then, the drifts chosen for the flexural behaviour are equal to twice those adopted for the shear, as observed in § 5.2.1. The assumed residual resistances are consistent with the values suggested by CNR-DT 212/2013 [25].

Table 32: "KS_1_1" specimen: values of drift and residual resistance

	Drift			Residual resistance ($V_{res,i} / V_{Rd,s}$)	
Damage	DL3	DL4	DL5	DL3 → DL4	DL4 → DL5
Shear	$d_{3,S}$ = 0.82%	$d_{4,S}$ = 1.0%	$d_{5,S}$ = 1.22%	0.63	0.41
Bending	$d_{3,B}$ = 1.64%	$d_{4,B}$ = 2.0%	$d_{5,B}$ = 2.44%	1.0	0.85

The results of the modelling in TREMURI are presented in Figure 5.7 with the experimental ones. The "KS_1_1" specimen was tested under the constant vertical load of 240 kN. The maximum shear force $V_{j,max}$ obtained from the numerical modelling, is equal to 232.53 kN, and it is very close to the expected one $V_{Rd,s}$ = 232.43 kN. The ML-BEAM element reaches the in-plane resistance with shear failure from DL2 until DL5, reproducing, in this way, the cracking pattern observed during the test.

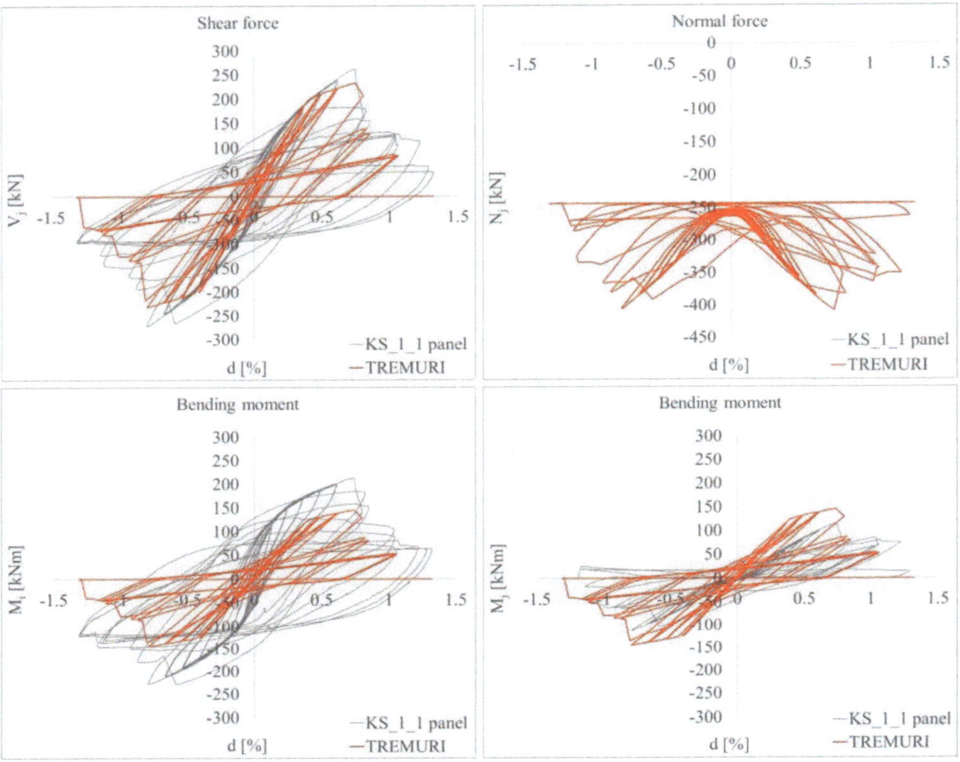

Figure 5.7: "KS_1_1" specimen: internal forces and moments

The strength domain of the panel in Figure 5.8 confirms this type of failure since the normal force N_j varies between 243 kN and 407 kN (Figure 5.7). In this range of values, no hybrid behaviour can occur. It is only possible at normal forces between 48 kN - 106 kN and 1636 kN - 1906 kN.

Furthermore, as shown in Figure 5.7, the bending moments, M_i and M_j, are under- and overestimated, respectively. The assumption made on the rotation at the top of the panel ($\varphi_j = 0$) causes the difference with the experimental values. However, this approximation does not influence the shear force values because the global equilibrium of the element is always correctly ensured.

Moreover, as described in § 3.6.3.2, the experimental data of the second calcium-silicate brick masonry panel strengthened with EQ-GRID, i.e. the "KS_1_2" specimen, were not completely reliable in the post-peak phase of the test. Therefore, only the first part of the hysteresis curve, until achieving the maximum shear force, is modelled in TREMURI. The masonry mechanical properties assumed for the "KS_1_2" panel are indicated in

Table 30. The strengthening effect of the EQ-GRID system is calculated with the approach proposed by CNR-DT 215/2018.

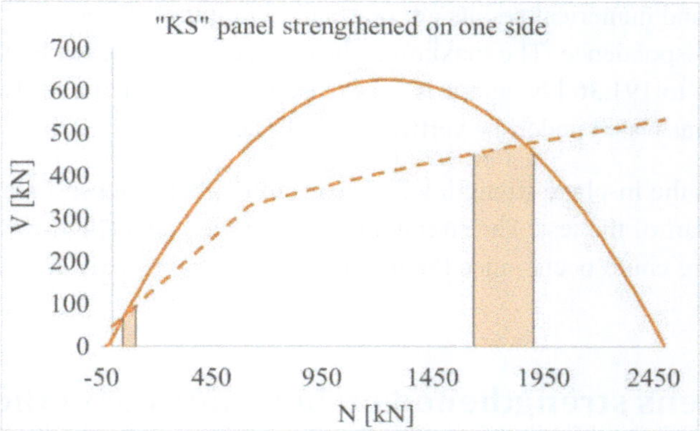

Figure 5.8: In-plane failure domain of the "KS" panels strengthened on one side

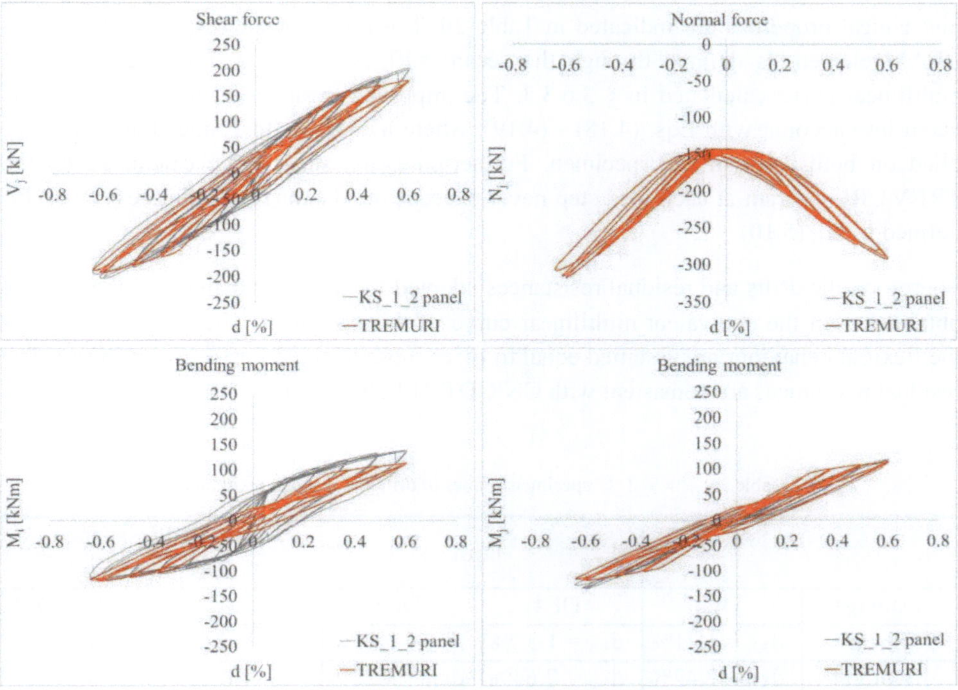

Figure 5.9: "KS_1_2" specimen: experimental and numerical values of internal forces and moments

The adopted values of drift and residual resistances are equal to those defined for the "KS_1_1" panel (Table 32) since the equivalent multilinear curve could not be determined in this case.

Therefore, the experimental and numerical results are presented in Figure 5.9, in which it is evident the very good correspondence. The maximum shear force $V_{j,max}$, reached by the ML-BEAM element, is equal to 191.36 kN, and it is very close to the expected one, $V_{Rd,s}$ = 191.29 kN (the specimen was tested under the vertical load of 120 kN).

Finally, the element achieves the in-plane strength with shear failure, but it doesn't reach the DL3, as during the first part of the test. The strength domain of the panel (Figure 5.8) confirms that no mixed failure could occur since the normal force N_j varies between 123 kN and 335 kN (Figure 5.9).

5.2.3 "KS" specimens strengthened on both sides with the EQ-GRID system

The modelling results of the "KS_2_1" specimen are shown in Figure 5.10 with the experimental ones. As for the samples reinforced on one side, the adopted masonry mechanical properties are indicated in Table 30. The linear elastic response of the ML-BEAM element is defined through the secant stiffness obtained from the equivalent multilinear curve calculated in § 3.6.3.3. The improved shear resistance of the panel is taken into account with Eqs. (4.18) - (4.19), where n_f is equal to 2 since the grid is applied on both sides of the specimen. Furthermore, the shear force calculated by the TREMURI program at each load step never exceeds the diagonal crushing resistance $V_{t,c}$ defined in Eq. (5.10).

Therefore, the drifts and residual resistances adopted for the shear damage (Table 33) are obtained from the equivalent multilinear curve of this specimen. The values chosen for the flexural behaviour are assumed equal to twice those adopted for the shear failure. The residual resistances are consistent with CNR-DT 212/2013 [25].

Table 33: "KS_2_1" specimen: values of drift and residual resistance

	Drift			Residual resistance ($V_{res,i}$ / $V_{Rd,s}$)	
Damage	DL3	DL4	DL5	DL3 → DL4	DL4 → DL5
Shear	$d_{3,S}$ = 1.21%	$d_{4,S}$ = 1.32%	$d_{5,S}$ = 1.61%	0.63	0.33
Bending	$d_{3,B}$ = 2.42%	$d_{4,B}$ = 2.64%	$d_{5,B}$ = 3.22%	1.0	0.85

As shown in Figure 5.10, the numerical results have a good match with the experimental ones since the ML-BEAM element reaches the in-plane strength with shear failure from DL2 until DL5. The hysteretic behaviour is also well reproduced.

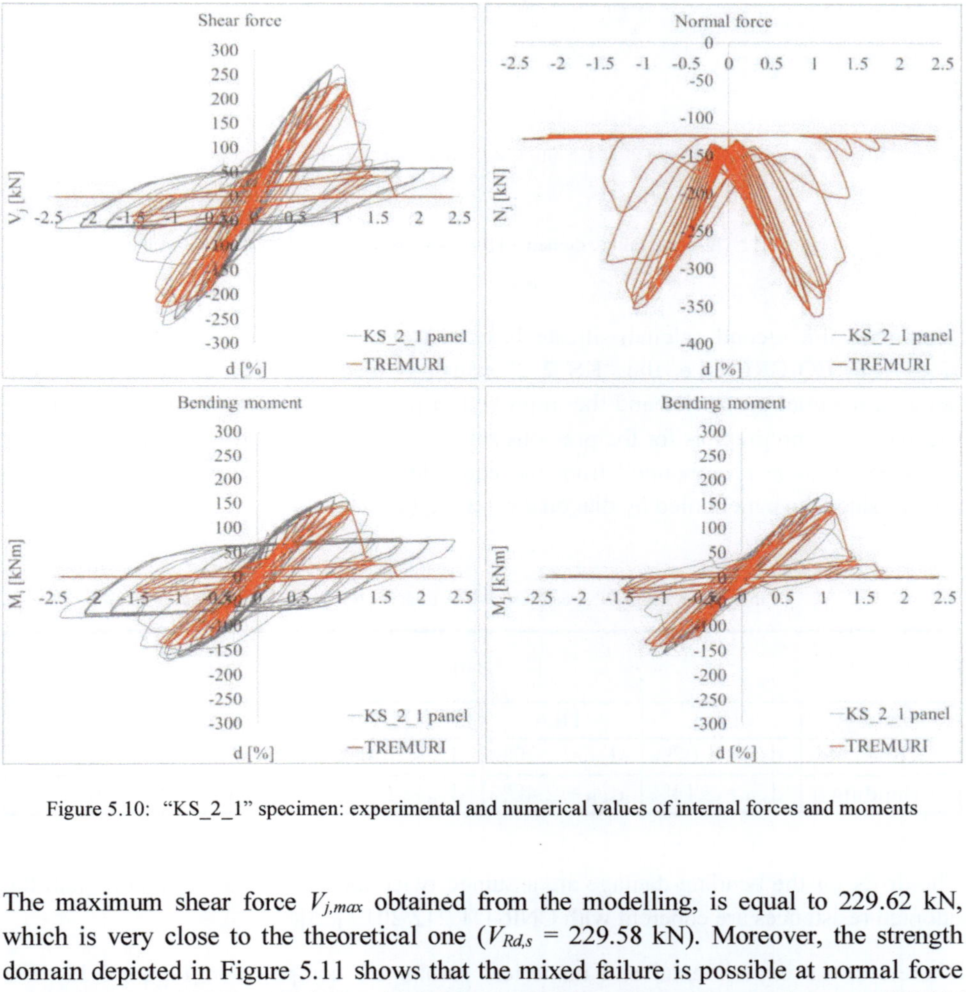

Figure 5.10: "KS_2_1" specimen: experimental and numerical values of internal forces and moments

The maximum shear force $V_{j,max}$ obtained from the modelling, is equal to 229.62 kN, which is very close to the theoretical one ($V_{Rd,s}$ = 229.58 kN). Moreover, the strength domain depicted in Figure 5.11 shows that the mixed failure is possible at normal force values between 45 kN - 120 kN and 1590 kN - 1880 kN. Since N_j varies between 123 kN and 363 kN (Figure 5.10), no mixed failure occurs. It is important to stress that the cross-section analysis described in § 4.2.3. is also performed by the program. In this case, the consideration of the increased in-plane flexural capacity is fundamental to reproduce the panel's experimental behaviour correctly.

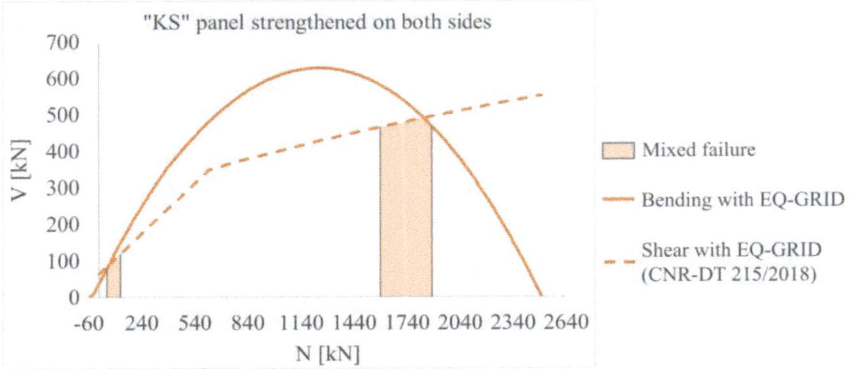

Figure 5.11: In-plane failure domain of the "KS" panels strengthened on both sides

Moreover, the second calcium-silicate brick masonry specimen strengthened on both sides with EQ-GRID, i.e. the "KS_2_2" panel, is also modelled in TREMURI. The adopted material properties and the improved in-plane load-bearing capacity are implemented in the program as for the previous sample. The drifts and residual resistances for the shear damage are obtained from the equivalent multilinear curve determined in § 3.6.3.3 since the panel failed by diagonal cracking (Table 34).

Table 34: "KS_2_2" specimen: values of drift and residual resistance

	Drift			Residual resistance ($V_{res,i}$ / $V_{Rd,s}$)	
Damage	DL3	DL4	DL5	DL3 → DL4	DL4 → DL5
Shear	$d_{3,S}$ = 1.07%	$d_{4,S}$ = 1.20%	$d_{5,S}$ = 1.46%	0.65	0.39
Bending	$d_{3,B}$ = 2.14%	$d_{4,B}$ = 2.40%	$d_{5,B}$ = 2.92%	1.0	0.85

The drifts for the bending damage are assumed twice those of the shear failure, and the adopted resistances are coherent with CNR-DT 212/2013 [25].

Therefore, the experimental and numerical results are compared in Figure 5.12. The maximum shear force, $V_{j,max}$, attained during the analysis is equal to 231.13 kN, and the expected value, $V_{Rd,s}$, is 231.09 kN (the "KS_2_2" was tested under the constant vertical load of 120 kN). Moreover, the ML-BEAM element achieves the in-plane resistance with shear failure from DL2 until DL5. No mixed failure occurs since the absolute values of N_j vary between 123 kN and 367 kN (Figure 5.12). Finally, also in this case, the consideration of the improved in-plane flexural capacity is fundamental to reproduce the panel's hysteretic behaviour correctly.

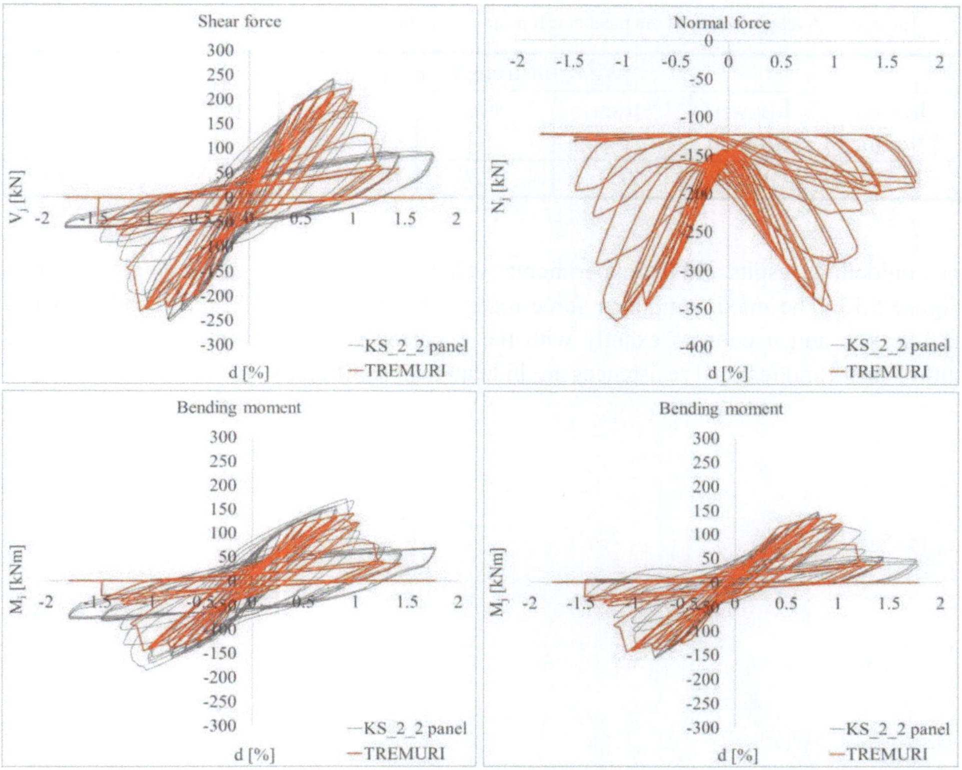

Figure 5.12: "KS_2_2" specimen: experimental and numerical values of internal forces and moments

5.3 Modelling results of the hollow clay brick masonry panels

5.3.1 "WZI" unstrengthened specimens

The cyclic lateral shear tests performed on the unreinforced "WZI" specimens are modelled with the Equivalent frame method implemented in the TREMURI program. The elastic properties of the ML-BEAM element are defined through the secant stiffness, k_{sec}, which is obtained from the calculation of the equivalent multilinear curve (§ 3.6.4.1). Moreover, Table 35 reports the adopted average values of masonry compressive strength $f_{m,u,WZI}$, initial shear strength $f_{vm0,WZI}$, local coefficient of friction μ_{WZI}, interlocking parameter ϕ_{WZI} and tensile strength of the brick $f_{bt,WZI}$.

Table 35: Average values of the mechanical properties of the unstrengthened "WZI" masonry panels

"WZI" unstrengthened panels						
$f_{m,u,WZI}$ [N/mm²]	$f_{vm0,WZI}$ [N/mm²]	μ_{WZI} [-]	ϕ_{WZI} [-]	$f_{vm0,red,WZI}$ [N/mm²]	$\mu_{red,WZI}$ [-]	$f_{bt,WZI}$ [N/mm²]
5.28	0.30	0.8	1.29	0.15	0.39	0.90

The modelling results and the experimental values of the "WZI_0_1" panel are shown in Figure 5.13. The maximum shear force reached by the ML-BEAM element is equal to 180.86 kN, and it concurs exactly with the expected value of resistance, V_{Rd}. The assumed drifts and residual resistances are indicated in Table 36.

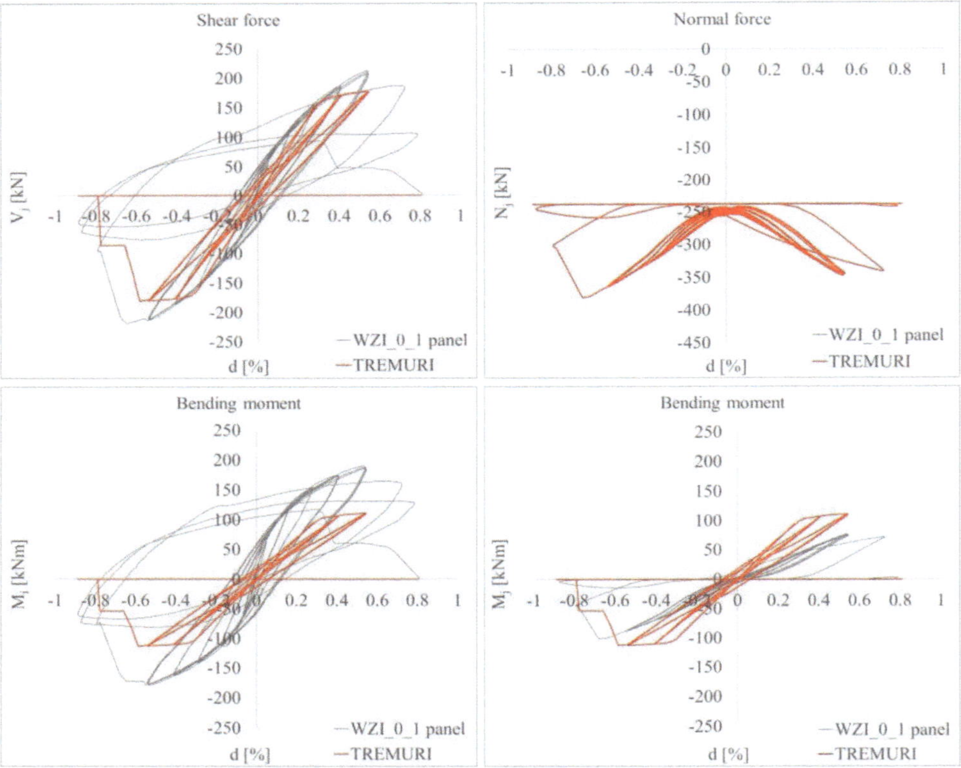

Figure 5.13: "WZI_0_1" specimen: experimental and numerical values of internal forces and moments

Table 36: "WZI_0_1" specimen: values of drift and residual resistance

	Drift			Residual resistance ($V_{res,i}$ / V_{Rd})	
Damage	DL3	DL4	DL5	DL3 → DL4	DL4 → DL5
Shear	$d_{3,S}$ = 0.60%	$d_{4,S}$ = 0.65%	$d_{5,S}$ = 0.73%	0.69	0.47
Bending	$d_{3,B}$ = 1.20%	$d_{4,B}$ = 1.30%	$d_{5,B}$ = 1.46%	1.0	0.85

Regarding the shear damage, the adopted values are obtained from the equivalent multi-linear curve determined in § 3.6.4.1 since the panel failed by diagonal cracking. Concerning the flexural behaviour, the chosen drifts are equal to twice those assumed for the shear failure. The residual resistances suggested by CNR-DT 212/2013 [25] are adopted.

The ML-BEAM element attains the in-plane strength showing shear damage from DL2 to DL5 and finding a good match with the cracking pattern observed during the test. The failure domain of the panel also confirms this result since the absolute values of N_j vary between 242 kN and 381 kN. As depicted in Figure 5.14, no mixed failure can occur in that range. It is only possible at normal force values between 74 kN - 136 kN and 885 kN - 1025 kN.

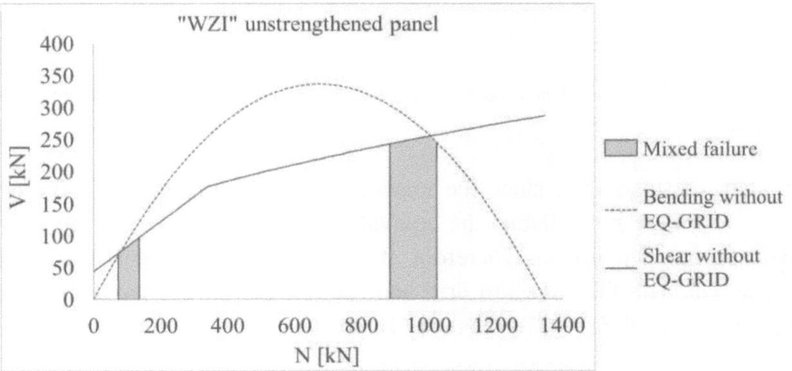

Figure 5.14: In-plane failure domain of the unstrengthened "WZI" specimens

Concerning the "WZI_0_2" panel, the diagrams of the internal forces and moments are shown in Figure 5.15.

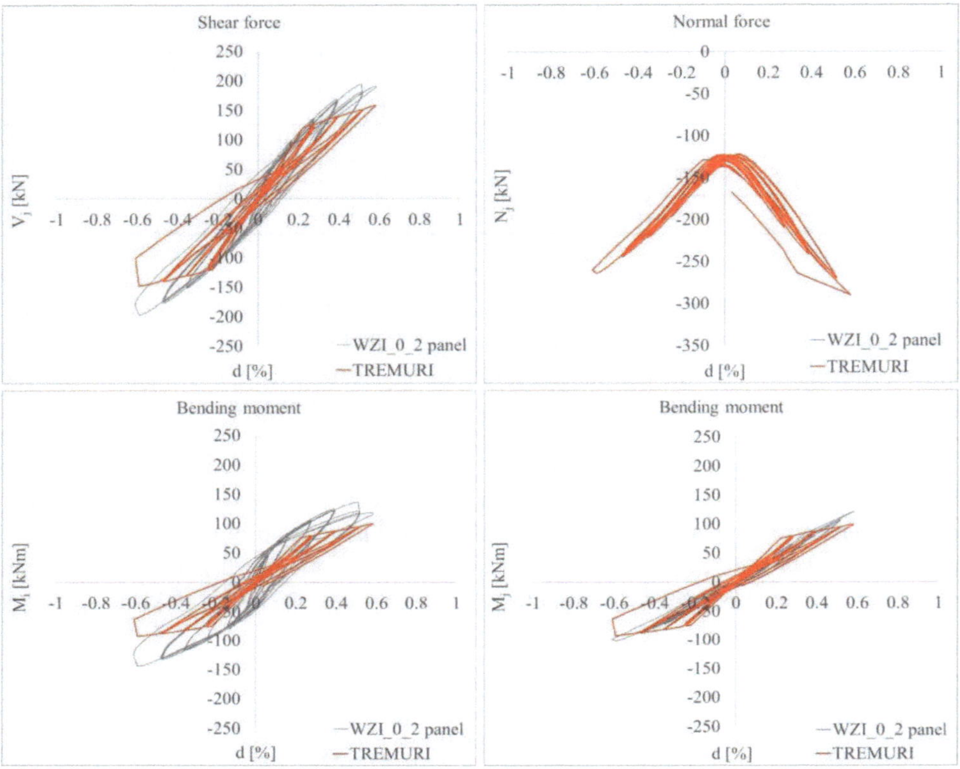

Figure 5.15: "WZI_0_2" specimen: experimental and numerical values of internal forces and moments

It is important to stress that, since the experimental values recorded in the post-peak phase of the test were not reliable, the equivalent multilinear curve of the "WZI_0_2" panel could be not determined. Therefore, the modelling is performed using only the available test data, and the values of drift and residual resistances are assumed equal to those defined for the "WZI_0_1" specimen (Table 36).

Moreover, as depicted in Figure 5.15, the maximum shear force, $V_{j,max}$, attained during the modelling is 158.56 kN, and it is very close to the expected one (V_{Rd} = 158.47 kN). It is worth noting that the absolute values of N_j vary between 122 kN and 290 kN (Figure 5.15). Therefore, the ML-BEAM element shows a hybrid behaviour in the range 122 kN - 136 kN. However, since these values were recorded during the transition from positive to negative displacements and vice versa, it is only temporary. In fact, the ML-BEAM element achieves the in-plane resistance with shear damage until DL3, as during the first part of the test (Figure 5.16).

Figure 5.16: "WZI_0_2" specimen: damage state at the end of the first part of the test (DL3)

5.3.2 "WZI" specimens strengthened on one side with the EQ-GRID system

The cyclic lateral shear tests carried out on the "WZI_1_1" and "WZI_1_2" specimens are also modelled in TREMURI. As described in § 5.1, the elastic secant stiffness, k_{sec}, is experimentally obtained by calculating the equivalent multilinear curve. The masonry mechanical properties are equal to those adopted for the unreinforced WZI specimens (Table 35). The improved load-bearing capacity is calculated with the design approach of the CNR-DT 215/2018, as for the "KS" specimens. Thus, the shear resistance $V_{Rd,s}$ is obtained through Eq. (4.19), in which n_f is equal to 1 since the grid is applied only on one side of the panels. Moreover, the shear force calculated by the program, V_j, is always lower than the diagonal crushing resistance $V_{t,c}$:

$$V_{t,c} = 0.25 \cdot f_{m,u,WZI} \cdot t \cdot d_f = 0.25 \cdot 5.28 \cdot 240 \cdot 1250 = 396kN \tag{5.11}$$

Where $f_{m,u,WZI}$ is the average masonry compressive strength (Table 35), t is the thickness of the wall (240 mm), d_f is the distance between the compressed edge of the panel and the fibre of the reinforcement that attains the highest tensile strain. As the EQ-GRID system is applied on the entire surface, d_f corresponds to the length of the specimen (1250 mm).

The drifts and residual resistances adopted for the modelling of the "WZI_1_1" specimen are summarized in Table 37. Since the panel showed diagonal cracks during the test, the values indicated for the shear damage are obtained from the equivalent multilinear curve (§ 3.6.4.2). Concerning the bending failure, the chosen drifts are equal to twice the values assumed for the shear damage, and the residual resistances are coherent with CNR-DT 212/2013 [25].

Table 37: "WZI_1_1" specimen: values of drift and residual resistance

		Drift		Residual resistance ($V_{res,i}$ / $V_{Rd,s}$)	
Damage	DL3	DL4	DL5	DL3 → DL4	DL4 → DL5
Shear	$d_{3,S}$ = 0.62%	$d_{4,S}$ = 0.75%	$d_{5,S}$ = 0.93%	0.72	0.44
Bending	$d_{3,B}$ = 1.24%	$d_{4,B}$ = 1.5%	$d_{5,B}$ = 1.86%	1.0	0.85

The numerical results of the "WZI_1_1" specimen are directly compared with the experimental diagrams of the internal forces and moments, as depicted in Figure 5.17.

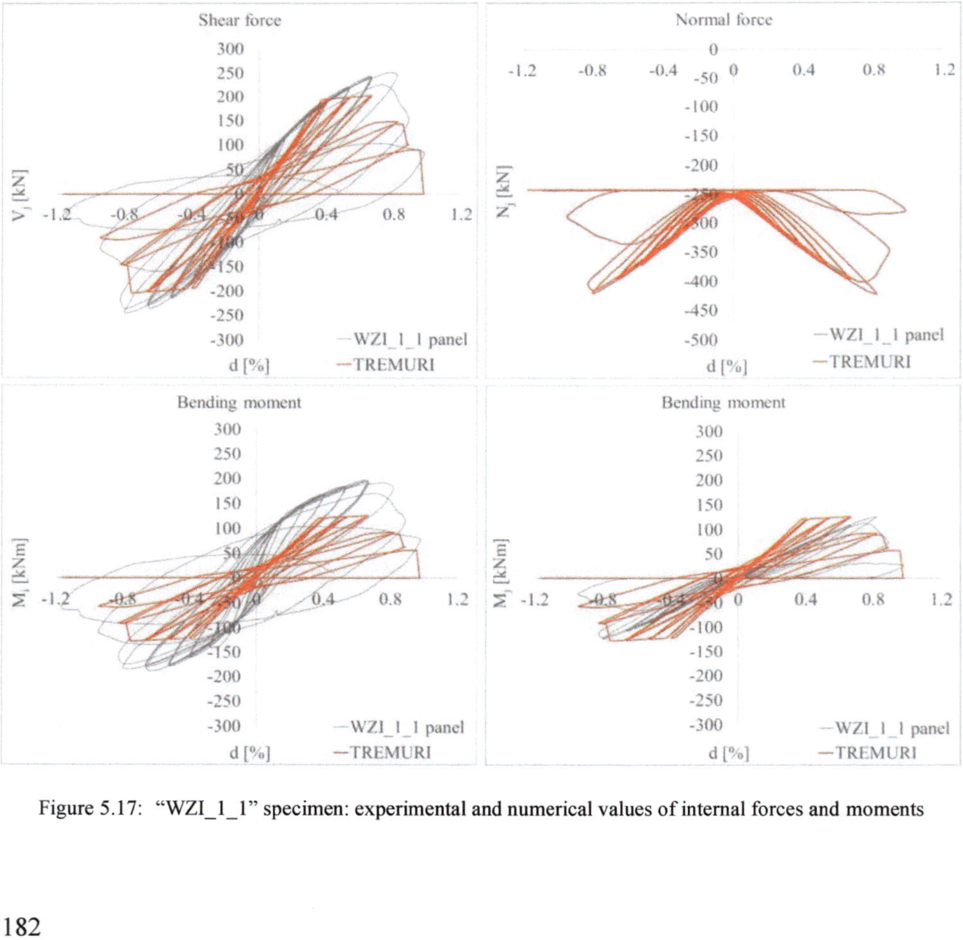

Figure 5.17: "WZI_1_1" specimen: experimental and numerical values of internal forces and moments

The sample was tested under the constant vertical load of 240 kN, and the normal force N_j varies between 242 kN and 422 kN. As shown by the failure domain of the panel (Figure 5.18), no mixed failure can occur in this range. The latter is only possible at normal force values between 73 kN - 163 kN and 833 kN - 998 kN. Therefore, the ML-BEAM element achieves the in-plane strength with shear failure from DL2 to DL5. The maximum attained shear force, $V_{j,max}$, is equal to 204.23 kN, and it is very close to the expected value ($V_{Rd,s}$ = 204.29 kN). As depicted in Figure 5.17, the results reproduce the hysteretic response of the panel very well and have a good match with the behaviour shown during the test.

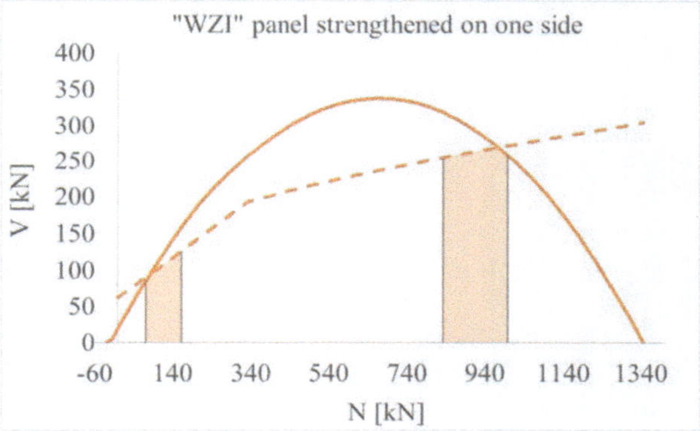

Figure 5.18: In-plane failure domain of the "WZI" specimens strengthened on one side

Concerning the "WZI_1_2" panel, the experimental and numerical diagrams are reported in Figure 5.19. Since the test data were not completely reliable in the post-peak phase, only the first part of the hysteresis curve, until the attainment of the maximum shear force, has been modelled with TREMURI.

The sample was tested under the constant vertical load of 120 kN, and the absolute values of N_j vary between 122 kN and 285 kN. Therefore, during the modelling with TREMURI, the mixed failure occurs at normal force values between 122 kN and 163 kN, as shown by the failure domain of the panel (Figure 5.18). The adopted values of drift and residual resistances are equal to those of the "WZI_1_1" panel (Table 37), as the equivalent multilinear curve could not be determined. The maximum shear force $V_{j,max}$ reached by the ML-BEAM element is equal to 173.92 kN, the expected value, $V_{Rd,s}$, is 173.89 kN. Finally, the element achieves the in-plane strength with shear failure, but it doesn't reach the DL3, as during the first part of the test.

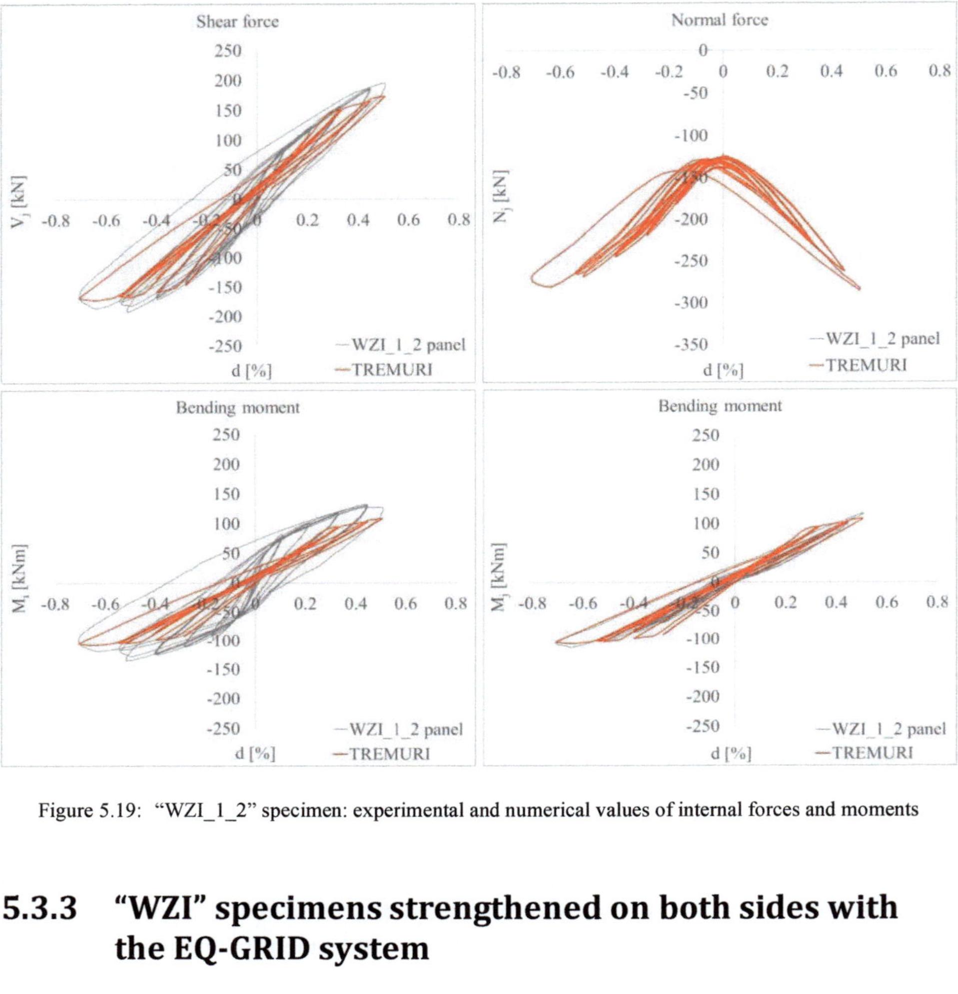

Figure 5.19: "WZI_1_2" specimen: experimental and numerical values of internal forces and moments

5.3.3 "WZI" specimens strengthened on both sides with the EQ-GRID system

The modelling results of the cyclic lateral shear tests performed on the "WZI_2_1" and "WZI_2_2" specimens are presented. The elastic properties of the ML-BEAM element are defined through the secant stiffness, k_{sec}, obtained from the equivalent multilinear curve determined in § 3.6.4.3. The masonry mechanical properties are described in Table 35. The in-plane shear resistance is calculated through Eqs. (4.18) and (4.19), in which n_f is equal to 2 since the textile is applied on both sides of the panels. Furthermore, the shear force V_j calculated at each load step is always lower than the diagonal crushing resistance $V_{t,c}$ (Eq. (5.11)) for both specimens. The in-plane flexural capacity is calculated through the cross-section analysis described in § 4.2.3. Therefore, the numerical and experimental results of the "WZI_2_1" panel are firstly presented (Figure 5.22). The maximum shear force $V_{j,max}$ attained by the ML-BEAM element, is equal to 215.51 kN, which is very close to the expected one calculated in § 4.2.4 ($V_{Rd,s}$ = 215.63 kN). The drifts and residual resistances adopted for the modelling are summarized in Table 38.

Table 38: "WZI_2_1" specimen: values of drift and residual resistance

	Drift			Residual resistance ($V_{res,i}$ / $V_{Rd,s}$)	
Damage	DL3	DL4	DL5	DL3 → DL4	DL4 → DL5
Shear	$d_{3,S}$ = 1.20%	$d_{4,S}$ = 1.40%	$d_{5,S}$ = 1.60%	0.55	0.35
Bending	$d_{3,B}$ = 2.40%	$d_{4,B}$ = 2.80%	$d_{5,B}$ = 3.20%	1.0	0.85

It is worth noting that the "WZI_2_1" sample achieved the in-plane resistance with the formation of diagonal cracks from corner to corner, but it also showed sub-vertical cracks across the thickness during the post-peak phase. As depicted in Figure 5.20, stress concentrations at the compressed toes caused the achievement of the masonry compressive strength. Therefore, the values indicated in Table 38 for the shear failure are lower than those obtained from the equivalent multilinear curve calculated in § 3.6.4.3 (Table 19).

Figure 5.20: Sub-vertical cracks in the "WZI" specimens strengthened on both sides

Moreover, it is important to stress that the experimental behaviour of the panel can be correctly reproduced thanks to the cross-section analysis performed by the program to calculate the in-plane flexural capacity. In this regard, Figure 5.21 shows the in-plane strength domain of the "WZI" panels, in which the orange and grey lines represent the bending and shear domain when the reinforcing system is applied on both sides of the specimens. The dotted line corresponds to the bending domain of the unreinforced panel.

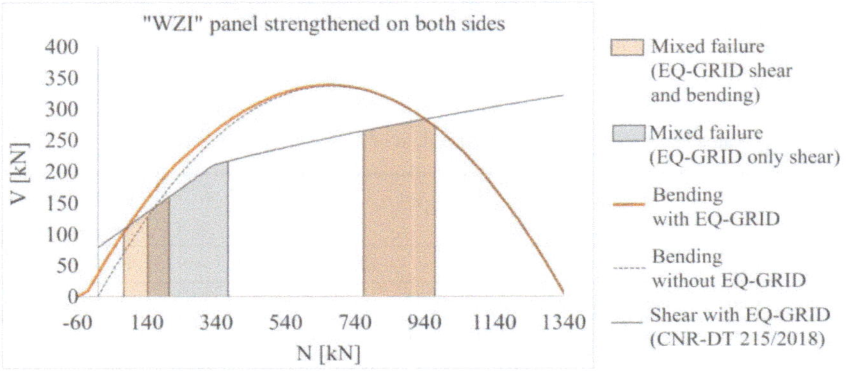

Figure 5.21: In-plane failure domain of the "WZI" specimens strengthened on both sides

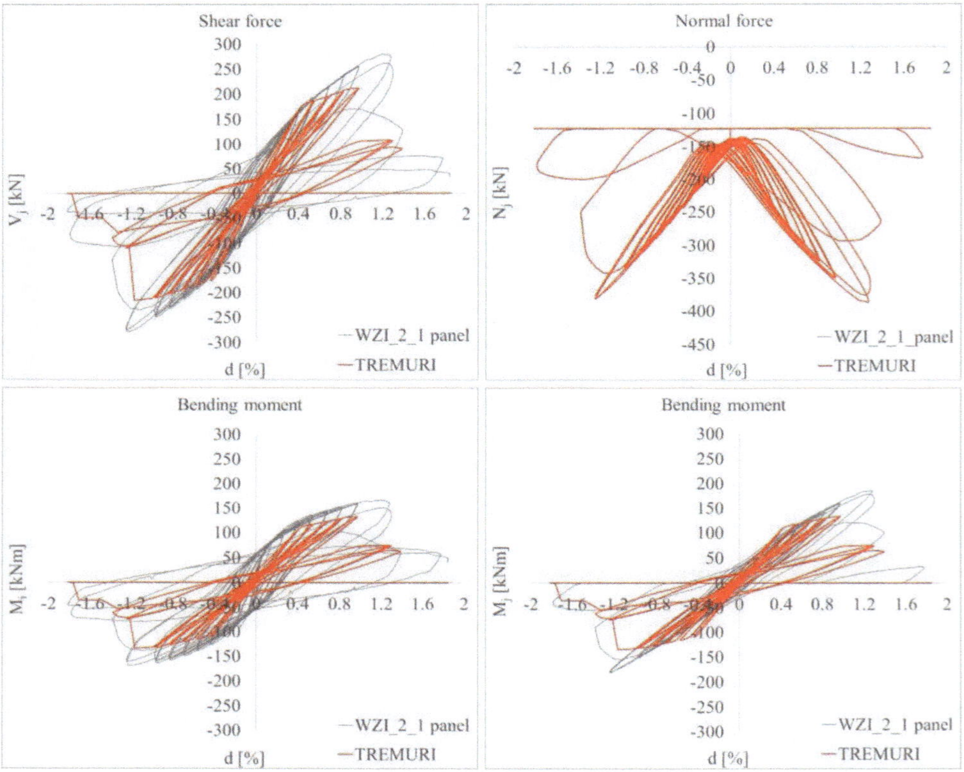

Figure 5.22: "WZI_2_1" specimen: experimental and numerical values of internal forces and moments

If the strengthening effect of the EQ-GRID system on the shear and flexural capacity is considered, the mixed failure can occur at normal force values between 75 kN and 205 kN. On the contrary, if the contribution of the EQ-GRID system to the flexural strength of the panel is neglected, the mixed failure can occur at normal force values between 143 kN and 375 kN.

Since the variation range of N_j is 122 kN - 387 kN (Figure 5.22), the ML-BEAM element would present a hybrid behaviour in most of the load steps if the improvement of flexural strength is neglected. Furthermore, during the final phase of the modelling, in which the normal force values are lower than 143 kN, the element would show only bending failure.

Therefore, the cyclic response of the sample can be correctly reproduced only by considering the contribution of EQ-GRID in the shear and flexural strength of the panel. In this case, the ML-BEAM element can attain the maximum horizontal force showing the shear failure. Then, only in the post-peak phase, when the acting normal force is lower than 205 kN, it can present the hybrid behaviour observed during the test.

Concerning the "WZI_2_2" panel, the experimental and numerical results are compared in Figure 5.23. The maximum shear force achieved during the modelling is equal to 213.19 kN, which is very close to that calculated in § 4.2.4 ($V_{Rd,s}$ = 213.30 kN). The adopted drifts and residual resistances are summarized in Table 39.

Table 39: "WZI_2_2" specimen: values of drift and residual resistance

	Drift			Residual resistance ($V_{res,i}$ / $V_{Rd,s}$)	
Damage	DL3	DL4	DL5	DL3 → DL4	DL4 → DL5
Shear	$d_{3,S}$ = 0.88%	$d_{4,S}$ = 0.93%	$d_{5,S}$ = 1.20%	0.70	0.45
Bending	$d_{3,B}$ = 1.76%	$d_{4,B}$ = 1.86%	$d_{5,B}$ = 2.40%	1.0	0.85

It is worth pointing out that the "WZI_2_2" specimen reached the in-plane shear resistance with the formation of diagonal cracks and presented sub-vertical cracks across the thickness during the post-peak phase. Therefore, as for the "WZI_2_1" panel, the experimental behaviour can be numerically reproduced only considering the increased shear and flexural strength due to EQ-GRID. In this regard, Figure 5.23 shows that the absolute values of N_j vary between 122 kN and 363 kN, and the mixed failure is possible in the range 122 kN - 205 kN, as during the test.

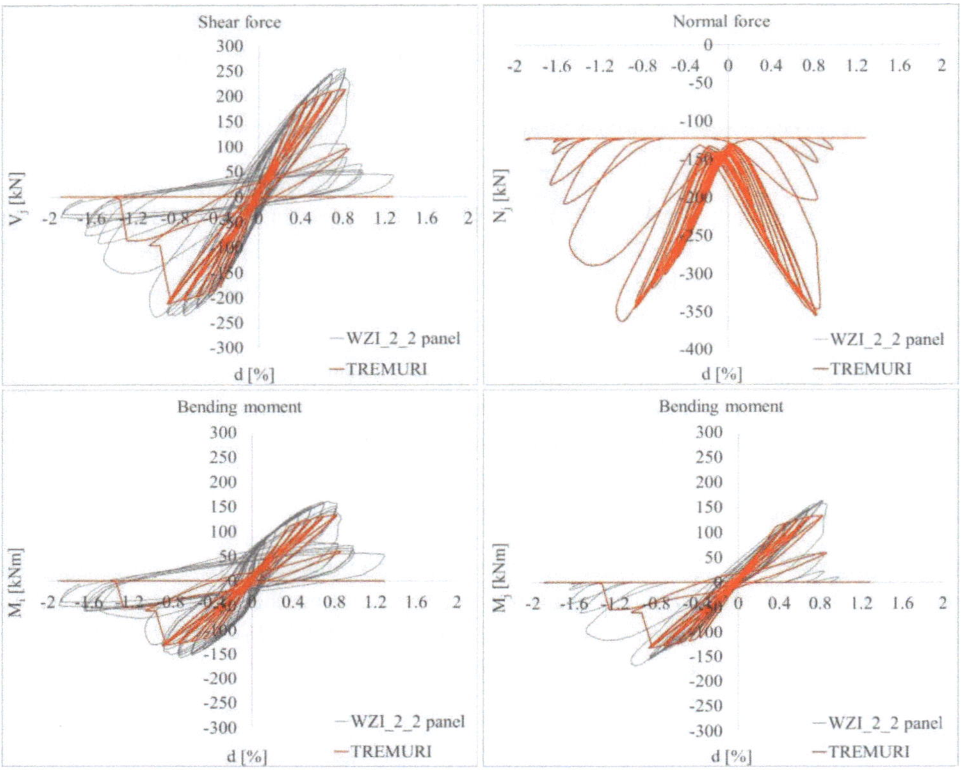

Figure 5.23: "WZI_2_2" specimen: experimental and numerical values of internal forces and moments

Chapter 6

Modelling of shaking table tests with the Equivalent frame method

The three-dimensional model of a hypothetical prototype masonry structure was tested on the seismic shaking table of the Dynamic Testing Laboratory at the Institute of Earthquake Engineering and Engineering Seismology (IZIIS) in Skopje from February to September 2013 [73]. Figure 6.1 shows the geometry of the three-dimensional BM model.

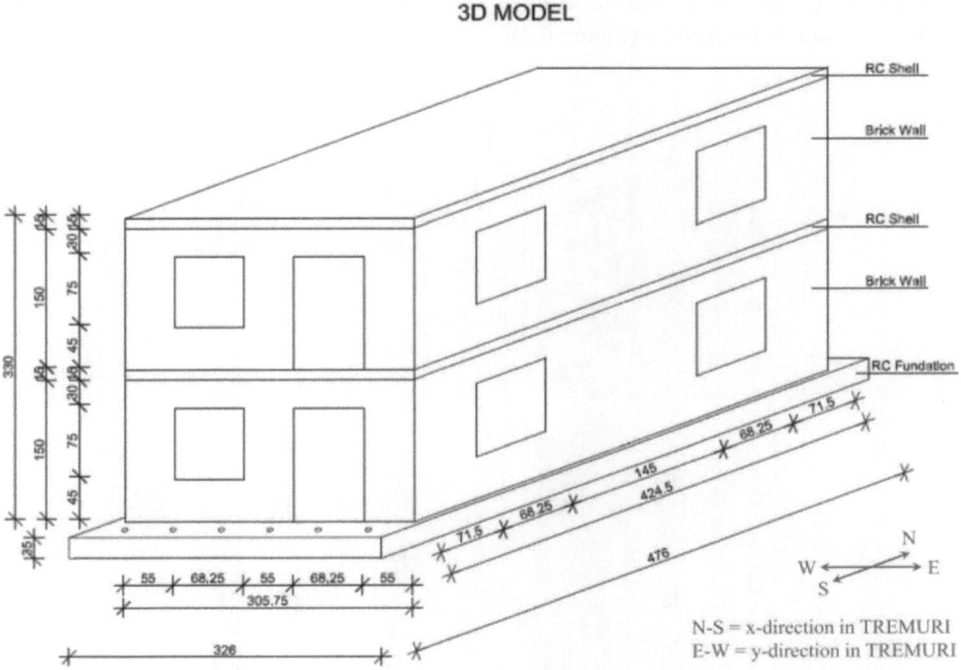

Figure 6.1: BM model: the unstrengthened three-dimensional model [73]

The experimental campaign was carried out to validate the EQ-GRID system as seismic retrofitting technology for traditional historic masonry structures. Therefore, a two-story brick masonry building with reinforced concrete floors was chosen as a prototype structure. It represented most of the existing residential and public masonry buildings of the second half of the XX century in the Republic of Macedonia and on the territory of the Balkan region.

The geometrical scale of the three-dimensional BM model was 1:2. It was chosen based on the following criteria:

- Dimensions of the shaking table (4.5 m x 4.5 m);
- Allowed total height and weight on the shaking table (10 m and 400 kN, respectively);
- Realistic reproduction of the possible failure mechanisms of masonry walls under seismic actions.

Thus, the total length, width and height of the building were 4.24 m, 3.06 m and 3.30 m, respectively (Figure 6.1). The structural system consisted of five bearing walls, four facades and one in the middle (Figure 6.2).

Figure 6.2: Construction of the BM model [73]

The walls were constructed in a running bond with a thickness of 12.5 cm (Figure 6.3). The adopted bricks were manufactured with 12.5 x 6 x 3.25 cm dimensions to reproduce the standard brick size of 25 x 12 x 6.5 cm. The mortar joints were prepared with a lime/cement/sand ratio equal to 1/1/3, river sand with fraction 0-2 cm as filler, and water. The thickness of the vertical and horizontal mortar joints was equal to 0.5 cm [73]. Therefore, the physical, chemical and mechanical masonry properties were similar to those of the hypothetical prototype structure.

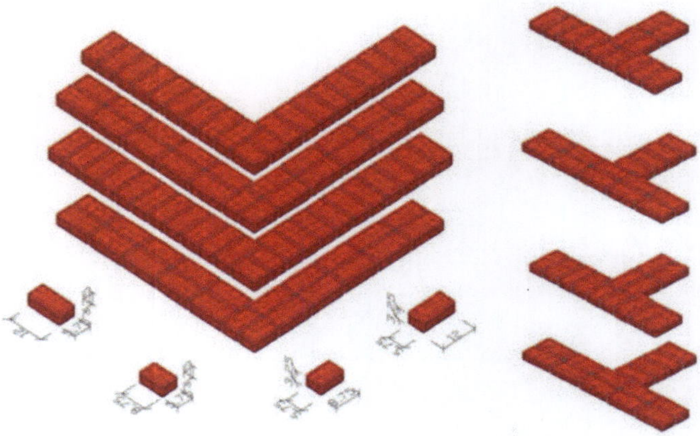

Figure 6.3: Masonry pattern of the BM model [73]

Table 40 shows the average compressive strength of the adopted solid clay brick and mortar. The values were experimentally obtained through compressive tests performed on bricks and mortar prisms.

Table 40: Average values of compression strength of the solid bricks and lime-cement mortar of the BM model [73]

Solid clay brick	Mortar (cement: clay: sand = 1 : 1 : 3)
f_b = 20,12 MPa	f_M = 15,52 MPa (28 days)

Moreover, a reinforced concrete foundation with dimensions of 3.26 x 4.76 x 0.30 m was built to transport and anchor the structure on the shaking table (26 anchors were used). After the construction phase, the BM model was left to dry for 30 days at the place where it was constructed. Finally, it was transported with a 90-tons auto crane on the shaking table [73].

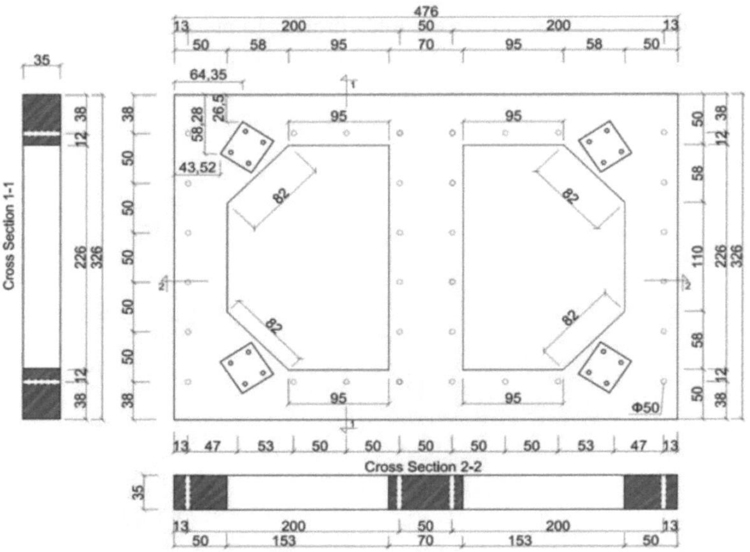

Figure 6.4: Foundation plan of the BM model [73]

The model was dynamically tested by applying two main types of earthquakes in its W-E direction: one local (Northridge, 1994) and two distant (El Centro, 1940 and Petrovac N-S, 1979). The time histories were scaled in compliance with the principles of modal analysis, and their application was made gradually, increasing the peak ground acceleration [73].

Table 41 shows the selected experimental tests starting from moderate to destructive intensities and the reached damage level.

Since the experimental campaign aimed to investigate the effectiveness of the EQ-GRID strengthening system by comparing the behaviour of the retrofitted structure with the original one, the testing program provided for the application of earthquake excitations with a gradual increase in intensity. In this way, the progressive development of cracks, the modification of the dynamic characteristics, the elasticity limit, i.e. occurrence of the first cracks, and the failure mechanisms could be monitored.

Table 41: Dynamic input excitation of the BM model [73]

Earthquake	Span [%]	a_{input} [g]	Damage BM model
El Centro	50	0.14	-
Petrovac	36	0.16	
Northridge	16	0.18	
El Centro	65	0.18	
Petrovac	40	0.18	
El Centro	75	0.21	Initial fine cracks
Petrovac	45	0.20	
Northridge	20	0.21	
El Centro	80	0.27	Further propagation of initial cracks
Northridge	25	0.23	
Petrovac	50	0.22	Damage development
Petrovac	70	0.32	
Petrovac	75	0.35	

The BM model was tested until observing considerable damages. As depicted in Figure 6.5, the North-East corner of the model resulted dislocated at the end of the dynamic tests. Therefore, it was first directly stabilized on the shaking table. Then, the structure was transported back to the place where it was constructed. The cracks were repaired through injection of lime-cement based mixtures. Finally, the EQ-GRID system was applied on the outer side of the walls.

Figure 6.5: Damage to the North-East corner of the BM model and transportation from the shaking table [73]

As illustrated in Figure 6.6, the grid was applied with 10-15 cm overlapping in the following order:

- at the window and door corners (diagonal textile strips),

- around the floor slabs,
- around the foundation slab,
- from the top of the walls to the foundation.

Figure 6.6: Retrofitting phases of the BM-SR model [73]

Figure 6.7: Retrofitting phases of the BM-SR model – anchoring detail [73]

Moreover, the grid was anchored into the foundation through steel L-profile and bolts at regular distances of 15-20 cm along the perimeter of the facade walls (Figure 6.7).

Finally, the model was left drying for 30 days at the construction place and then positioned on the shaking table, where it was appropriately anchored through 26 anchors (Figure 6.8).

Figure 6.8: Transportation of the BM-SR model and positioning on the shaking table [73]

The BM-SR model was subjected to the same dynamic tests' series of the unstrengthened specimen to directly compare the experimental results and evaluate the EQ-GRID system's performance. Table 42 shows the list of the scaled earthquakes with the structural damage reached by both structures. Due to the presence of the system, the BM-SR model was tested under higher intensities of input excitation. It is worth noting that the maximum acceleration applied at the base of the BM-SR model (Petrovac 260) is about 3.45 times higher than the last one assigned to the unstrengthened structure (Petrovac 75).

Therefore, both BM and BM-SR specimens are modelled with the equivalent frame method. The frame-type representation of the structure is depicted in Figure 6.9, in which piers, spandrels, r. c. beams and rigid nodes assemble each wall. The concrete slabs are simulated as rigid diaphragms. It is important to stress that the r. c. beams are introduced in the modelling to consider that the masonry spandrels are interrupted by the concrete slabs. Therefore, the beams' dimensions are assumed equal to the wall and slab thickness (b x h = 12.5 x 15 cm).

Concerning the masonry panels, the ML-BEAM element is adopted for both BM and BM-SR models. The initial stiffness k_{el} is calibrated through modal analyses since the natural frequencies of the BM and BM-SR models were measured before and after testing by ambient vibration and random excitation techniques.

Table 42: Dynamic input excitation of the BM-SR and BM models [73]

<table>
<tr><th>Earthquake</th><th>Span [%]</th><th>a_{input} [g]</th><th>Damage BM-SR model</th><th>Damage BM model</th></tr>
<tr><td>El Centro</td><td>50</td><td>0.14</td><td rowspan="5">-</td><td rowspan="5">-</td></tr>
<tr><td>Petrovac</td><td>36</td><td>0.16</td></tr>
<tr><td>Northridge</td><td>16</td><td>0.18</td></tr>
<tr><td>El Centro</td><td>65</td><td>0.18</td></tr>
<tr><td>Petrovac</td><td>40</td><td>0.18</td></tr>
<tr><td>El Centro</td><td>75</td><td>0.21</td><td rowspan="3">-</td><td rowspan="3">Initial fine cracks</td></tr>
<tr><td>Petrovac</td><td>45</td><td>0.20</td></tr>
<tr><td>Northridge</td><td>20</td><td>0.21</td></tr>
<tr><td>El Centro</td><td>80</td><td>0.27</td><td rowspan="2">-</td><td rowspan="2">Further propagation of initial cracks</td></tr>
<tr><td>Northridge</td><td>25</td><td>0.23</td></tr>
<tr><td>Petrovac</td><td>50</td><td>0.22</td><td rowspan="4">Initial fine cracks</td><td rowspan="3">Damage development</td></tr>
<tr><td>Petrovac</td><td>70</td><td>0.32</td></tr>
<tr><td>Petrovac</td><td>75</td><td>0.35</td></tr>
<tr><td>Petrovac</td><td>100</td><td>0.51</td><td rowspan="8">-</td></tr>
<tr><td>Petrovac</td><td>120</td><td>0.60</td><td rowspan="3">Further propagation of initial cracks</td></tr>
<tr><td>El Centro</td><td>100</td><td>0.31</td></tr>
<tr><td>Petrovac</td><td>150</td><td>0.82</td></tr>
<tr><td>Petrovac</td><td>180</td><td>0.92</td><td rowspan="4">Damage development</td></tr>
<tr><td>Petrovac</td><td>220</td><td>1.04</td></tr>
<tr><td>Petrovac</td><td>260</td><td>1.22</td></tr>
<tr><td>Petrovac</td><td>250</td><td>1.21</td></tr>
</table>

The progressive stiffness degradation in the elastic phase is considered by assigning two proper ratios. The first one is between the initial and secant stiffness, and it is assumed equal to 1.6. The second one, k_0 (Figure 1.41), is defined between the shear at the end of the initial elastic phase and the shear strength attained by the element. It is assumed 0.7 for the piers and 0.5 for spandrels. The experimental evidence presented in [84], [85] and [30] confirms these assumptions.

The maximum strength of the panels is calculated as the minimum obtained from the chosen failure criteria. The Mann and Müller theory [23] is adopted in the case of shear failure. Eqs. (1.18), (1.19) for the piers and Eqs. (1.29), (1.30) for the spandrels are implemented in the TREMURI program. The increase in strength due to the EQ-GRID system is computed after CNR-DT 215/2018, as shown in § 4.2.3.

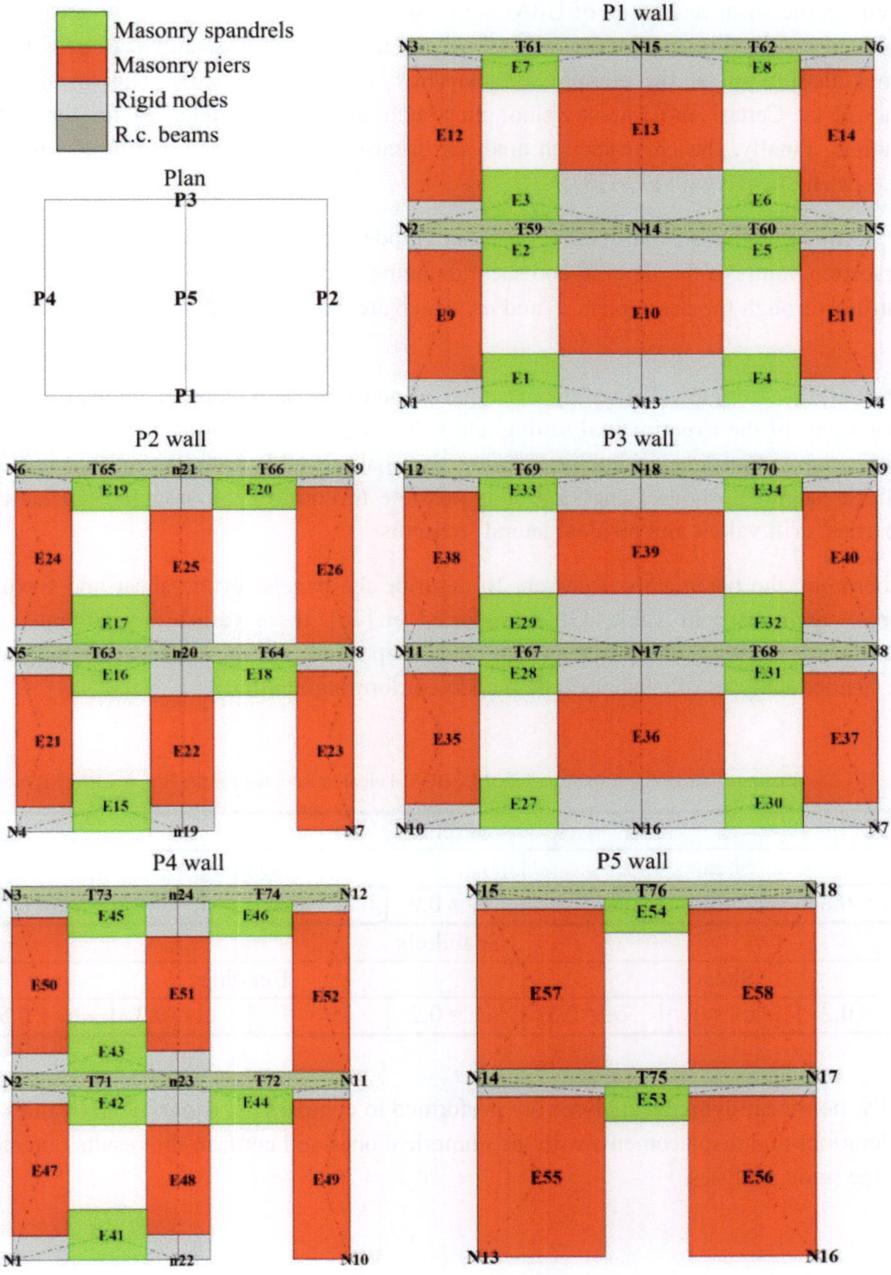

Figure 6.9: Frame-type representation of the walls of the BM and BM-SR Model in TREMURI

Moreover, the shear resistance of URM piers associated with the flexural failure mode is calculated, neglecting the tensile strength of the masonry and assuming a stress block normal distribution at the compressed toe (Eqs. (1.25) and (1.26)). The formulation proposed by Cattari and Lagomarsino, illustrated in § 1.4, is adopted for the URM spandrels. Finally, the cross-section analysis discussed in § 4.2.3 is performed for the panels strengthened with EQ-GRID.

It is worth remembering that the hybrid failure mode is possible in two areas close to the intersection points of the flexural and shear domains. As depicted in Figure 1.40, they are identified through the parameters a_1 and a_2, which are assumed 0.95 and 1.25, respectively.

Moreover, the strength values of the masonry elements are defined based on the results of the first part of the experimental testing and values typical for solid clay brick masonry. By comparing the damage attained during the shaking table tests with the results of nonlinear static (pushover) analyses, it is possible to confirm the assumed mechanical properties, drift values and residual lateral strengths.

Furthermore, the parameters necessary to describe the panels' cyclic shear and flexural response are shown in Table 43. As reported in [46], these values are calibrated on experimental results ([30], [86]), and they are adopted for the non-linear modelling and time-history analyses of masonry Italian code-conforming buildings.

Table 43: Parameters of the cyclic response of the ML-BEAM element adopted for the BM and BM-SR model

Piers						
Shear			Bending			
$c_1 = 0.8$	$c_2 = 0.8$	$c_3 = 0$	$c_1 = 0.9$	$c_2 = 0.8$	$c_3 = 0.6$	$c_4 = 0.5$
Spandrels						
Shear			Bending			
$c_1 = 0.2$	$c_2 = 0$	$c_3 = 0.3$	$c_1 = 0.2$	$c_2 = 0$	$c_3 = 0.3$	$c_4 = 0.8$

Finally, nonlinear dynamic analyses are performed to compare the experimental values of accelerations and displacements with the numerical ones and confirm the results obtained with the static analyses.

6.1 The unstrengthened BM model

6.1.1 Modal analysis of the BM model

The modal analysis allows identifying the mode shapes, natural frequencies/periods and modal masses of a structure in its original condition. As shown in Figure 6.10, the seismic masses are assumed to be lumped at each model's level. It is worth noting that only the dead loads need to be considered, as no accidental loads were present during the test.

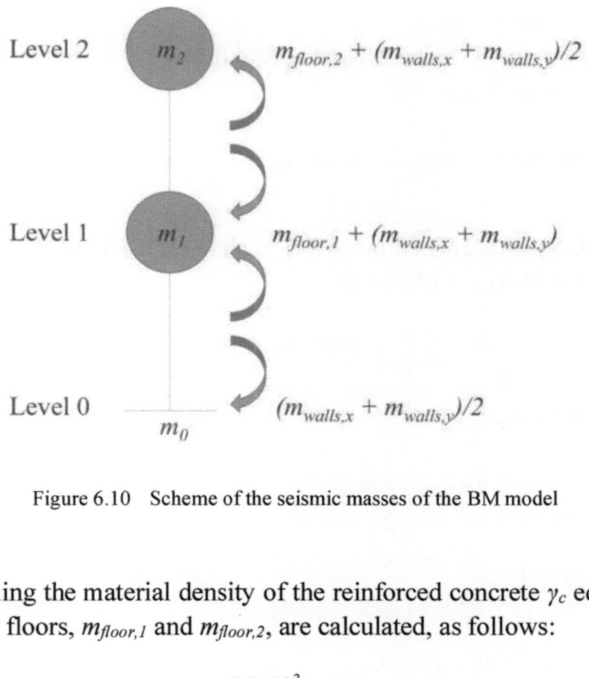

Figure 6.10 Scheme of the seismic masses of the BM model

Therefore, assuming the material density of the reinforced concrete γ_c equal to 25 kN/m³, the masses of the floors, $m_{floor,1}$ and $m_{floor,2}$, are calculated, as follows:

$$m_{floor,1} = \gamma_c \cdot A_{floor,1} \cdot h_{floor,1} = \frac{25 \cdot 10^3}{9.81} \cdot (4.245 \cdot 3.058) \cdot 0.25 = \\ = 8269.11 kg \tag{6.1}$$

$$m_{floor,2} = \gamma_c \cdot A_{floor,2} \cdot h_{floor,2} = \frac{25 \cdot 10^3}{9.81} \cdot (4.245 \cdot 3.058) \cdot 0.15 = \\ = 4961.26 kg \tag{6.2}$$

Where $A_{floor,1}$ and $A_{floor,2}$, $h_{floor,1}$ and $h_{floor,2}$ are the areas and thicknesses of the first and second floors. It is important to stress that an additional RC plate was constructed on the first floor to create an additional load, as this was not connected to the walls. For this reason, the thickness of the first slab, $h_{floor,1}$, is equal to 0.25 m.

The masses of the walls in x- and y-direction, $m_{walls,x}$ and $m_{walls,y}$, are determined assuming the material density of the masonry equal to 1950 kg/m³ [73]. The thickness and the clear interstorey height of each wall are equal to 0.125 m and 1.50 m, respectively.

$$\begin{aligned} m_{walls,x} &= \gamma_{masonry} \cdot V_{walls,x} = 1950 \cdot (2 \cdot 4.245 - 4 \cdot 0.75 \cdot 0.683) \cdot 0.125 \cdot 1.50 \\ m_{walls,x} &= 2605.08kg \end{aligned} \tag{6.3}$$

$$\begin{aligned} m_{walls,y} &= \gamma_{masonry} \cdot V_{walls,y} \\ &= 1950 \cdot (3 \cdot 2.81 \cdot 1.50 - (2 \cdot 0.683 \cdot 0.75 + 3 \cdot 0.725 \cdot 1.2)) \cdot 0.125 \\ m_{walls,y} &= 2193.75kg \end{aligned} \tag{6.4}$$

$V_{walls,x}$ and $V_{walls,y}$ are the volumes of the masonry walls placed in x- an y-direction of the model.

Therefore, assigning the calculated masses to each level (Figure 6.10), the lumped seismic masses, m_0, m_1 and m_2, are obtained:

$$\begin{aligned} m_0 &= \left(\frac{m_{walls,x} + m_{walls,y}}{2}\right) = \left(\frac{2605.08 + 2193.75}{2}\right) = \\ &= 2399.41kg \end{aligned} \tag{6.5}$$

$$\begin{aligned} m_1 &= m_{floor,1} + 2 \cdot \left(\frac{m_{walls,x} + m_{walls,y}}{2}\right) = \\ &= 8269.11 + 2 \cdot \left(\frac{2605.08 + 2193.75}{2}\right) = 13067.87kg \end{aligned} \tag{6.6}$$

$$\begin{aligned} m_2 &= m_{floor,2} + \left(\frac{m_{walls,x} + m_{walls,y}}{2}\right) = 4961.26 + \left(\frac{2605.08 + 2193.75}{2}\right) = \\ &= 7360.84kg \end{aligned} \tag{6.7}$$

Finally, the total mass of the model, m_{tot}, is equal to:

$$m_{tot} = m_0 + m_1 + m_2 = 22828.12kg \tag{6.8}$$

Table 44 shows the values calculated by the TREMURI program. They are very close to those determined with Eqs. (6.5) - (6.8).

Table 44: Seismic masses of the BM model calculated by TREMURI

Level	Nodes of the level	Mass of the level
0	19; 22; 1; 4; 7; 10; 13; 16	$m_0 = 2472.72$ kg
1	20; 23; 2; 5; 8; 11; 14; 17	$m_1 = 12684.06$ kg
2	21; 24; 3; 6; 9; 12; 15; 18	$m_2 = 7446.35$ kg
$m_{tot} = 22603.13$ kg		

Therefore, the natural frequencies *f*, periods *T* and modal masses m_x, m_y of the first ten mode shapes obtained with the TREMURI program, are reported in Table 45.

Table 45: Natural frequencies *f*, periods *T* and modal masses m_x, m_y of the BM model calculated by TREMURI

Mode shape	f [Hz]	T [s]	m_x [kg]	m_y [kg]
1	10.71	0.093	0.001	17850.13
2	13.74	0.073	163.87	0.066
3	14.98	0.067	16813.18	0
4	28.10	0.036	0	2263.57
5	35.72	0.028	6.12	0.012
6	40.10	0.025	3073.90	0
7	49.06	0.020	0.005	0.84
8	50.26	0.020	24.31	0
9	51.45	0.019	0.002	3.23
10	52.56	0.019	27.29	0

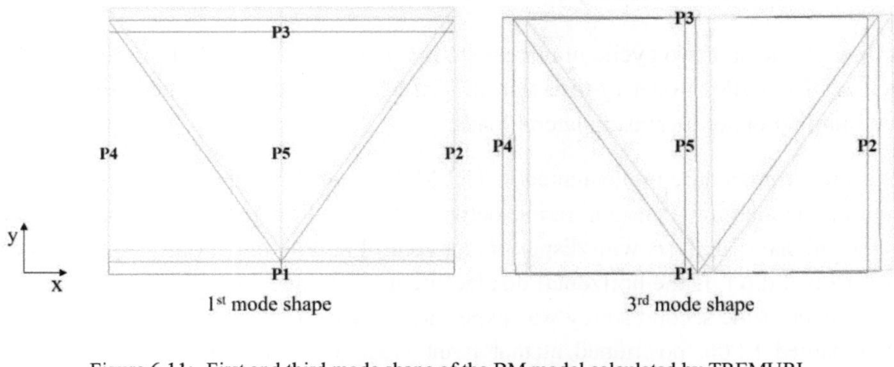

Figure 6.11: First and third mode shape of the BM model calculated by TREMURI

The first vibration mode (Figure 6.11) is translational along the East-West direction of the structure (y-direction in TREMURI) with a frequency of 10.71 Hz and modal mass equal to 79% of the total model's mass, m_{tot}. On the contrary, the third mode shape (Figure 6.11) is translational along the North-South direction (x-direction in TREMURI) with a frequency of 14.98 Hz and modal mass equal to 74% of the total model's mass, m_{tot}. Since the natural frequencies of the BM model were measured during the first part of the tests by the ambient vibration and random excitation techniques, they can be compared with the fundamental frequency calculated by TREMURI. In this way, it is possible to calibrate the elastic properties of the ML-BEAM element for the masonry piers and spandrels. Table 46 shows the values adopted for the elastic moduli perpendicular and parallel to the bed joints adopted. As suggested in the literature and most codes for masonry buildings, the shear modulus G is assumed equal to $0.4E$.

Table 46: Natural frequencies of the BM model and assumed elastic moduli

Experimental (before testing)			TREMURI			
Direction	f	f	f	E	$E_{//}$	G
E-W (y-direction)	10.95 Hz (ambient vibration)	10.57 Hz (average)	10.71 Hz (1st mode shape)	2150 N/mm²	1075 N/mm²	860 N/mm²
	10.18 Hz (random excitation)					

6.1.2 Pushover analyses of the BM model

Four monotonic and two cyclic nonlinear static analyses are carried out in the East-West direction of the BM model (y-direction in TREMURI) to compute the response of the tested building under increasing lateral loads.

The pushover procedure implemented in TREMURI transforms the problem of pushing a structure, maintaining constant ratios between the applied forces, into an equivalent incremental static analysis with displacement control at only one d.o.f. [14]. In this case, the controlled d.o.f. is the horizontal displacement u_y of node 18 of the P3 wall since the displacement of the second storey was experimentally measured through a linear potentiometer named LP06, positioned at that point [73]. Since the values recorded by this sensor are absolute displacements and the nodes at the base of the numerical model are fixed, the difference (LP06 - LP01) [73] is considered for the modelling [1].

The nonlinear static analyses can be carried out with different load patterns. The results are generally expressed in terms of a base shear versus control node displacement relationship, i.e. the so-called pushover curve. The two most commonly adopted force distributions are the uniform and the triangular inverse. The first one is proportional to the mass of the levels, while the second one is proportional to the mass and height of the levels. Both force distributions may be considered as boundary conditions for seismic analyses. The actual dynamic response of the structure can be assumed to be within these two solutions, and the real failure mode is generally predicted by one of the two distributions [87].

Since the pushover curve may be considered an envelope of the dynamic response of a structure, the monotonic pushover analyses' results are first discussed. In this way, it is possible to individuate the best force distribution to interpret the failure mechanism observed during the test and calibrate the material properties and drift values of the masonry elements (piers and spandrels).

Table 47: Mechanical properties of the solid clay brick masonry

Material properties	NTC 2018 (explanatory Circular)	TREMURI
Compressive strength $f_{m,u}$	2.6 ÷ 4.3 N/mm² (min. and max. values)	5.0 N/mm²
Compressive strength $f_{mh,u}$	-	2.5 N/mm²
Young's modulus E	1200 ÷ 1800 N/mm² (min. and max. values)	2150 N/mm²
Young's modulus $E_{//}$	-	1075 N/mm²
Shear modulus G	400 ÷ 600 N/mm² (min. and max. values)	860 N/mm²
Initial shear strength f_{vmo}	0.13 ÷ 0.27 N/mm² (min. and max. values)	0.13 N/mm²
Friction coefficient μ	0.577	0.577
Brick tensile strength f_{bt}	$0.1 f_b$	0.20 N/mm²
Interlocking ϕ	$\phi = \frac{2\Delta y}{\Delta x}$	1.18

1 The linear potentiometer LP01 was positioned on the foundation of the BM model to measure its absolute displacement. The complete arrangement of the adopted sensors is indicated in [73].

Table 47 shows the average mechanical properties suggested by the explanatory Circular of the Italian Building Code NTC 2018 [22] for existing solid clay brick masonry and those adopted for the modelling with TREMURI. It is important to stress that minimum and maximum average values are given since the masonry quality is usually not a constant factor for the existing buildings.

The average compressive strength perpendicular to the bed joints, $f_{m,u}$, is obtained from three compressive tests carried out on masonry panels during the first part of the experimental characterization [73]. It is worth noting that it is 1.2 times higher than the maximum indicated by the Italian Building Code in Table 47. Moreover, the value assumed for the Young's modulus perpendicular to the bed joints, E, is also 1.2 times higher than the maximum one. Since the tests performed on the bricks and mortar prisms in the first part of the experimental testing showed that the masonry quality of the BM model is good (Table 40), the adopted values are consistent with the type of masonry.

Finally, the initial shear strength f_{vmo}, the friction coefficient μ and the tensile brick strength f_{bt} are assumed equal to 0.13 N/mm², 0.577 and 0.20 N/mm², as suggested by the Italian Building Code. It is important to stress that only the piers of the middle wall under the first floor (elements 55 and 56 of the P5 wall in TREMURI) failed with diagonal cracks during the shaking table test. If initial shear strength values higher than 0.13 N/mm² are adopted, the bending-rocking behaviour becomes relevant for these elements, and the numerical model can't reproduce the experimental results well.

6.1.2.1 Monotonic pushover analyses of the BM model

The load pattern proportional to the masses allows examining the response of a building under extensive damage. Force redistributions among the levels are prevented, and actions at lower storeys cannot shift to higher ones [87].

In the case of the BM model, the seismic masses m_1 and m_2 are quite different. Therefore, the force distribution is not uniform along with the building height and the load multiplier λ_i assumes unit value at the first level. Thus, the vector $[\lambda]$ is calculated as follows:

$$[\lambda]=\begin{bmatrix}\lambda_2\\ \lambda_1\end{bmatrix}=\begin{bmatrix}\dfrac{m_2}{m_0+m_1+m_2}\\ \dfrac{m_1}{m_0+m_1+m_2}\end{bmatrix}=\begin{bmatrix}\dfrac{7446.35}{2472.72+12684.06+7446.35}\\ \dfrac{12684.06}{2472.72+12684.06+7446.35}\end{bmatrix}= \tag{6.9}$$

$$=\begin{bmatrix}0.33\\ 0.56\end{bmatrix}\Rightarrow[\lambda]=\begin{bmatrix}\lambda_2=0.59\\ \lambda_1=1.00\end{bmatrix}$$

With this type of distribution, the final damage observed during the test is well represented by the modelling in TREMURI. It is worth pointing out that the comparison between the experimental and numerical model is made based on the experimental observations reported in [73] and evaluating the shear-drift curve of the masonry elements. In this way, the drifts of the ML-BEAM elements can be well-calibrated, and a clear overview of the final state of damage can be obtained. In this regard, Table 48 summarizes the values of drift and residual resistance chosen for the masonry piers and spandrels.

It is worth noting that experimental evidence, e.g. [30], [31], [45], has shown that the spandrels generally achieve drifts significantly higher than those adopted for piers. In particular, the values indicated in Table 48 are consistent with those suggested by the Italian technical standard CNR-DT 212/2013 [25] for masonry spandrels supported by concrete lintel and tie beam.

Table 48: Drift values of the masonry piers and spandrels for the BM model

Masonry piers					
	Drift			Residual resistance ($V_{res,i} / V_{Rd}$)	
Damage	DL3	DL4	DL5	DL3 → DL4	DL4 → DL5
Shear	$d_{3,S} = 0.6\%$	$d_{4,S} = 1.1\%$	$d_{5,S} = 1.6\%$	0.7	0.4
Bending	$d_{3,B} = 1.2\%$	$d_{4,B} = 2.2\%$	$d_{5,B} = 3.2\%$	1.0	0.85
Masonry spandrels with concrete lintel and tie beam					
	Drift			Residual resistance ($V_{res,i} / V_{Rd}$)	
Damage	DL3	DL4	DL5	DL3 → DL4	DL4 → DL5
Shear	$d_{3,S} = 1.2\%$	$d_{4,S} = 2.0\%$	$d_{5,S} = 2.6\%$	1.0	0.8
Bending	$d_{3,B} = 1.2\%$	$d_{4,B} = 2.0\%$	$d_{5,B} = 2.6\%$	1.0	0.8

Moreover, Figure 6.12 shows the pushover curves of two monotonic non-linear static analyses with load pattern proportional to the masses (one performed in the positive y-direction of the model and the other in the negative one). The attained maximum and minimum base shear forces are equal to 51.66 kN and -50.95 kN, respectively. Hereafter, the results of the nonlinear static analysis performed in the +y direction are in detail discussed. The horizontal displacement u_y of node 18 is chosen as controlled d.o.f., whose limit value is the maximum relative displacement (LP06 - LP01 = 17.073 mm) measured during the last earthquake applied to the BM model (Petrovac 75).

Moreover, it is important to specify that the P2, P4, P5 walls are stressed in-plane since they are in the same direction of the seismic input. On the contrary, the P1, P3 walls are positioned in the x-direction of the model and are stressed out-of-plane. Therefore, the results of the P2 wall are firstly presented. In this regard, the experimental and numerical damage patterns are depicted in Figure 6.13, in which the most damaged elements are the 21 and 22, as experimentally for the BM Model.

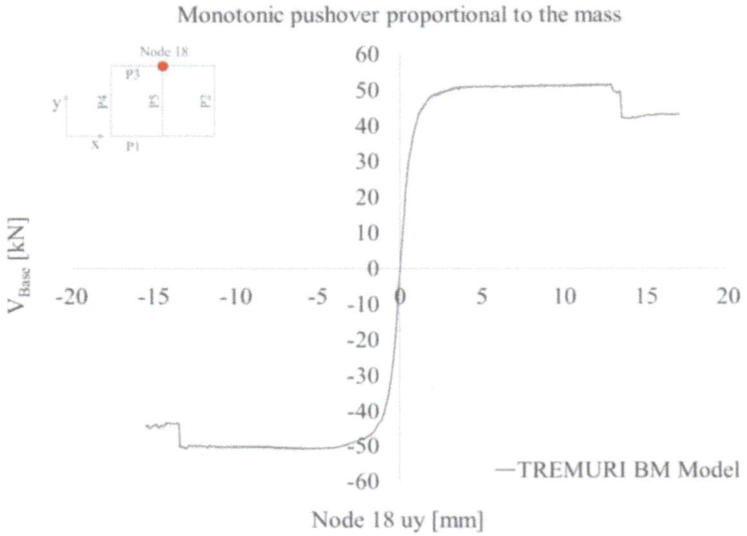

Figure 6.12: Monotonic pushover curves of the BM model with load pattern proportional to the masses

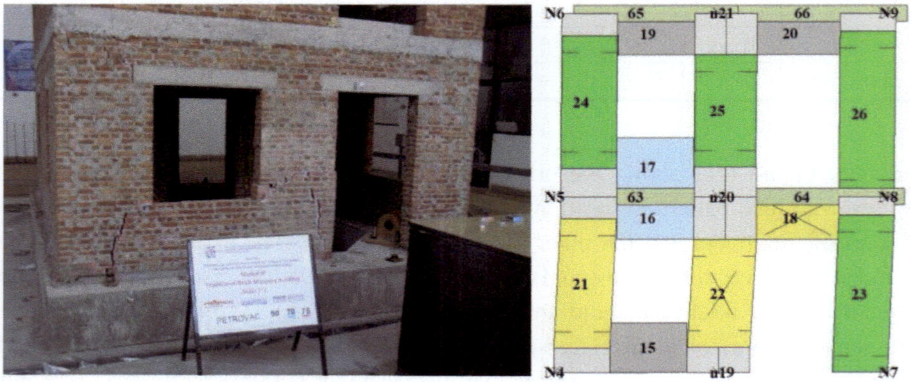

Figure 6.13: P2 wall of the BM model: experimental [73] and numerical final DLs (monotonic pushover analysis in the +y direction with load pattern proportional to the masses)

In particular, pier 21 shows a bending behaviour, while pier 22 presents a mixed shear-flexural failure. Furthermore, pier 23 reaches the bending resistance, but it does not exceed the DL3. The maximum drift value attained by this latter is about 1.0% (Figure 6.14), and it is lower than the imposed limit of 1.2%. This result confirms the experimental evidence that pier 23 is less damaged than the 21 and 22. Moreover, the upper elements are only slightly stressed, as experimentally observed after the shaking table test (no evident cracks were present [73]).

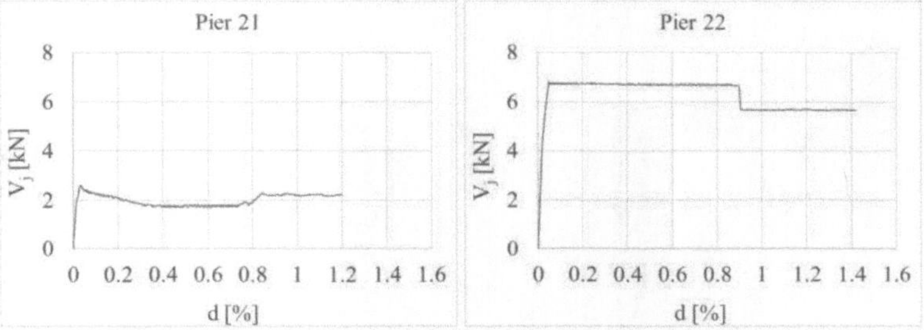

Figure 6.14: BM model: shear-drift diagrams of the masonry piers of the P2 wall (monotonic pushover in the +y direction with load pattern proportional to the masses)

As shown in Figure 6.14, the maximum drift reached by the elements 24, 25, 26 is equal to 0.36%. Therefore, it is possible to assume that the upper piers are not visible cracked, and the damage is concentrated in the bottom ones. Finally, the masonry spandrels and the r. c. beams are not damaged except for spandrel 18 that achieves the shear strength (spandrels 16 and 17 show only a degradation of the initial elastic stiffness k_{el}).

Concerning the masonry elements of the P4 wall, they show the same shear-drift diagrams of the P2 since they are identical. Therefore, Figure 6.15 illustrates only the experimental and numerical final damage pattern of the P4 wall.

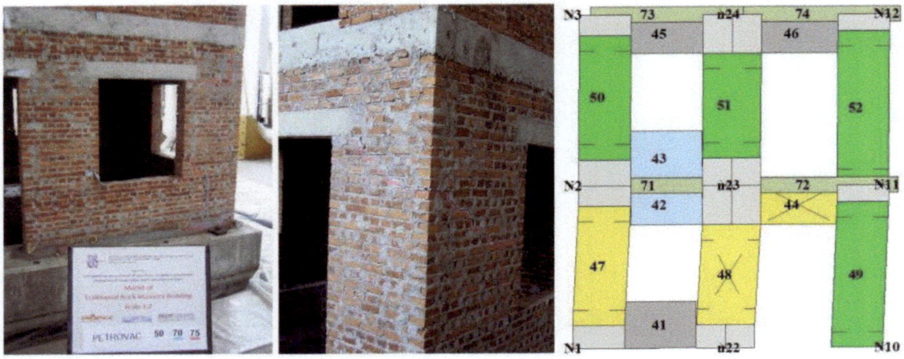

Figure 6.15: P4 wall of the BM model: experimental [73] and numerical final DLs (monotonic pushover analysis in the +y direction with load pattern proportional to the masses)

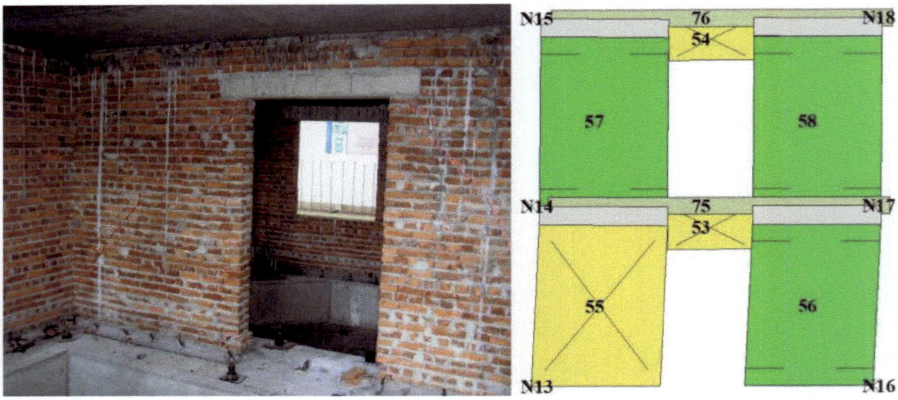

Figure 6.16: P5 wall of the BM model: experimental [73] and numerical final DLs (monotonic pushover analysis in the +y direction with load pattern proportional to the masses)

Moreover, the final failure mechanism showed by the middle wall P5 is reported in Figure 6.16. The masonry pier 55 reaches the shear resistance with diagonal cracking, confirming the experimental damage, and the maximum drift value is lower than 1.0%

(Figure 6.17). Therefore, it has exceeded the DL3 since $d_{3,S}$ is equal to 0.6%, but it doesn't attain the DL4 ($d_{4,S}$ = 1.1%). Moreover, pier 56 achieves the bending resistance, and the maximum drift is 1.1%, which means that it is largely in the non-linear range but just before the DL3 ($d_{3,B}$ = 1.2%). It is worth noting that this type of damage was experimentally observed. As depicted in Figure 6.16, piers 55 and 56 also show horizontal cracks at the bottom corners.

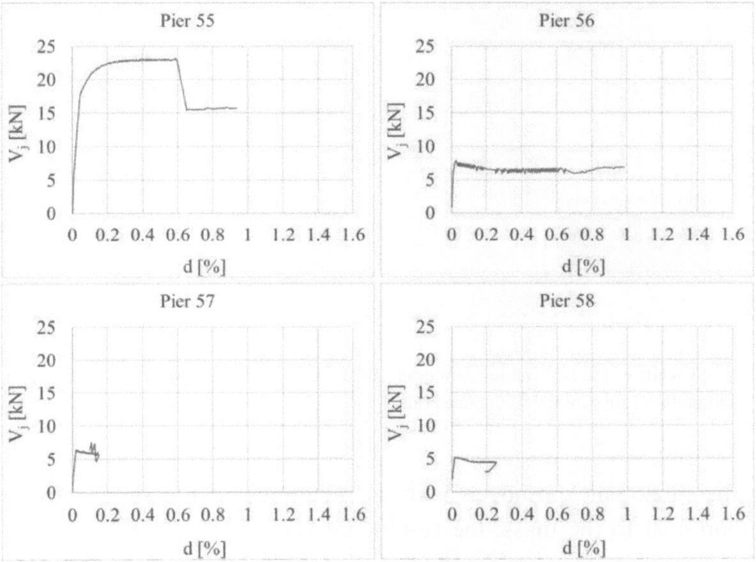

Figure 6.17: BM model: shear-drift diagrams of the masonry piers of the P5 wall (monotonic pushover in the +y direction with load pattern proportional to the masses)

On the contrary, the upper elements 57 and 58 are only slightly stressed. They achieve the bending resistance and are far from the DL3 since the maximum drift value is equal to 0.25%, attained by pier 58 (Figure 6.17). Therefore, it is possible to assume that these elements are not cracked, as observed after the shaking table test. Finally, the masonry spandrels 53 and 54 achieve the shear strength, and the reinforced concrete beams are not damaged.

Concerning the P1 and P3 walls, they are positioned in the x-direction of the model and, for this reason, loaded out-of-plane. Figure 6.18 shows the final DLs obtained with TREMURI. The elements 1, 4, 13 of the P1 wall remain linear elastic; the masonry spandrels 2, 3, 5, 6, 7, 8 and piers 12, 14 present only a degradation of the initial stiffness k_{el}. Piers 9, 11 attain the bending resistance remaining, however, at the beginning of the plastic phase. It is worth noting that pier 10 is subjected to a vertical tensile force since the seismic input is applied in the direction perpendicular to the wall, and the building has a stiff floor. In this case, the program deactivates the element, as the masonry tensile strength is neglected, and the unreinforced pier cannot be subjected to normal tensile forces. Finally, all the elements of the P3 wall remain in the linear elastic phase.

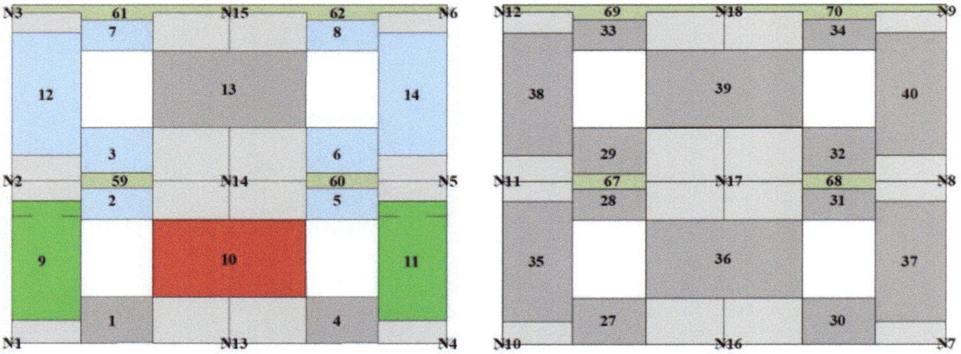

Figure 6.18: P1, P3 walls of the BM model: numerical final DLs (monotonic pushover in the +y direction with load pattern proportional to the masses)

Regarding the monotonic pushover analysis performed in the -y direction with load pattern proportional to the mass, the results are very similar to those obtained in the positive one. Therefore, they are reported in Chapter 8. In this case, the maximum horizontal displacement u_y imposed on the control node is -15.467 mm, and it is slightly lower than the positive value (17.073 mm).

The second load pattern adopted for the non-linear static analyses is the triangular inverse one, where the force distribution is practically proportional to the first mode shape. It allows representing the structural dynamic amplification by increasing the action on the higher storeys, and the load multiplier vector $[\lambda]$ is proportional to the mass and the height of the levels (Eq. (6.10)).

Compared to the previous force distribution, the load multiplier λ_i assumes unit value at the second level of the BM model. Thus, the damage is expected to be concentrated in the upper masonry elements.

The pushover curves obtained by imposing the maximum and minimum relative displacement (LP06 - LP01), i.e. 17.073 mm and -15.467 mm, as horizontal displacement limit of the control node, are shown in Figure 6.19. The maximum and minimum base shear forces are equal to 49.03 kN and -48.83 kN, respectively.

$$[\lambda]=\begin{bmatrix}\lambda_2\\ \lambda_1\end{bmatrix}=\begin{bmatrix}\dfrac{m_2\cdot(h_1+h_2)}{m_0\cdot h_0+m_1\cdot h_1+m_2\cdot(h_1+h_2)}\\ \dfrac{m_1\cdot h_1}{m_0\cdot h_0+m_1\cdot h_1+m_2\cdot(h_1+h_2)}\end{bmatrix}=$$

$$=\begin{bmatrix}\dfrac{7446.35\cdot 3.23}{2472.72\cdot 0+12684.06\cdot 1.58+7446.35\cdot 3.23}\\ \dfrac{12684.06\cdot 1.58}{2472.72\cdot 0+12684.06\cdot 1.58+7446.35\cdot 3.23}\end{bmatrix}= \tag{6.10}$$

$$=\begin{bmatrix}0.55\\ 0.45\end{bmatrix}\Rightarrow[\lambda]=\begin{bmatrix}\lambda_2=1.00\\ \lambda_1=0.82\end{bmatrix}$$

Moreover, the results of the non-linear static analysis carried out in the positive direction are first discussed, and the damage state obtained with TREMURI is directly compared with the experimental one.

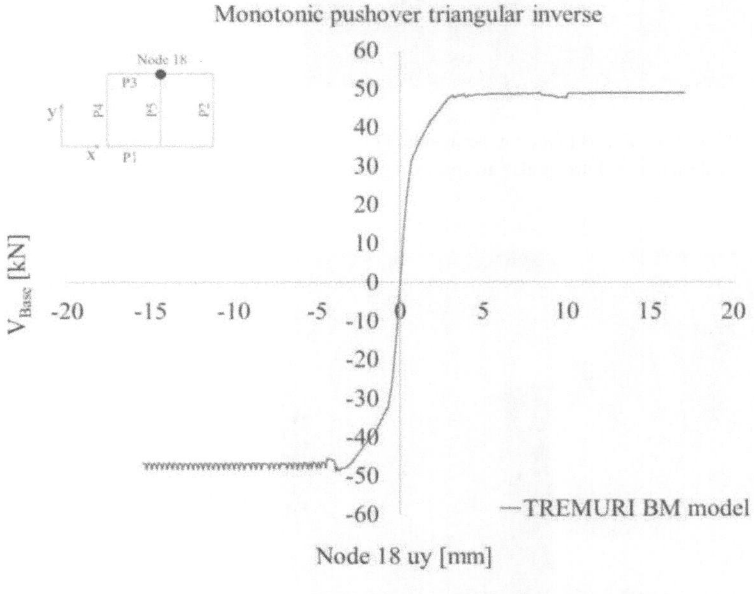

Figure 6.19: Monotonic pushover curves of the BM model with triangular inverse load pattern

The final damage showed by the P2 wall is depicted in Figure 6.20, in which the upper masonry piers are more damaged than the lower ones. The shear-drift curves of the elements (Figure 6.22) confirm that piers 21, 22, 23 attain the bending resistance (DL2) and are far from the DL3; thus, they should not present any visible cracks. On the contrary, the elements 24, 25 exceed the DL3 with a maximum drift equal to about 1.3% and 1.5%, respectively. Pier 26 is close to the DL3, with a maximum drift of 1.05%. The masonry spandrels and reinforced concrete beams remain linear elastic, except for spandrel 18 that achieves the shear strength. As shown in Figure 6.20, elements 16 and 17 present only a degradation of the initial elastic stiffness k_{el}.

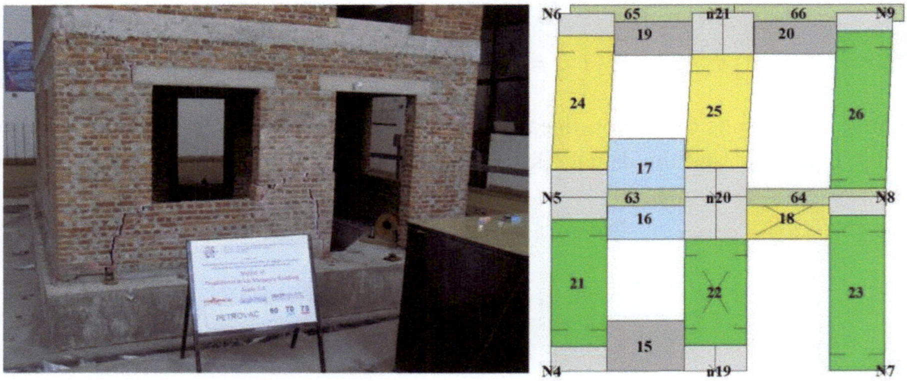

Figure 6.20: P2 wall of the BM model: experimental [73] and numerical final DLs (monotonic pushover in the +y direction with triangular inverse load pattern)

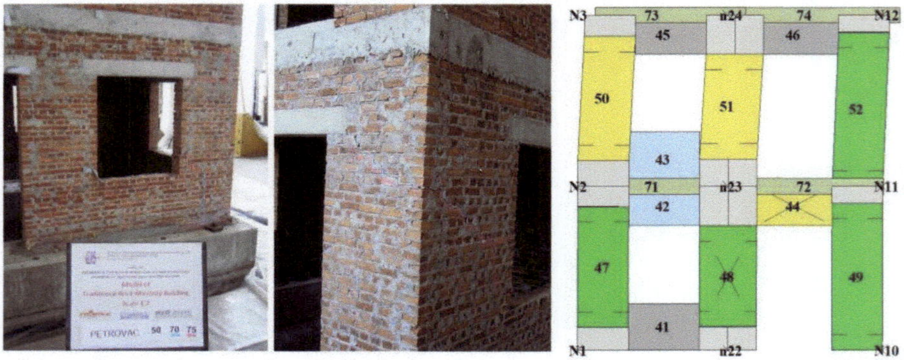

Figure 6.21: P4 wall of the BM model: experimental [73] and numerical final DLs (monotonic pushover in the +y direction with triangular inverse load pattern)

The masonry piers of the P4 wall are characterized by the same shear-drift curves depicted in Figure 6.22. Thus, only the final damage state is illustrated in Figure 6.21.

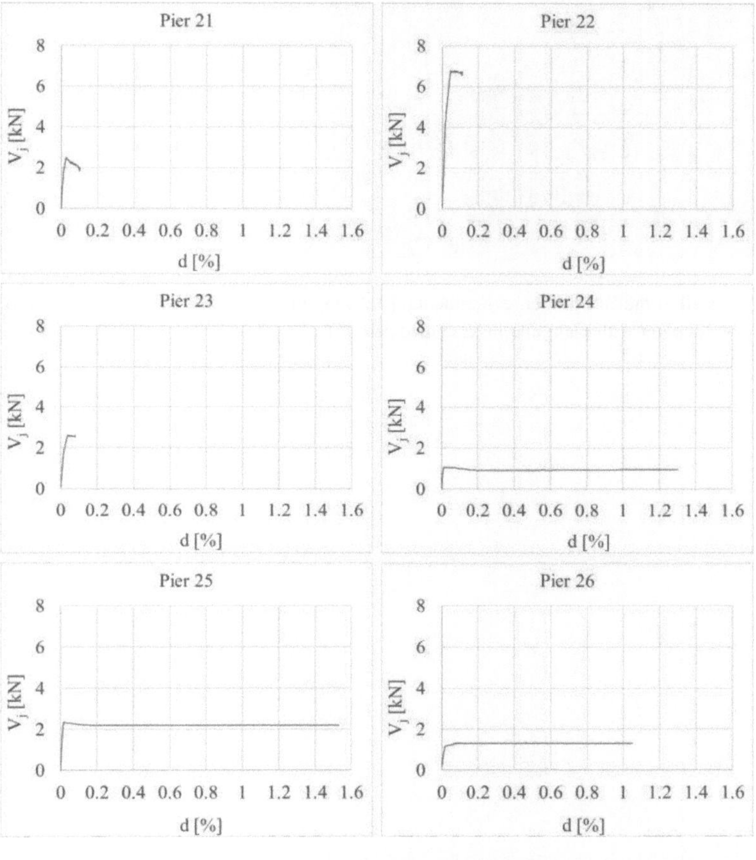

Figure 6.22: BM model: shear-drift diagrams of the masonry piers of the P2 wall (monotonic pushover in the +y direction with triangular inverse load pattern)

Furthermore, the upper masonry piers of the P5 wall are also more damaged than those of the lower level. As shown in Figure 6.24, pier 55 behaves linear elastic with only a degradation of the initial elastic stiffness k_{el}. The element 56 attains the bending strength (DL2), and it is very far from the DL3; thus, it is reasonable to assume that they are not cracked. The elements 57, 58 achieve the in-plane resistance and have a maximum drift equal to 0.96% and 1.05%, respectively. The masonry spandrels reach the shear strength, and the reinforced concrete beams are undamaged.

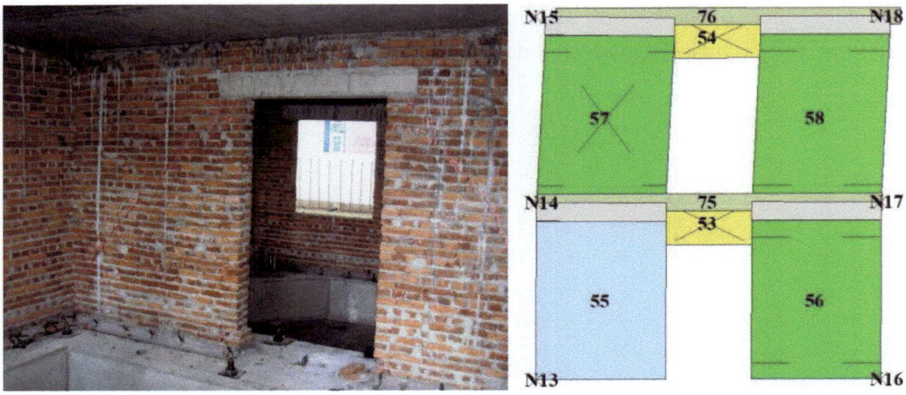

Figure 6.23: P5 wall of the BM model: experimental [73] and numerical final DLs (monotonic pushover in the +y direction with triangular inverse load pattern)

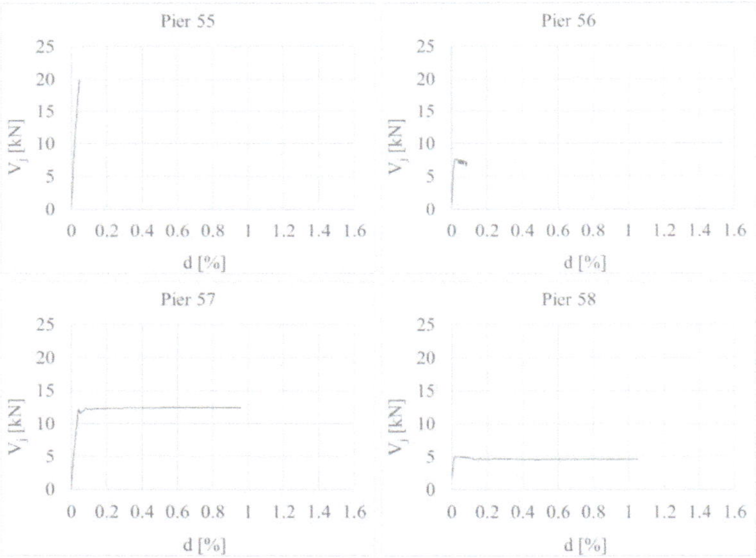

Figure 6.24: BM model: shear-drift diagrams of the masonry piers of the P5 wall (monotonic pushover in the +y direction with triangular inverse load pattern)

Finally, the out-of-plane walls, i.e. the P1 and P3, remain linear elastic. The elements 7, 8, 12, 14 have a degradation of the initial elastic stiffness and the piers 10, 13 are subjected to normal tensile forces (Figure 6.25).

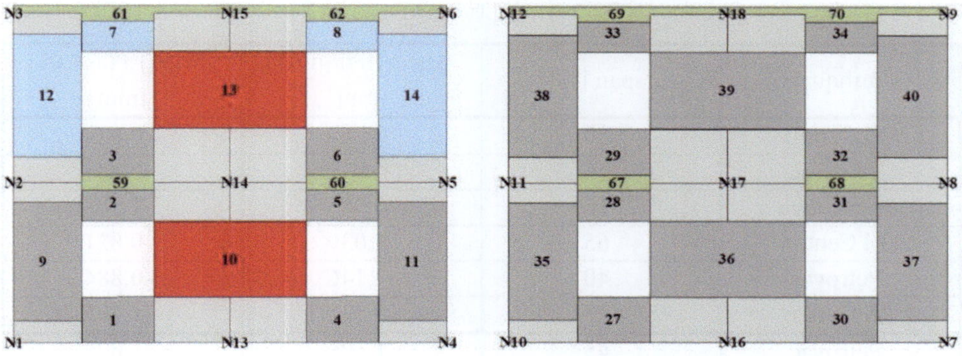

Figure 6.25: P1, P3 walls of the BM model: numerical final DLs (monotonic pushover in the +y direction with triangular inverse load pattern)

The results of the nonlinear static analysis performed in the -y direction with the triangular inverse force distribution are similar to those obtained in the positive one. For this reason, they are reported in Chapter 8.

Therefore, it is possible to conclude that the results obtained from the monotonic pushover analyses with force distribution proportional to the masses reproduce the failure mechanism of the BM model very well. The experimental damage is concentrated in the bottom masonry piers. The shear failure is dominant in the piers of the middle wall P5 under the first storey, like in the model in TREMURI. Thus, it is reasonable to assume that the adopted values of strength (Table 47), drift, and residual resistance (Table 48) are well calibrated.

On the contrary, the triangular inverse load distribution increases the action on the upper masonry elements. Therefore, the results cannot reproduce the same final damage state of the BM model. In this case, the upper piers of the model in TREMURI are more damaged than the bottom ones, which are only slightly stressed. Since no visible cracks were observed in the upper masonry elements and the bottom piers of the P5 wall failed with diagonal shear cracks [73], the final damage state obtained with the triangular inverse load pattern doesn't correspond to the experimental one.

6.1.2.2 Cyclic pushover analyses of the BM model

Two cyclic pushover analyses are performed through sequential monotonic pushover analyses to compare the damage evolution of the numerical model with the experimental one. The maximum and minimum horizontal displacements (LP06 - LP01) recorded during each test are applied as displacement limits u_y at the control node of the model (i.e. node 18). Table 49 summarizes the sequence of the applied earthquakes.

Table 49: Maximum and minimum displacement (LP06-LP01) measured during the test on the BM model

Earthquake	span [%]	Max (LP06-L01) [mm]	Min (LP06-L01) [mm]
El Centro	50	0.769	-0.621
Petrovac	36	0.979	-0.841
Northridge	16	0.985	-1.208
El Centro	65	1.039	-0.821
Petrovac	40	1.146	-0.884
El Centro	75	1.542	-1.405
Petrovac	45	1.298	-1.078
Northridge	20	1.518	-1.797
El Centro	80	5.382	-4.433
Northridge	25	1.910	-2.313
Petrovac	50	3.929	-3.493
Petrovac	70	9.887	-11.444
Petrovac	75	17.073	-15.467

The first adopted load pattern is proportional to the masses. Figure 6.26 shows the obtained cyclic pushover curve. As reported in [73], the first visible horizontal cracks occurred between the foundation slab and the out-of-plane walls (i.e. the P1, P3 walls in TREMURI) during the application of the Petrovac 45 earthquake while the rest of the masonry was without any visible crack. Therefore, the structure was estimated to be in a state of initial non-linearity.

This experimental observation finds a very good match with the results of the cyclic pushover analysis. As shown in Figure 6.26, the model is at the end of the elastic phase of the pushover curve when the load step Petrovac 45 is applied.

Then, the next seismic inputs (i.e. Northridge 20, El Centro 80 and Northridge 25) caused fine diagonal cracks in the North and South window corners (P2 and P4 walls) while the existing ones became wider. Finally, only the time history of the Petrovac earthquake was applied (Petrovac 50, 70 and 75) up to damage the BM model considerably. The experimental failure occurred in the lower zone of the structure with crack openings equal to 0.5 - 2.5 cm, and no visible cracks were observed at the upper storey. The mainly damaged elements were the piers under the first storey that were stressed in the same direction of the dynamic excitation, i.e. the piers of the P2 (Figure 6.27), P4 (Figure 6.29), and P5 walls (Figure 6.30). On the other hand, the P1 and P3 facades suffered minimal damage, as in the modelling with TREMURI (Figure 6.31).

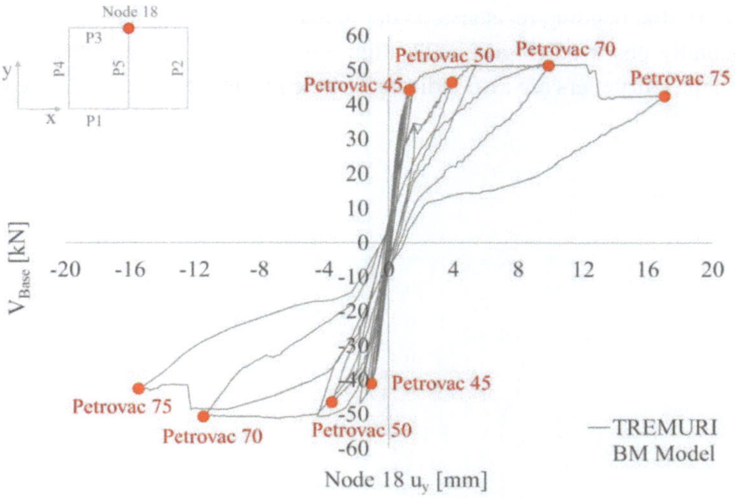

Figure 6.26: BM model: cyclic pushover curve with load pattern proportional to the masses

Therefore, it is possible to conclude that the seismic response of the BM model is very well represented by the cyclic pushover analysis with force distribution proportional to the masses. The curve shows that the structure is widely in the plastic range when the inputs Petrovac 70 and 75 are applied (Figure 6.26), as during the tests. Moreover, a strength degradation occurs after Petrovac 70, and larger dissipative cycles develop due to the attainment of the shear failure DL3 by the piers 55 and 56 (Figure 6.30).

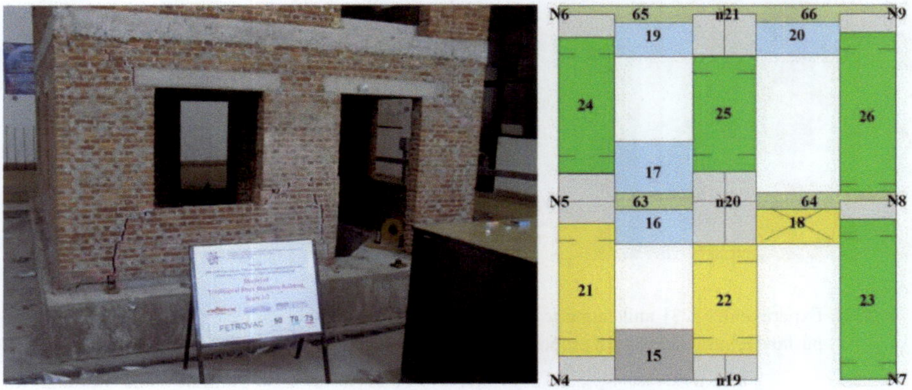

Figure 6.27: Experimental [73] and numerical final damage state of the P2 wall of the BM model (cyclic pushover with load pattern proportional to the masses)

On the contrary, the bending resistance is dominant in the lower piers of the P2, P4 walls, as experimentally observed (Figure 6.27, Figure 6.29). In this regard, the percentages of the damaged masonry piers are also indicated on the pushover curve in Figure 6.28.

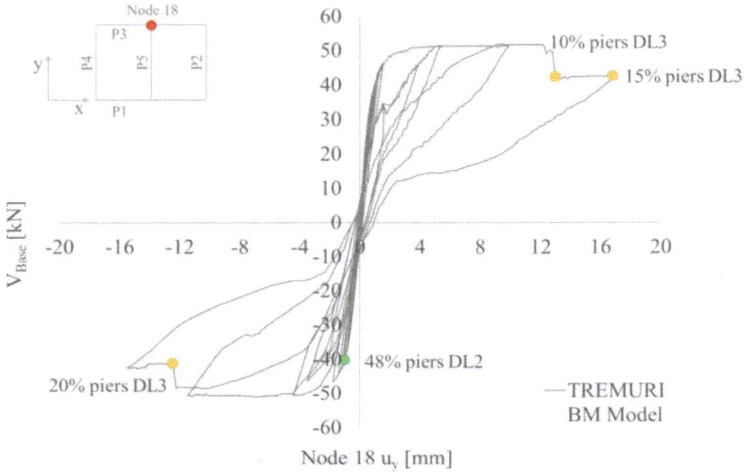

Figure 6.28: Cyclic pushover curve of the BM model with load pattern proportional to the masses: DLs of the masonry piers

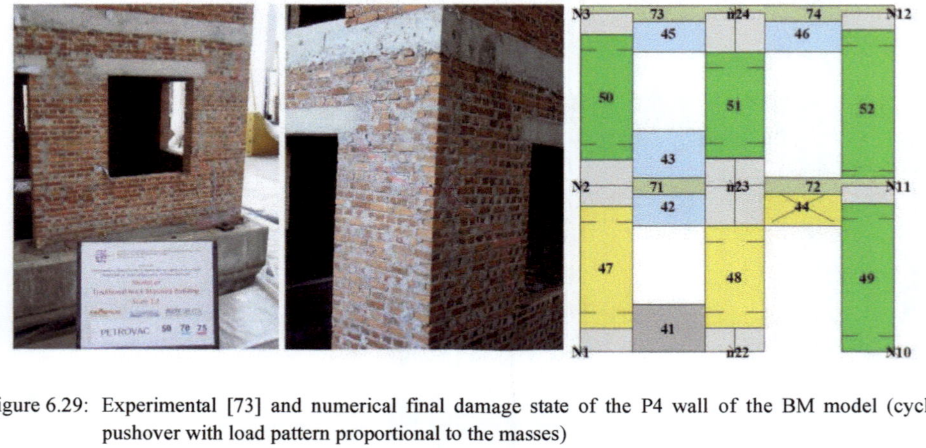

Figure 6.29: Experimental [73] and numerical final damage state of the P4 wall of the BM model (cyclic pushover with load pattern proportional to the masses)

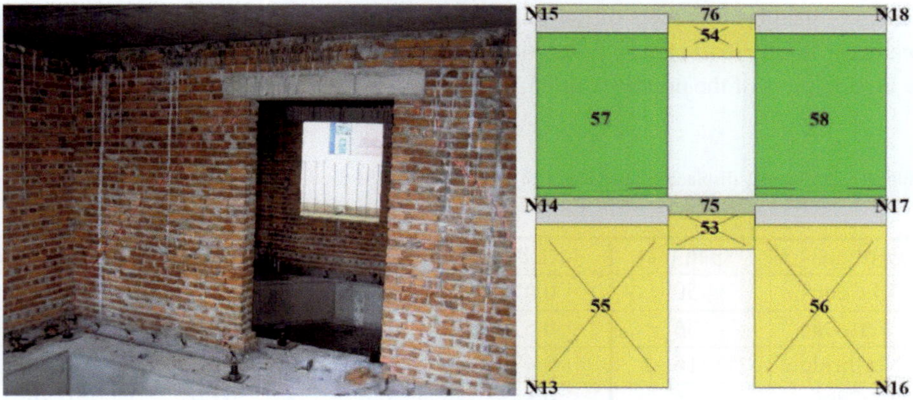

Figure 6.30: Experimental [73] and numerical final damage state of the P5 wall of the BM model (cyclic pushover with load pattern proportional to the masses)

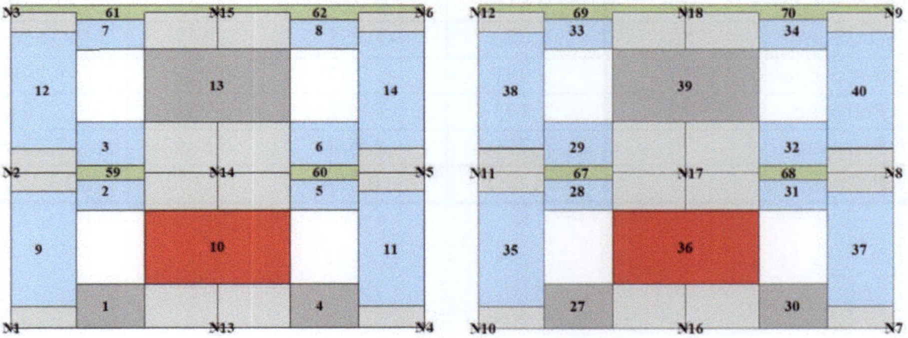

Figure 6.31: Numerical final damage state of the P1, P3 walls of the BM model (cyclic pushover with load pattern proportional to the masses)

Moreover, the absolute horizontal displacement of the first storey was experimentally measured through the linear potentiometer LP04, whose position corresponds to node 17 of the P3 wall in TREMURI. Since the floors are modelled as rigid diaphragms, the deformation of the structure can be compared with the experimental one checking the displacement of node 17. Therefore, the experimental and numerical values are summarized in Table 50, in which it is evident the very good match between the results.

Concerning the results of the pushover analysis with the triangular inverse force distribution, these are shown in Figure 6.32 through the base shear versus horizontal displacement of the control node relationship. It is worth noting that the capacity curve presents narrower dissipative cycles with the typical "flag" shape due to the prevailing bending failures of the masonry piers of the walls loaded in-plane. In fact, compared to the results obtained with the load pattern proportional to the masses, no diagonal cracking occurs in the bottom piers of the middle wall P5.

Table 50: Horizontal displacement of the first storey of the BM model in the y-direction (cyclic pushover analysis with load pattern proportional to the masses)

Earthquake	span [%]	(LP04 - LP01) [mm]		TREMURI node 17 u_y [mm]	
El Centro	50	0.480	-0.297	0.437	-0.355
Petrovac	36	0.579	-0.470	0.571	-0.491
Northridge	16	0.479	-0.674	0.575	-0.709
El Centro	65	0.545	-0.445	0.603	-0.362
Petrovac	40	0.720	-0.462	0.641	-0.507
El Centro	75	0.788	-0.788	0.838	-0.760
Petrovac	45	0.731	-0.567	0.707	-0.611
Northridge	20	0.818	-0.932	0.812	-0.989
El Centro	80	2.772	-2.255	3.542	-2.666
Northridge	25	0.997	-1.267	0.979	-1.298
Petrovac	50	2.143	-1.962	2.738	-2.204
Petrovac	70	6.796	-8.508	6.868	-7.700
Petrovac	75	14.023	-10.584	14.320	-13.127

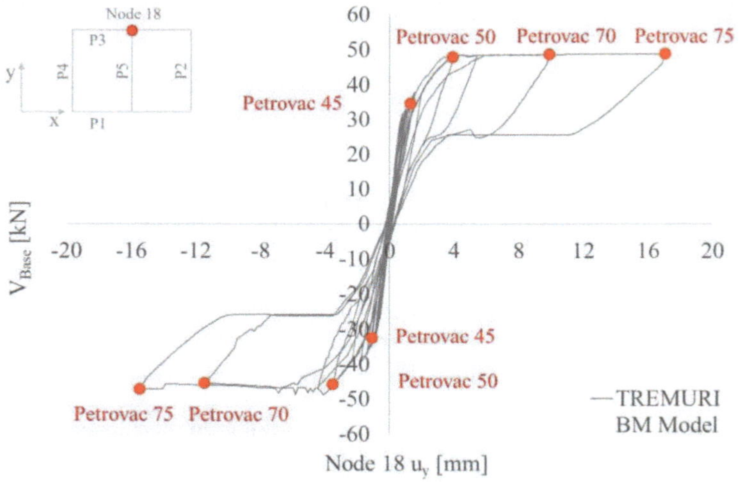

Figure 6.32: Cyclic pushover curve of the BM model obtained with the triangular inverse load pattern

Moreover, in Figure 6.33 are indicated the percentages of the damaged masonry piers on the cyclic pushover curve. In this case, only 10% of the piers attain the DL3, i.e. elements 24, 25 of the P2 wall (Figure 6.34) and 50, 51 of the P4 wall (Figure 6.35).

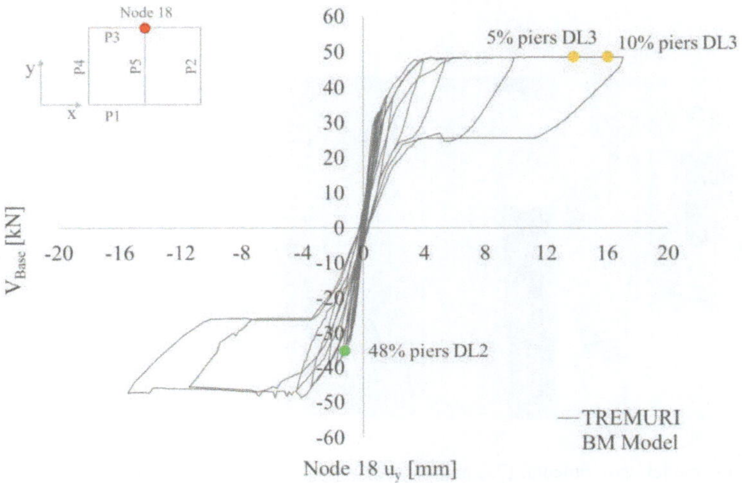

Figure 6.33: Cyclic pushover curve of the BM model obtained with the triangular inverse load pattern: DLs of the masonry piers

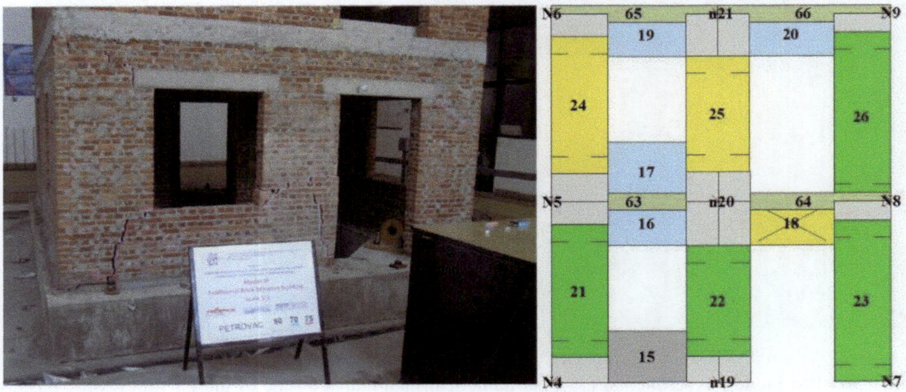

Figure 6.34: BM model: experimental [73] and numerical final DLs of the P2 wall (cyclic pushover with triangular inverse load pattern)

The experimental and numerical final damage state of the BM model is reported from Figure 6.34 to Figure 6.37. Since the triangular inverse force distribution can represent the structural dynamic amplification, the action is generally increased on the higher storey. Therefore, the damage is consequently concentrated in the upper piers of the model.

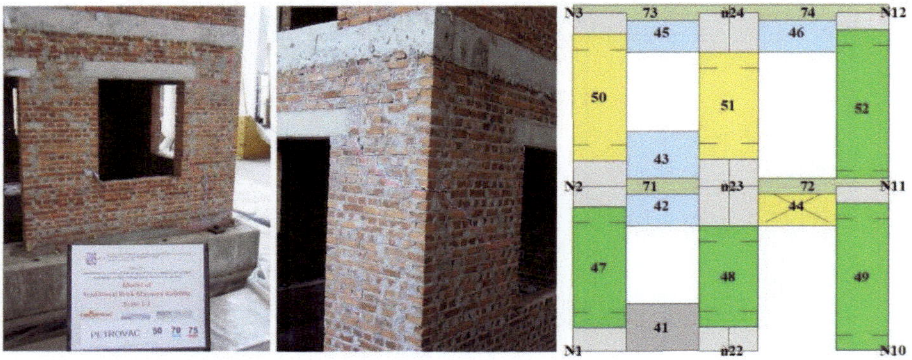

Figure 6.35: BM model: experimental [73] and numerical final damage state of the P4 wall (cyclic pushover with triangular inverse load pattern)

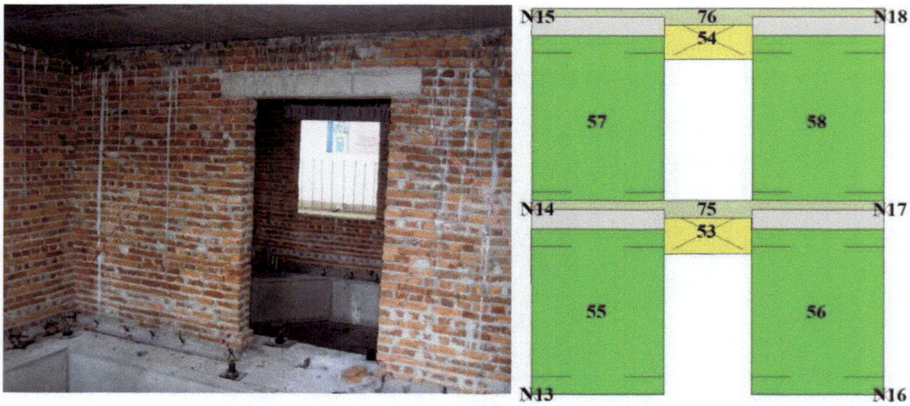

Figure 6.36: BM model: experimental [73] and numerical final damage state of the P5 wall (cyclic pushover with triangular inverse load pattern)

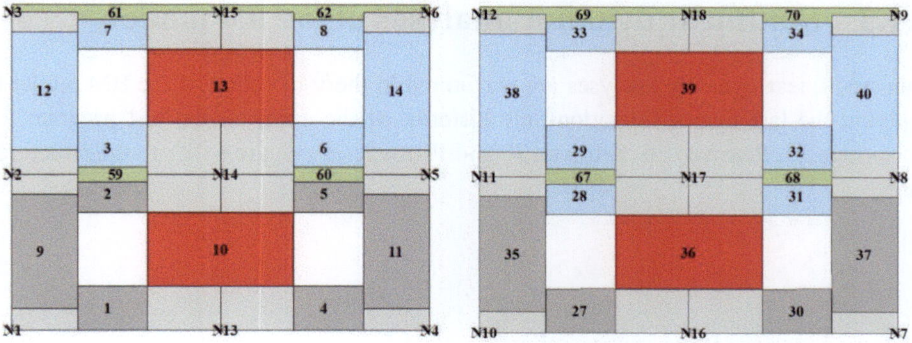

Figure 6.37: BM model: numerical final damage state the P1, P3 walls (cyclic pushover with triangular inverse load pattern)

Finally, the horizontal displacement u_y of node 17 is also compared to the experimental one (LP04 – LP01). The results are shown in Table 51 and confirm that the deformation of the model in TREMURI is very different from the experimental one. In particular, during the last sequence of the Petrovac earthquakes, the structure is widely in the non-linear range, but the horizontal displacement u_y of the first storey is lower than the experimental values since the bottom masonry elements are not plasticized like the upper ones.

Table 51: Horizontal displacement of the first storey of the BM model in the y-direction (cyclic pushover analysis with triangular inverse load pattern)

Earthquake	span [%]	(LP04 - LP01) [mm]		TREMURI node 17 u_y [mm]	
El Centro	50	0.480	-0.297	0.347	-0.315
Petrovac	36	0.579	-0.470	0.433	-0.423
Northridge	16	0.479	-0.674	0.450	-0.510
El Centro	65	0.545	-0.445	0.457	-0.411
Petrovac	40	0.720	-0.462	0.490	-0.422
El Centro	75	0.788	-0.788	0.602	-0.544
Petrovac	45	0.731	-0.567	0.512	-0.454
Northridge	20	0.818	-0.932	0.580	-0.692
El Centro	80	2.772	-2.255	1.343	-1.351
Northridge	25	0.997	-1.267	0.628	-0.730
Petrovac	50	2.143	-1.962	1.288	-1.174
Petrovac	70	6.796	-8.508	1.326	-1.334
Petrovac	75	14.023	-10.584	1.347	-1.307

6.1.3 Nonlinear dynamic analyses of the BM model

Four non-linear dynamic analyses are performed in the y-direction of the BM model by applying the last four acceleration-time histories of the experimental test program, i.e. Northridge 25, Petrovac 50, Petrovac 70 and Petrovac 75 (Figure 6.38), to the structure's base.

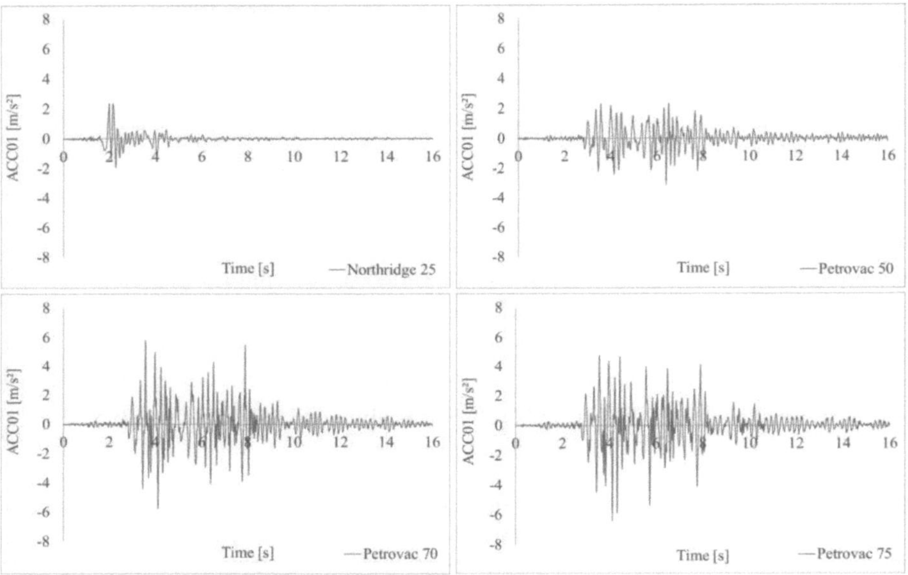

Figure 6.38: Acceleration-time histories applied to the base of the BM model in TREMURI

It is important to stress that the material properties of the model and the parameters assumed for the cyclic nonlinear behaviour of the ML-BEAM elements (piers and spandrels) are the same as those defined in § 6.1.1 and § 6.1.2 for the modal and pushover analyses.

Furthermore, the Rayleigh viscous damping is adopted by the TREMURI program. Since the latter is proportional to the mass and stiffness of the structure through the coefficients α_1 and β_1, it is necessary to assign these two parameters in the input phase of the modelling. Usually, they can be calculated assuming that two relevant modes of the structure, i^{th} and j^{th}, have the same damping ratio ξ, as indicated in Eq. (6.11):

$$\begin{cases} \alpha_1 = \xi \dfrac{2\omega_i \omega_j}{\omega_i + \omega_j} \\ \beta_1 = \xi \dfrac{2}{\omega_i + \omega_j} \end{cases} \tag{6.11}$$

It is worth noting that the appropriate selection of two relevant modes is important. A common approach is to choose the first mode and the last significant one that allows reaching at least 85% of mass participation to the structure's response in the analysis direction [88].

In this case, the mode shapes in the y-direction useful for achieving at least 85% of the total seismic mass are the first and the fourth. According to the modal analysis results shown in Table 45, 89% of mass participation is reached by considering these two modes.

Moreover, the elastic damping ratio ξ is assumed equal to 3% to correctly reproduce the structure's inherent damping and energy dissipation [46]. As shown in Figure 6.39, the elastic damping ratio can be assumed nearly constant in the range between the 1st and 4th frequency.

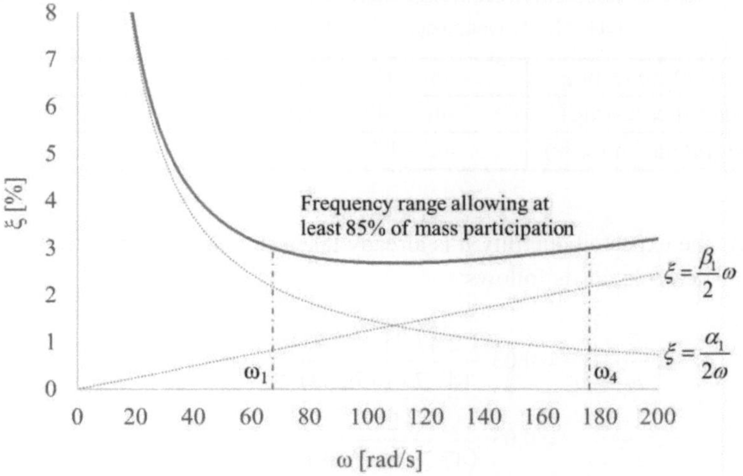

Figure 6.39: Rayleigh damping ratio and frequency range that allows reaching at least 85% of the BM model's mass participation

It is worth noting that, since the Rayleigh damping is a combination of mass and stiffness proportional terms, it may lead to physically not plausible responses under certain conditions. The latter can occur during a nonlinear dynamic analysis when the damping force coming from the stiffness-proportional part is computed with the initial stiffness. In this case, the damping force can become unrealistically large compared to the restoring one calculated in the inelastic response phase, and the results may be non-conservative [88].

Therefore, a possible solution to this problem is to calculate a lower first frequency ω_1^* in consideration of an expected displacement ductility μ, as proposed by [89]:

$$\omega_1^* = \frac{\omega_1}{\sqrt{\mu}} \tag{6.12}$$

According to this approach, the potential error induced by an initial overestimation of the elastic damping can also be reduced since it usually becomes greater more the structural response is in the inelastic phase.

Moreover, as reported in [73], the natural frequencies of the BM model were measured before and after testing with the ambient vibration technique to check the stiffness degradation of the structure. Since the first natural frequency in the East-West direction of the model decreased from 10.95 Hz to 7.6 Hz, it is possible to adopt this latter experimental value as first frequency ω_1^*.

Table 52: Natural frequencies and periods of the BM model

1st mode (before testing)	$\omega_1 = 67.31$ rad/s	$T_1 = 0.095$ s
1st mode (after testing)	$\omega_1^* = 47.75$ rad/s	$T_1^* = 0.13$ s
4th mode (modal analysis)	$\omega_4 = 176.58$ rad/s	$T_4 = 0.036$ s

In this way, the expected ductility μ is already taken into account, and the coefficients α_1 and β_1 can be calculated, as follows:

$$\begin{cases} \alpha_1 = \xi \dfrac{2\omega_1^* \omega_4}{\omega_1^* + \omega_4} = 0.03 \cdot \dfrac{(2 \cdot 47.75 \cdot 176.58)}{(47.75 + 176.58)} = 2.255 \\ \beta_1 = \xi \dfrac{2}{\omega_1^* + \omega_4} = 0.03 \cdot \dfrac{2}{(47.75 + 176.58)} = 0.000267 \end{cases} \tag{6.13}$$

It is worth noting that considering ω_1^* as 1st frequency means assuming a lower damping ratio ξ^* in the fundamental mode ω_1, as shown by Eq. (6.14) and Figure 6.40.

$$\xi^* = \frac{\alpha_1}{2\omega_1} + \frac{\beta_1}{2}\omega_1 = \frac{2.255}{2 \cdot 67.31} + \frac{0.000267}{2} \cdot 67.31 = 2.6\% \tag{6.14}$$

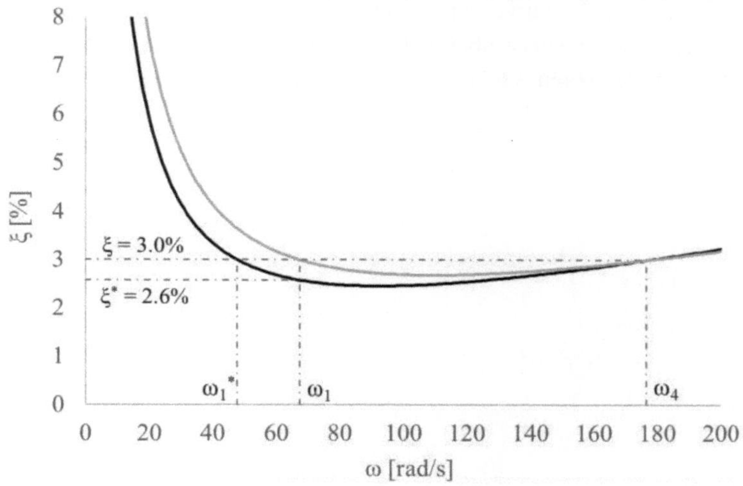

Figure 6.40: Rayleigh damping ratios ξ and ξ^* of the BM model

Therefore, according to the available test data, it is possible to compare the numerical and experimental results in terms of displacement and acceleration. The results of the last performed nonlinear dynamic analysis Petrovac 75 are depicted in Figure 6.41 (those of the analyses Northridge 25, Petrovac 50, Petrovac 70 are reported in Chapter 8, Figure 8.13 - Figure 8.15).

In detail, Figure 6.41 shows the horizontal relative displacements u_y of the second and first storey, (LP06 – LP01) and (LP04 – LP01), compared with those calculated by the TREMURI program in nodes 18 and 17. Furthermore, it illustrates the absolute accelerations recorded by the sensors ACC04, ACC05, ACC06, ACC07[2] compared with those calculated by the program in the corresponding nodes of the model (i.e. 12, 9, 17, 18, respectively).

Thus, it is evident the very good agreement between the numerical and experimental results. The nonlinear dynamic analyses confirm the material properties of the model and the parameters of the cyclic nonlinear behaviour of the ML-BEAM elements (piers and spandrels) adopted in § 6.1.1 and § 6.1.2 to perform modal and pushover analyses.

Moreover, the base shear force-displacement diagram resulting from the dynamic analyses is shown in Figure 6.42. The maximum and minimum shear force values are equal to 52,06 kN and -54,92 kN. They are achieved during the application of the seismic input Petrovac 70.

[2] The complete arrangement of the adopted sensors is illustrated in [73].

It is worth noting that the curve pattern and the attained strength values are consistent with the cyclic pushover curve obtained from the load distribution proportional to the masses of the structure (Figure 6.26), confirming, in this way, the same structural behaviour.

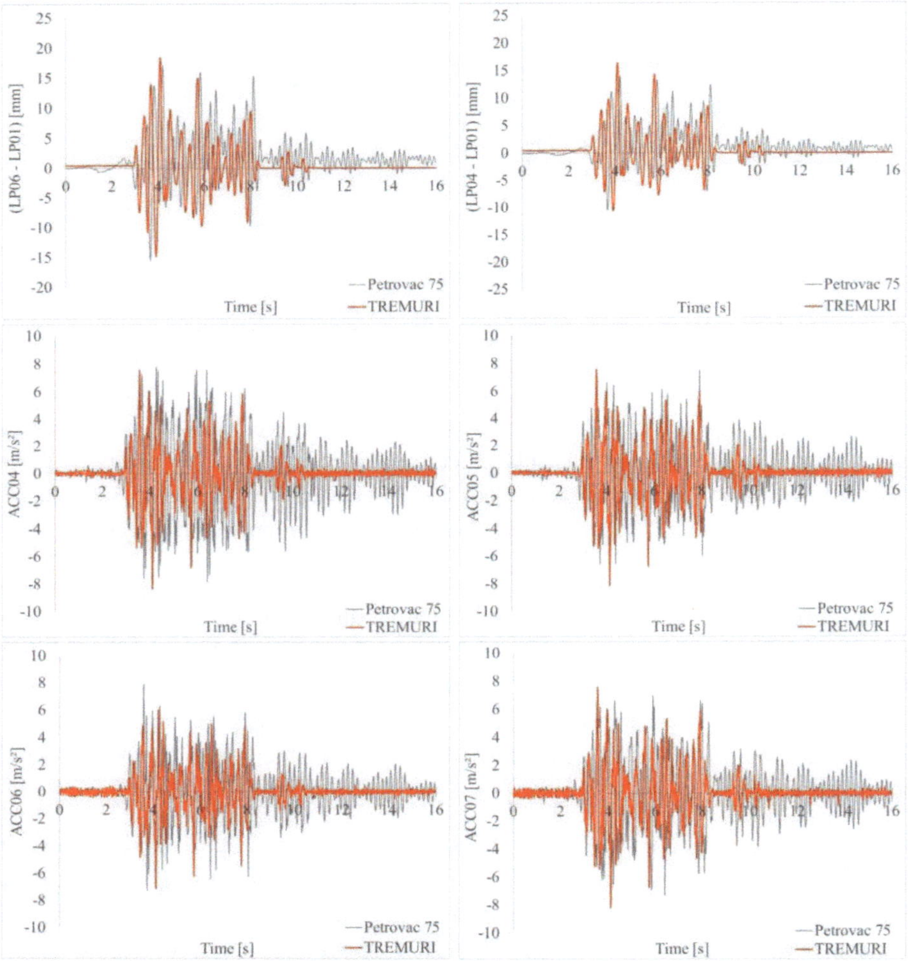

Figure 6.41: Results of the last nonlinear dynamic analysis Petrovac 75 performed on the BM model with TREMURI

Finally, the last damage state obtained from the dynamic analyses is compared to the experimental one (Figure 6.43 - Figure 6.46). As observed after the shaking table tests, the most damaged piers were the lower ones belonging to the P2, P4 and P5 walls, while the upper elements were completely undamaged. After the dynamic analyses, the piers 55 and 56 of the P5 wall achieve the DL4 by shear failure. Elements 21 and 22 (P2 wall), 47 and 48 (P4 wall) attain the flexural strength until DL4 or show a hybrid behaviour. Finally, some P1 and P3 wall elements are subjected to traction (Figure 6.45), which is is also expected since they are loaded out-of-plane.

Therefore, according to the obtained results, it is possible to conclude that the structural performance of the BM model is very well reproduced by the Equivalent frame method implemented in the TREMURI program. The values of strength (Table 47), drift, and residual resistance (Table 48) adopted for the static analyses are confirmed.

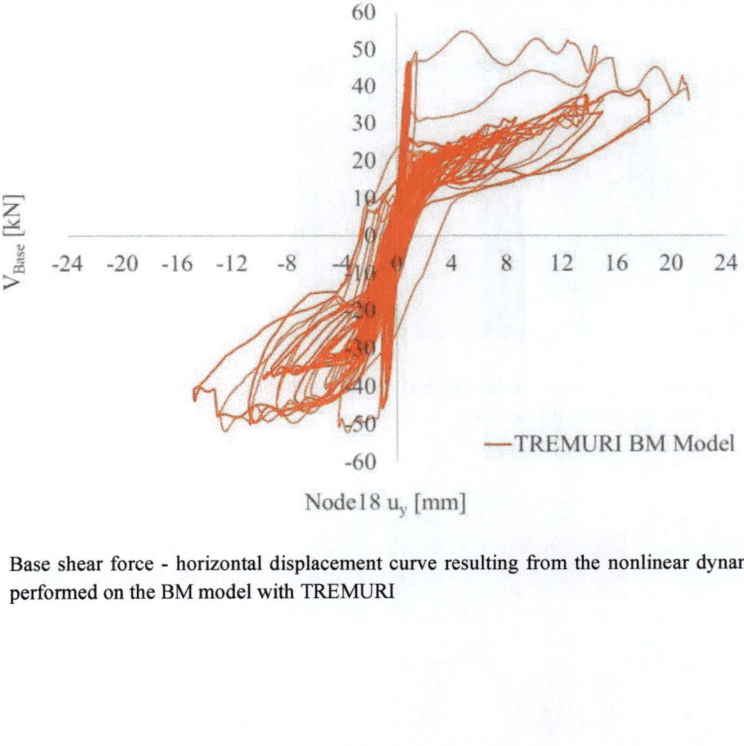

Figure 6.42: Base shear force - horizontal displacement curve resulting from the nonlinear dynamic analyses performed on the BM model with TREMURI

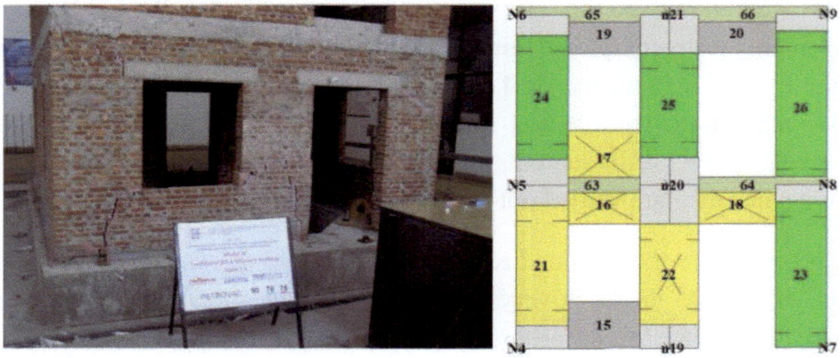

Figure 6.43: Experimental [73] and numerical final damage state of the P2 wall of the BM model (nonlinear dynamic analysis Petrovac 75)

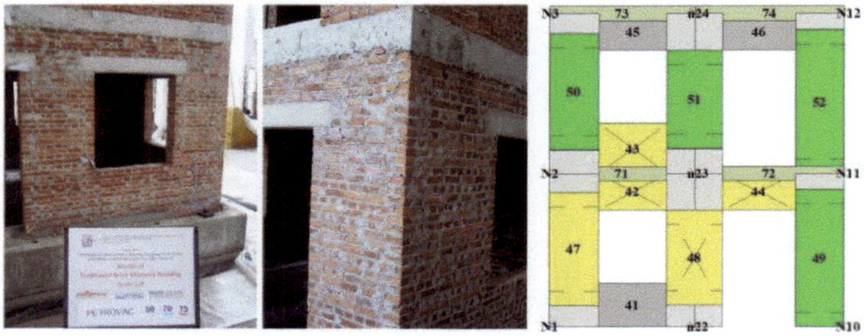

Figure 6.44: Experimental [73] and numerical final damage state of the P4 wall of the BM model (nonlinear dynamic analysis Petrovac 75)

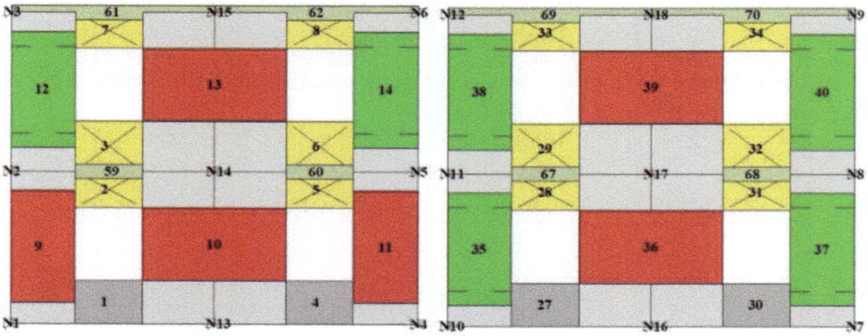

Figure 6.45: Numerical final damage state of the P1 and P3 walls of the BM model (nonlinear dynamic analysis Petrovac 75)

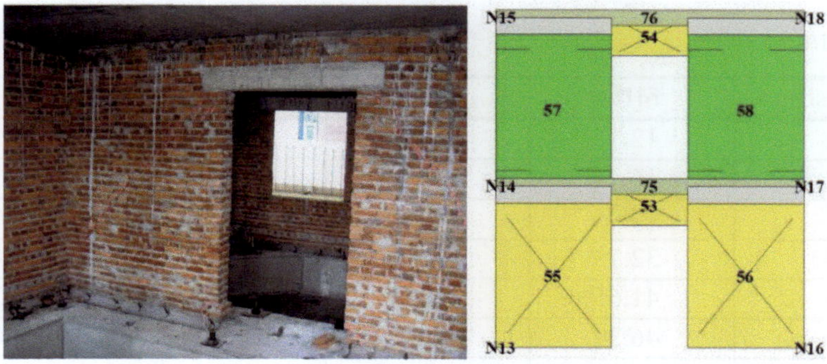

Figure 6.46: Experimental [73] and numerical final damage state of the P5 wall of the BM model (nonlinear dynamic analysis Petrovac 75)

6.2 The BM-SR model strengthened with the EQ-GRID system

6.2.1 Modal analysis of the BM-SR model

The natural frequencies of the BM-SR model were experimentally measured with the ambient vibration and random excitation techniques before applying the seismic inputs, as for the non-retrofitted specimen. Therefore, the modal analysis is performed to identify the mode shapes, frequencies/periods and participating masses of the BM-SR model. It is important to note that the EQ-GRID system doesn't increase the original model's mass since the total thickness of the applied system is only 8 mm. Thus, the retrofitted specimen is characterized by the same seismic masses calculated in § 6.1.1 for the unstrengthened structure.

The modal analysis results are illustrated in Table 53, in which the natural frequencies f, periods T and modal masses m_x, m_y of the first ten mode shapes, are reported. Furthermore, Table 54 shows the elastic moduli assumed for the masonry elements. The values are calibrated comparing the 1st frequency calculated by the program with the experimental average one.It is worth noting that the strengthening system slightly improved the stiffness of the structure since the initial frequency in the E-W direction increased from 10.57 Hz (without the EQ-GRID) to 12.63 Hz (with EQ-GRID). Therefore, the values adopted for the elastic and shear moduli are about 30% greater than those assumed for the BM model.

Table 53: Natural frequencies *f*, periods *T* and modal masses m_x, m_y of the BM-SR model calculated with TREMURI

Mode shape	f [Hz]	T [s]	m_x [kg]	m_y [kg]
1	12.77	0.078	0.001	18070.02
2	16.44	0.061	163.82	0.06
3	17.95	0.056	17054.78	0
4	32.73	0.031	0	2043.93
5	41.65	0.024	8.059	0.012
6	46.39	0.022	2838.71	0
7	56.80	0.018	0.004	1.058
8	58.47	0.017	18.775	0.001
9	59.43	0.017	0.003	4.387
10	61.16	0.016	29.07	0

Table 54: Natural frequencies of the BM-SR model and assumed elastic moduli for the modal analysis

<table>
<tr><th colspan="3">Experimental (before testing)</th><th colspan="4">TREMURI</th></tr>
<tr><th>Direction</th><th>f</th><th>f</th><th>f</th><th>E</th><th>$E_{//}$</th><th>G</th></tr>
<tr><td rowspan="2">E-W (y-direction)</td><td>13.77 Hz (ambient vibration)</td><td rowspan="2">12.63 Hz (average)</td><td rowspan="2">12.77 Hz (1st mode shape)</td><td rowspan="2">2800 N/mm²</td><td rowspan="2">1400 N/mm²</td><td rowspan="2">1120 N/mm²</td></tr>
<tr><td>11.48 Hz (random excitation)</td></tr>
</table>

Finally, Figure 6.47 illustrates the first and third mode shapes obtained from the modal analysis. The first one is translational along the East-West direction of the structure (y-direction in TREMURI) with a frequency equal to 12.77 Hz and 80% mass participation. The third one is translational along the North-South direction (x-direction in TREMURI) and has a frequency equal to 17.95 Hz with 75% mass participation.

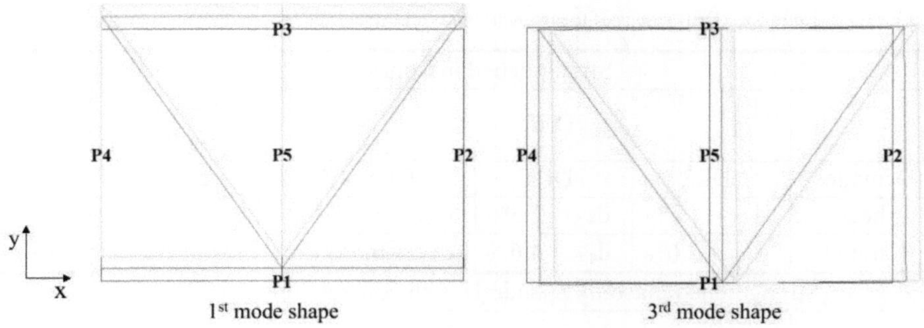

Figure 6.47: First and third mode shape of the BM-SR model obtained from the modal analysis

6.2.2 Pushover analyses of the BM-SR model

Four monotonic and two cyclic nonlinear static analyses are performed in the East-West direction of the BM-SR model (y-direction in TREMURI) to compute the response of the retrofitted masonry structure under increasing lateral loads.

As reported in [73], the experimental horizontal displacement of the second storey was measured through a linear potentiometer named LP04. Therefore, the controlled d.o.f. assumed for the analyses is the same as the unstrengthened model, i.e. the horizontal displacement u_y of node 18 of the P3 wall. It is worth noting that, since the values recorded by the sensor LP04 are absolute displacements, the difference (LP04 - LP01) is considered for the modelling[3].

The average masonry material properties are the same as those adopted for the unstrengthened structure (Table 47). The in-plane resistance increase due to the EQ-GRID system is computed after CNR-DT 215/2018, as shown in § 4.2.3.

6.2.2.1 Monotonic pushover analyses of the BM-SR model

The numerical results of the monotonic pushover analyses performed with the load pattern proportional to the masses are presented in the following. Since the EQ-GRID system doesn't increase the original model's mass, the load multipliers λ_i calculated in Eq. (6.9) for the unreinforced structure are still adopted.

Moreover, the values of drift and residual resistance assumed for the strengthened masonry piers and spandrels are indicated in Table 55.

---

3 The linear potentiometer LP01 measured the absolute displacement of the BM-SR model's foundation. The complete description of the adopted sensors is given in [73].

Table 55: Drift values of the masonry piers and spandrels for the BM-SR model

Strengthened masonry piers					
	Drift			Residual resistance ($V_{res,i}$ / $V_{Rd,s}$)	
Damage	DL3	DL4	DL5	DL3 → DL4	DL4 → DL5
Shear	$d_{3,S}$ = 1.3%	$d_{4,S}$ = 2.3%	$d_{5,S}$ = 3.4%	0.7	0.4
Bending	$d_{3,B}$ = 2.6%	$d_{4,B}$ = 4.6%	$d_{5,B}$ = 6.8%	1.0	0.85
Strengthened masonry spandrels with concrete lintel and tie beam					
	Drift			Residual resistance ($V_{res,i}$ / $V_{Rd,s}$)	
Damage	DL3	DL4	DL5	DL3 → DL4	DL4 → DL5
Shear	$d_{3,S}$ = 2.5%	$d_{4,S}$ = 4.2%	$d_{5,S}$ = 5.5%	1.0	0.8
Bending	$d_{3,B}$ = 2.5%	$d_{4,B}$ = 4.2%	$d_{5,B}$ = 5.5%	1.0	0.8

Figure 6.48 shows the pushover curves obtained from two monotonic non-linear static analyses with load pattern proportional to the masses (one performed in the positive y-direction of the model and the other in the negative one).

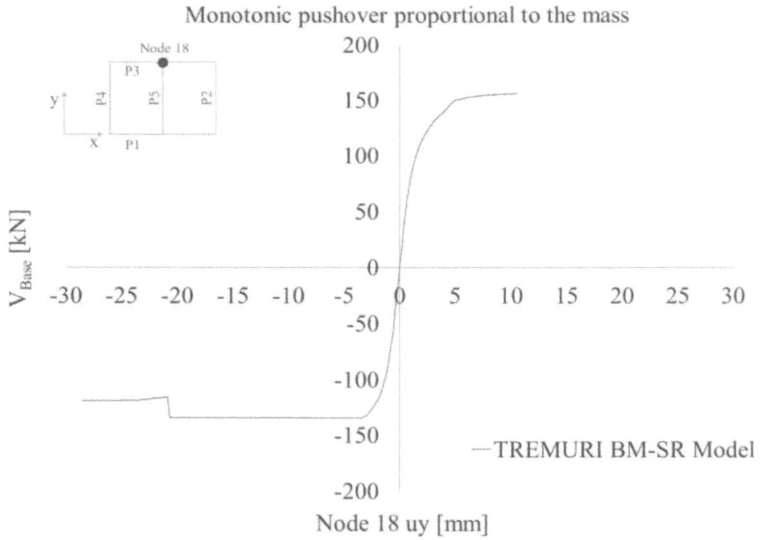

Figure 6.48: Monotonic pushover curves of the BM-SR model with load pattern proportional to the masses

The analyses are performed applying at the control node the maximum and minimum experimental value of the horizontal displacement (LP04 - LP01) measured during the last applied earthquake (Petrovac 250), that is 10.54 mm and -28.51 mm.

As depicted in Figure 6.48, the attained maximum and minimum base shear forces are equal to 156.0 kN and -134.8 kN, respectively. Therefore, thanks to the retrofitting with EQ-GRID, the global resistance of the BM-SR model is up to three times higher than the one achieved by the original structure (Figure 6.12). Moreover, since the positive and negative maximum values of the horizontal displacement (LP04 - LP01) are quite different, the nonlinear static analysis results performed in the most stressed direction (-y) are discussed in the following. Those obtained in the positive one are reported in Chapter 8.

Figure 6.49 shows the experimental final damage state of the corner between the P1 and P4 walls, where horizontal flexural cracks are evident in the lower piers. The DLs obtained from the pushover analysis are also depicted. The numerical results confirm that most of the piers and spandrels of the P4 wall attain the bending strength. The shear force-drift diagrams (Figure 6.50) show that the most damaged elements are piers 47 and 48, as experimentally observed (Figure 6.51). In particular, pier 48 reaches DL4 with a maximum drift of about 2.8%, while pier 47 doesn't exceed DL3 since its maximum drift value is equal to 2.3%.

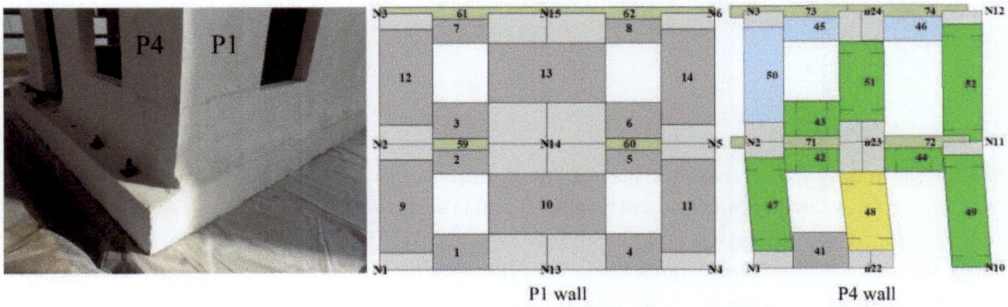

Figure 6.49: P4 - P1 corner of the BM-SR model: experimental [73] and numerical final DLs (monotonic pushover analysis in the -y direction with load pattern proportional to the masses)

Concerning the upper elements, they are only slightly stressed. As shown in Figure 6.50, the maximum drift achieved by elements 51 and 52 is 0.046%, while pier 50 has only a degradation of the initial elastic stiffness k_{el}. Moreover, the masonry spandrels and the r. c. beams are not damaged. In this regard, elements 42, 43, 44 attain the flexural strength, but they are far from the DL3 limit, and spandrels 45, 46 show only a degradation of the initial elastic stiffness.

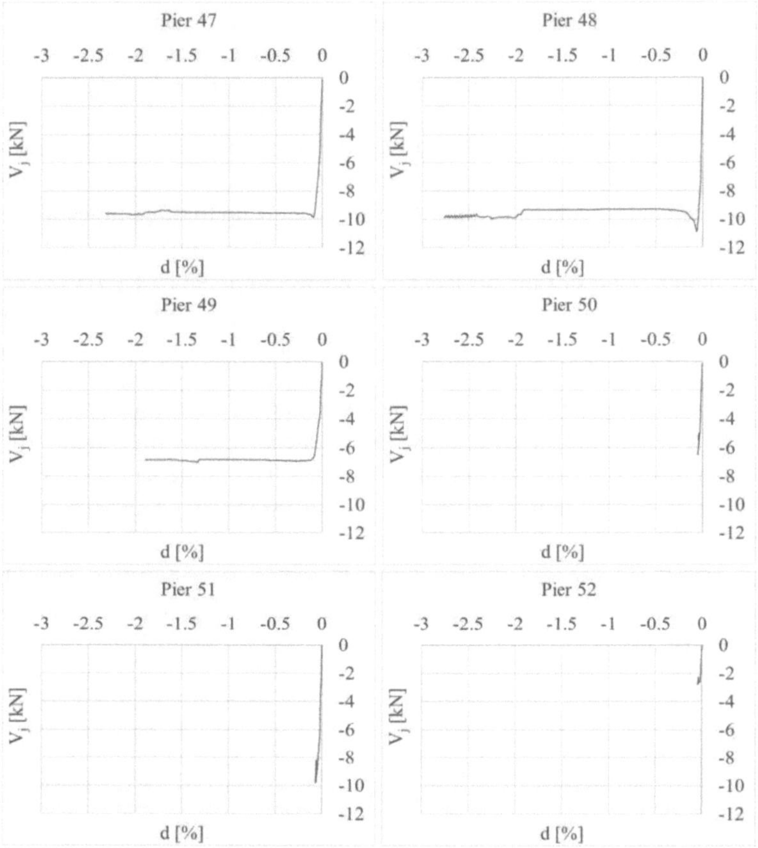

Figure 6.50: BM-SR model: shear-drift diagrams of the masonry piers of the P4 wall (monotonic pushover in the -y direction with load pattern proportional to the masses)

Therefore, it is possible to conclude that the upper elements are not cracked, and the damage is concentrated in the bottom ones, as experimentally observed.

Furthermore, Figure 6.52 illustrates the final damage state of the corner between the P2 and P3 walls and the DLs obtained from the pushover analysis. It is worth noting that the numerical failure attained by the P2 and P3 walls is identical to the P1 and P4 ones. Since the shear-drift diagrams of the P2 wall are very similar to those of the P4, they are not reported.

Finally, the numerical failure mechanism of the middle wall P5 is depicted in Figure 6.53. The masonry pier 56 reaches the shear resistance with diagonal cracking until DL4.

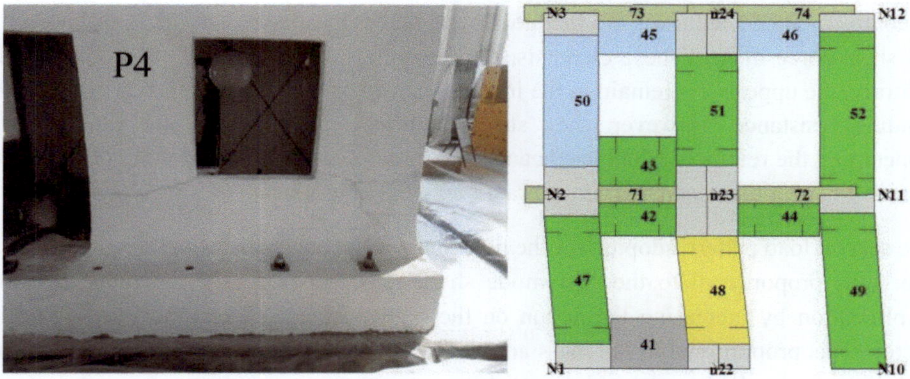

Figure 6.51: P4 wall of the BM-SR model: experimental [73] and numerical final DLs (monotonic pushover analysis in the -y direction with load pattern proportional to the masses)

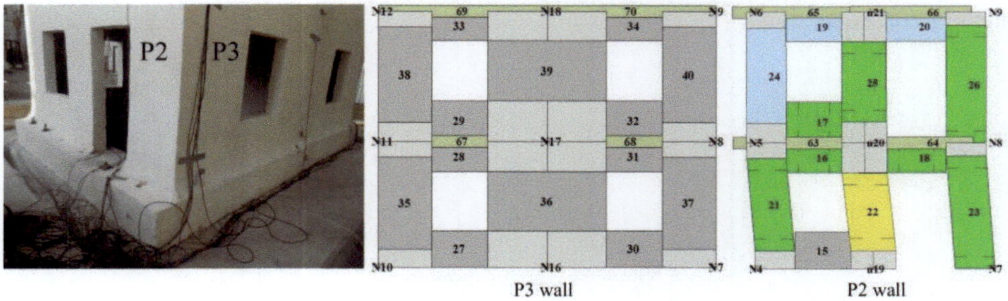

Figure 6.52: P2 - P3 corner of the BM-SR model: experimental [73] and numerical final DLs (monotonic pushover analysis in the -y direction with load pattern proportional to the masses)

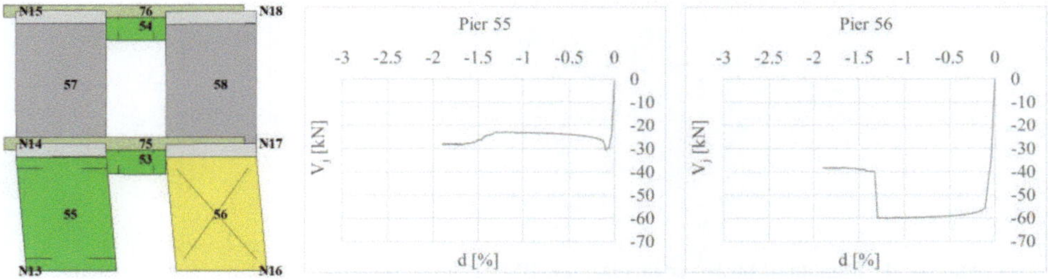

Figure 6.53: P5 wall of the BM-SR model: numerical final DLs and shear-drift diagrams of the masonry piers 55 and 56 (monotonic pushover analysis in the -y direction with load pattern proportional to the masses)

In contrast, pier 55 achieves the flexural strength and doesn't exceed DL3. As shown by the shear force-drift of these elements, the maximum drifts are equal to 1.9%. On the contrary, the upper piers remain in the initial elastic phase, and spandrels 53, 54 attain the bending resistance. However, they stay just at the beginning of the plastic state. Concerning the reinforced concrete beams, they are not damaged. Therefore, it is possible to conclude that the damage is concentrated only in the lower piers 55 and 56.

The second load pattern adopted for the non-linear static analyses is the triangular inverse one. It is proportional to the first mode shape and represents the structural dynamic amplification by increasing the action on the higher storeys. Thus, the load multiplier vector [λ] is proportional to the mass and the height of the levels and is the same as that calculated for the BM model with Eq. (6.10).

The pushover curves are obtained by imposing the maximum and minimum relative displacement (LP04 - LP01), i.e. 10.54 mm and -28.51 mm, to the control node. It is important to stress that the monotonic nonlinear static analyses are performed separately and not consequently.

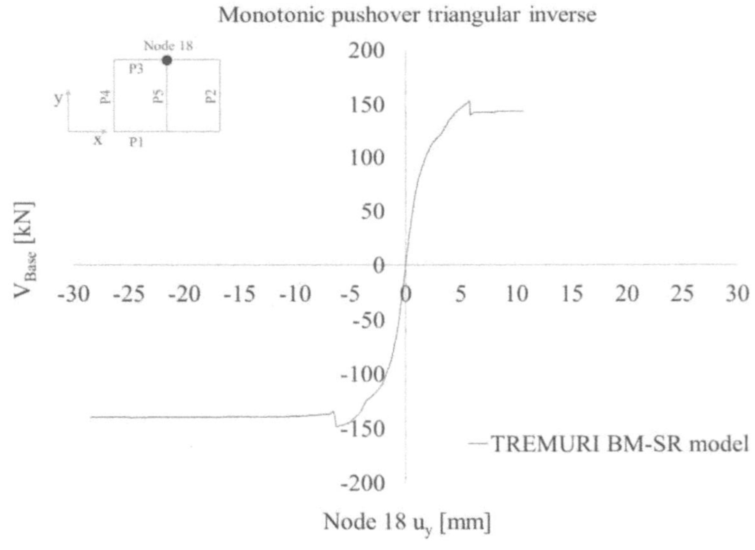

Figure 6.54: Monotonic pushover curves of the BM-SR model with triangular inverse load pattern

As depicted in Figure 6.54, the achieved maximum and minimum base shear forces are 153.41 kN and -148.45 kN. Therefore, the global resistance of the BM-SR model is also with this type of distribution about three times higher than the one reached by the original structure. It is worth noting that the development of the pushover curves in Figure 6.54 is similar to that obtained with the mass-proportional load distribution (Figure 6.48). However, in this case, the slight strength drops are due to the tensile failure of some masonry elements.

Therefore, the results of the pushover analysis carried out in the negative direction are discussed in the following. Those obtained in the positive one are reported in Chapter 8.

Figure 6.55 shows the flexural failure of the corner between the P1 and P4 walls. The damage state obtained with TREMURI indicates that the masonry elements, except spandrel 46, achieve the bending strength and present the damage state between DL2 and DL3.

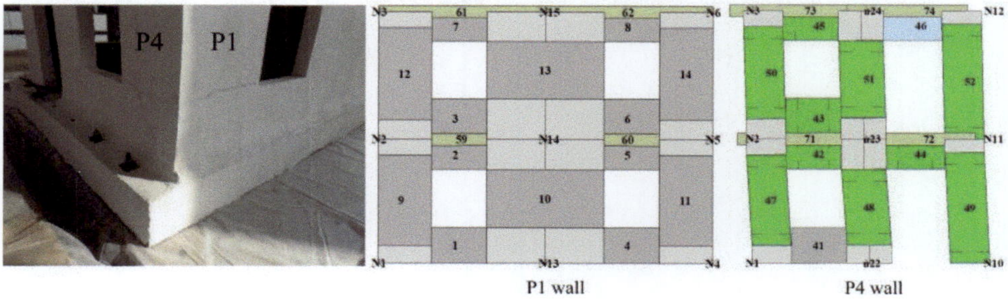

Figure 6.55: P4 - P1 corner of the BM-SR model: experimental [73] and numerical final DLs (monotonic pushover in the -y direction with triangular inverse load pattern)

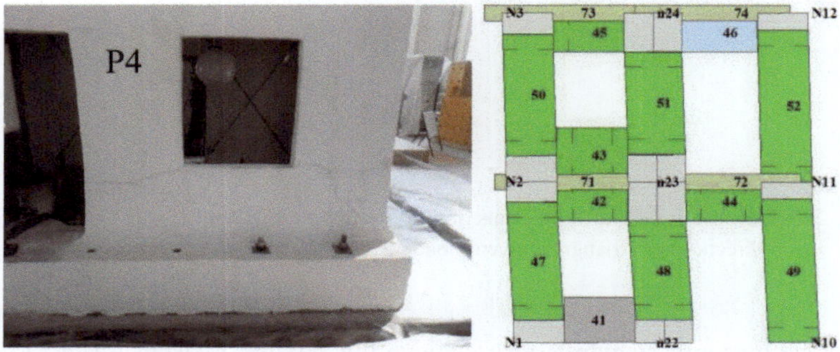

Figure 6.56: P4 wall of the BM-SR model: experimental [73] and numerical final DLs (monotonic pushover in the -y direction with triangular inverse load pattern)

Therefore, the shear-drift curves of the P4 wall's piers are also depicted in Figure 6.57. They confirm that the most stressed elements are the lower ones, 47, 48, 49, while the upper ones, 50, 51, 52, attain lower drift values. However, since they are in the plastic response phase, it is impossible to exclude that they are not cracked.

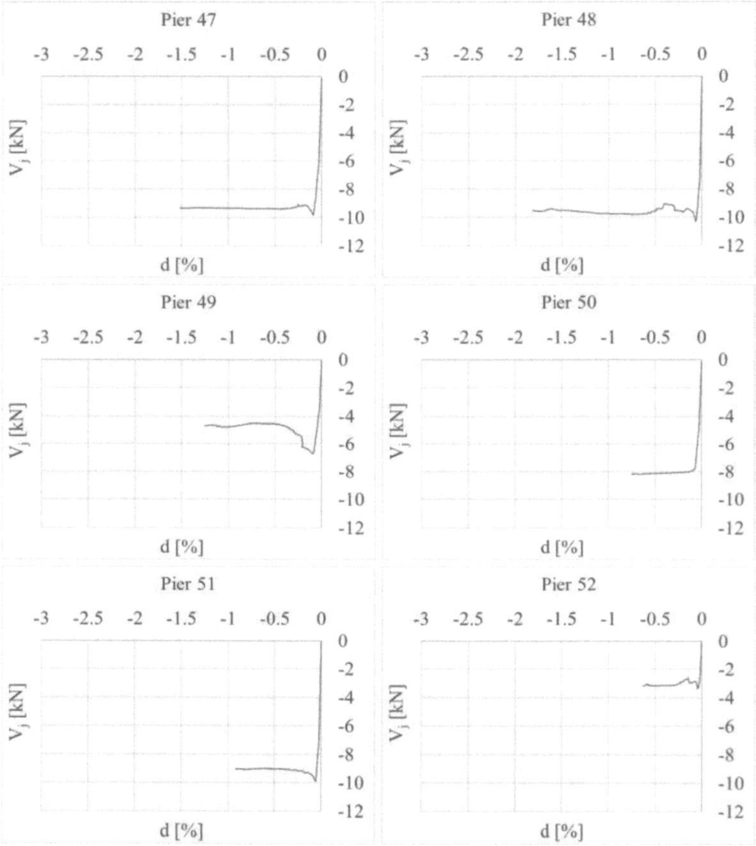

Figure 6.57: BM-SR model: shear-drift diagrams of the masonry piers of the P4 wall (monotonic pushover in the -y direction with triangular inverse load pattern)

Moreover, the masonry spandrels also achieve the flexural resistance (DL2). However, they remain at the beginning of the plasticity with a maximum drift equal to 0.12% for spandrels 42, 43, 44 and 0.05% for spandrel 45. On the contrary, all the elements of the P1 wall and the reinforced concrete beams remain in the elastic range.

Furthermore, Figure 6.58 illustrates the experimental damage of the corner between the P2 and P3 walls and the final damage state obtained from the pushover analysis. The shear-drift diagrams of the masonry elements of the P2 wall are very similar to those of the P4. Therefore, they are not further reported.

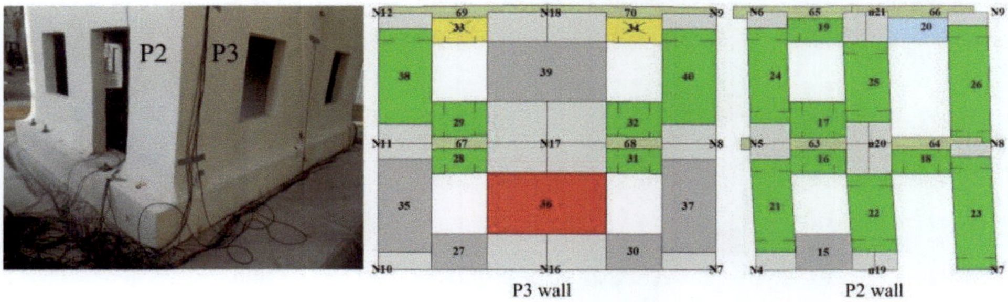

Figure 6.58: P2 - P3 corner of the BM-SR model: experimental [73] and numerical final DLs (monotonic pushover in the -y direction with triangular inverse load pattern)

It is worth noting that some masonry elements of the P3 wall attain the bending strength, while piers 35, 37 and spandrels 27, 30 remain in the elastic phase (Figure 6.58). Moreover, pier 36 fails by tension since the normal force acting at its nodes achieves the element's ultimate tensile value, as confirmed by the in-plane strength domain (Figure 6.59).

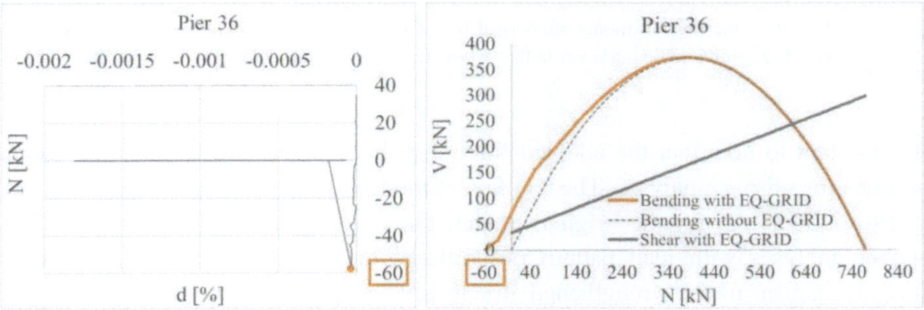

Figure 6.59: Normal force - drift diagram of pier 36 (monotonic pushover in the -y direction with triangular inverse load pattern) and in-plane failure domain of the element

Concerning the spandrels 33 and 34, they reach DL3 showing a mixed shear-flexural failure. It is worth pointing out that this type of damage is not coherent with the experimental results since the upper structural elements were not cracked at the end of the tests.

Furthermore, Figure 6.60 shows the damage state and the shear force-drift curves obtained from the pushover analysis for the P5 wall. The lower masonry piers are more damaged than the upper ones since the elements 55, 56 achieve the in-plane shear resistance (DL2) and have a maximum drift equal to 1.21% and 0.92%, respectively. Finally, the masonry spandrels fail by tension, and the reinforced concrete beams are undamaged.

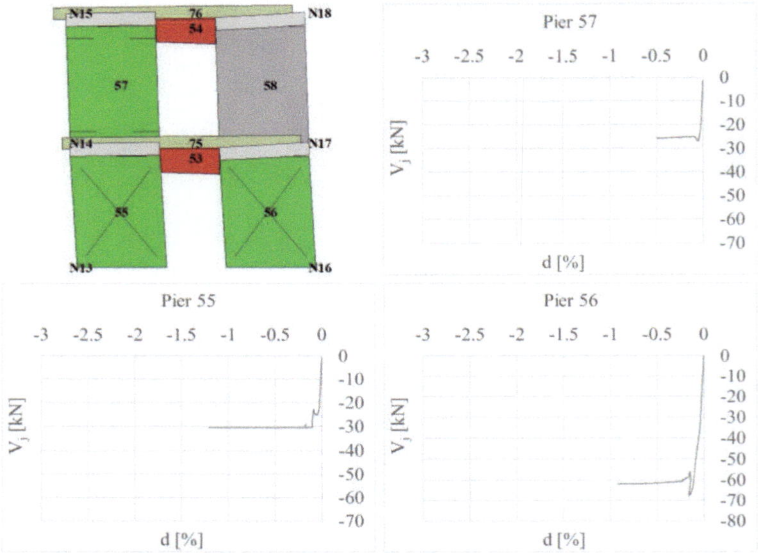

Figure 6.60: P5 wall of the BM-SR model: numerical final DLs and shear-drift diagrams of the masonry piers 55, 56, 57 (monotonic pushover in the -y direction with triangular inverse load pattern)

It is important to note that the adopted force distributions may be considered boundary conditions for seismic analyses. The real failure mode is generally predicted by one of the two distributions. As for the original model, the results obtained from the monotonic pushover analyses with load pattern proportional to the masses better reproduce the failure mechanism of the strengthened structure. The experimental damage was concentrated in the bottom masonry piers with a dominant flexural failure, while the upper elements were completely undamaged, as for the numerical model. The obtained results confirm that the response of the building under extensive damage can be reproduced by adopting the load pattern proportional to the masses since the force redistributions among the levels are prevented.

Finally, the assumed values of strength (Table 47), drift, and residual resistance (Table 55) may be considered well calibrated. The calculation of the EQ-GRID system according to CNR-DT 215/2018 enables reproducing the observed behaviour of the strengthened masonry elements, particularly for the ones in which the flexural damage was evident.

6.2.2.2 Cyclic pushover analyses of the BM-SR model

The cyclic pushover analyses are performed through sequential monotonic nonlinear static analyses. The maximum and minimum horizontal displacements (LP04 - LP01) measured during each test (Table 56) are applied as displacement limits u_y at the control node of the model. In this way, the damage evolution of the numerical model can be better evaluated and compared with that observed during the dynamic tests.

Table 56: Maximum and minimum displacement (LP04-LP01) measured during the test on the BM-SR model

Earthquake	span [%]	Max (LP04-L01) [mm]	Min (LP04-L01) [mm]
El Centro	50	0.588	-0.733
Petrovac	36	0.961	-1.447
Northridge	16	0.385	-0.555
El Centro	65	0.543	-1.016
Petrovac	40	1.018	-0.642
El Centro	75	0.677	-1.196
Petrovac	45	0.830	-0.708
Northridge	20	0.618	-0.521
El Centro	80	0.783	-1.213
Northridge	25	0.771	-0.790
Petrovac	50	0.940	-0.862
Petrovac	70	1.300	-2.316
Petrovac	75	1.520	-1.641
Petrovac	100	2.431	-2.860
Petrovac	120	2.937	-2.986
El Centro	100	1.221	-1.354
Petrovac	150	4.260	-4.004
Petrovac	180	5.749	-11.427
Petrovac	220	7.686	-20.122
Petrovac	260	10.116	-26.443
Petrovac	250	10.539	-28.514

As reported in [73], the experimental response of the retrofitted model was very different from that of the original one. The increased elasticity limit and slight reduction of displacements at the top were distinctive in the first part of the tests. In this regard, the unstrengthened model was estimated to be in a state of initial non-linearity during the application of the Petrovac 45 earthquake (a_{input} = 0.20g). The first visible horizontal cracks occurred between the foundation slab and the out-of-plane walls. On the contrary, no damage was present in the retrofitted model at the same excitation level. Therefore, the structure was estimated to be still in the elastic phase of the seismic response.

Furthermore, the BM model showed considerable nonlinear damage at the application of Petrovac 75 (a_{input} = 0.35g), which was considered similar to the ultimate state of load-bearing capacity. In comparison, the retrofitted structure was assumed to be just at the end of the elastic phase since the first fine horizontal cracks occurred along the bottom of the windows on the first floor (about 60 cm above the foundation). Moreover, with Petrovac 100 (a_{input} = 0.51g), the BM-SR model was estimated to be at the beginning of the nonlinear state.

As depicted in Figure 6.61, the experimental observations reported in [73] are confirmed by the cyclic pushover analysis with the force distribution proportional to the masses. As shown by the pushover curve, the numerical model behaves linear elastic during the application of Petrovac 45 and is at the end of this phase at the load step Petrovac 75. Moreover, it is at the beginning of the plastic phase with Petrovac 100.

Furthermore, in [73] is also reported that during the last applied seismic input Petrovac 250 (a_{input} = 1.22g), the initial cracks further propagated, and additional ones occurred along the contact line of the walls with the foundation. However, the structural stability was still preserved, and the damage was considered repairable. Although the stiffness and load-bearing capacity decreased, the BM-SR model showed structural integrity at the end of the dynamic tests, and it was far from failure. The last excitation level applied to the retrofitted model was about three times higher than the last one adopted for the unstrengthened structure. However, the damage level showed by the retrofitted model was considerably smaller. The cyclic pushover analysis also confirms these experimental observations. As depicted in Figure 6.61, the structure is in the plastic phase between Petrovac 180 and Petrovac 250. Moreover, the curve shows large dissipative cycles in the negative direction due to the dominant shear behaviour of piers 55 and 56 (Figure 6.64). Then, the strength drop that occurs during Petrovac 220 corresponds to the achievement of the DL3 by 5% of the masonry piers (Figure 6.62). Furthermore, the flexural behaviour is dominant in the masonry elements of the P2, P4 walls, as experimentally observed (Figure 6.63, Figure 6.64, Figure 6.65). In detail, the most damaged piers are the ones under the first storey in the same direction of the dynamic excitation, i.e. elements 47, 48, 49, 21, 22, 23 and 55, 56, while the two out of plane walls, P1 and P3, behave linear elastic.

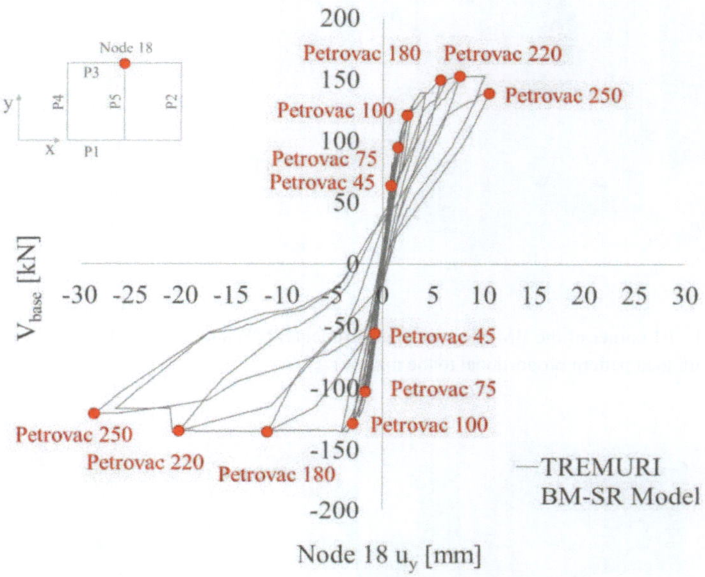

Figure 6.61: BM-SR model: cyclic pushover curve with load pattern proportional to the masses

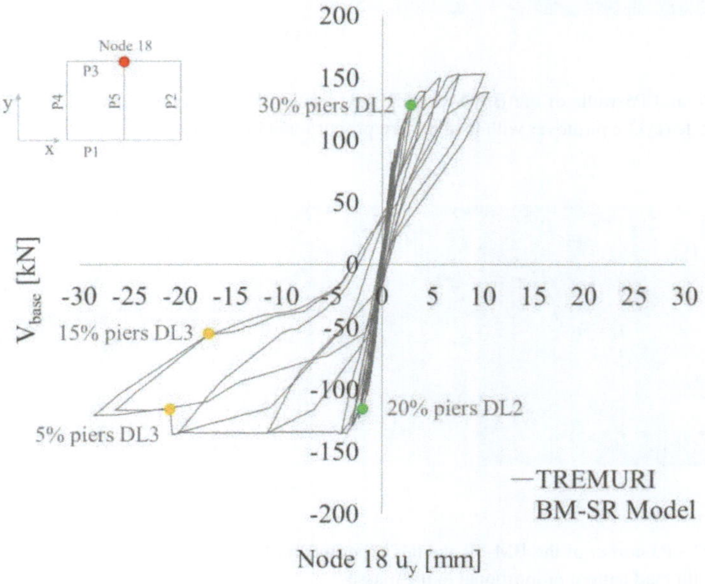

Figure 6.62: BM-SR model: cyclic pushover curve with load pattern proportional to the masses: DLs of the masonry piers

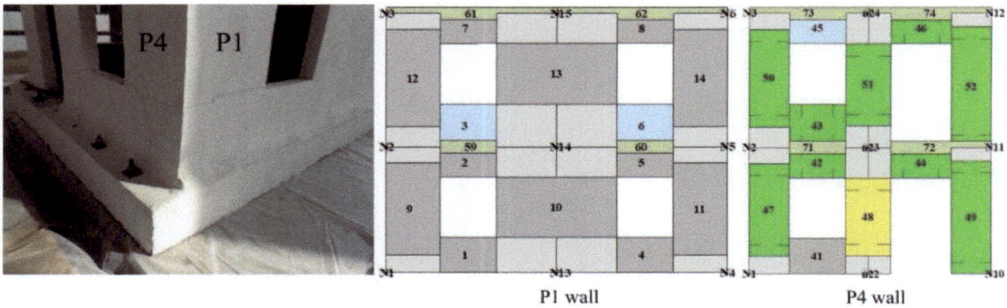

Figure 6.63: P4 - P1 corner of the BM-SR model: experimental [73] and numerical final DLs (cyclic pushover with load pattern proportional to the masses)

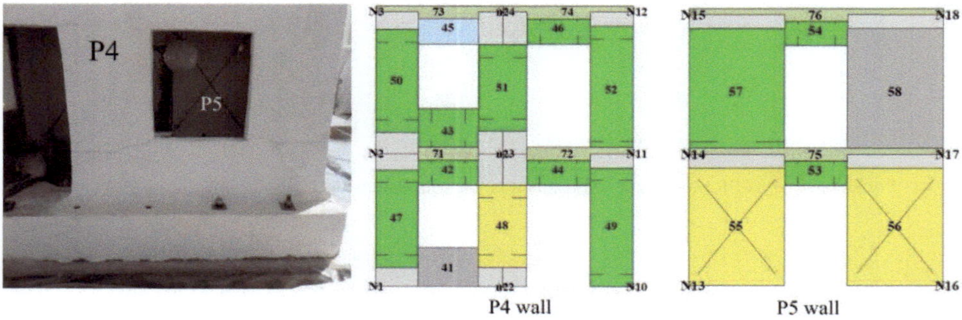

Figure 6.64: P4 and P5 walls of the BM-SR model: experimental [73] and numerical final DLs final damage state (cyclic pushover with load pattern proportional to the masses)

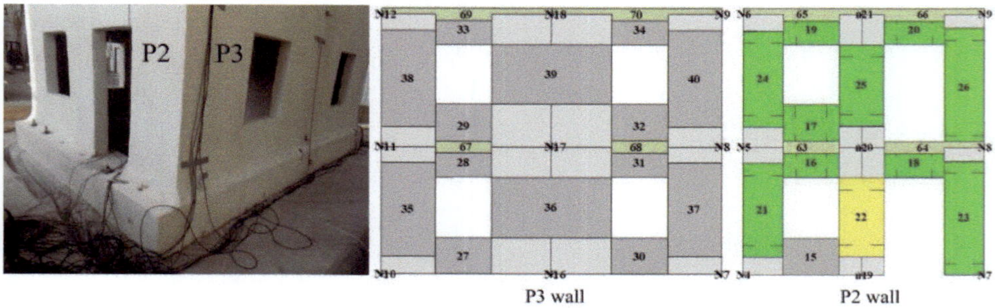

Figure 6.65: P2 - P3 corner of the BM-SR model: experimental [73] and numerical final DLs (cyclic pushover with load pattern proportional to the masses)

Finally, no visible cracks were experimentally observed at the upper storey as in the modelling with TREMURI. It is worth noting that, although some masonry elements of the second storey attain the in-plane strength, they remain at the beginning of the plastic phase since the plastic deformations are concentrated in the lower part of the structure. Therefore, it is possible to assume that they are not visible cracked.

Table 57: Horizontal displacement of the first storey of the BM-SR model in the y-direction (cyclic pushover analysis with load pattern proportional to the masses)

Earthquake	span [%]	(LP03 - LP01) [mm]		TREMURI node 17 u_y [mm]	
El Centro	50	0.423	-0.38	0.33	-0.415
Petrovac	36	0.438	-0.32	0.541	-0.84
Northridge	16	0.275	-0.271	0.227	-0.331
El Centro	65	0.579	-0.562	0.32	-0.604
Petrovac	40	0.629	-0.306	0.597	-0.385
El Centro	75	0.801	-0.613	0.397	-0.702
Petrovac	45	0.65	-0.463	0.488	-0.425
Northridge	20	0.449	-0.298	0.363	-0.313
El Centro	80	0.793	-0.679	0.46	-0.711
Northridge	25	0.556	-0.511	0.454	-0.473
Petrovac	50	0.751	-0.534	0.552	-0.516
Petrovac	70	1.22	-0.973	0.774	-1.383
Petrovac	75	1.113	-1.199	0.883	-0.978
Petrovac	100	1.981	-1.936	1.421	-1.791
Petrovac	120	2.378	-2.061	1.743	-1.851
El Centro	100	1.261	-0.845	0.683	-0.731
Petrovac	150	3.146	-3.575	2.74	-2.661
Petrovac	180	4.162	-11.044	3.934	-9.892
Petrovac	220	5.179	-19.742	5.661	-18.634
Petrovac	260	6.708	-26.033	8.087	-25.118
Petrovac	250	7.401	-27.983	8.809	-27.269

Moreover, the absolute horizontal displacement of the first storey was experimentally measured through a linear potentiometer (named LP03 in [73]), whose position corresponds to node 17 of the P3 wall in TREMURI. Thus, the deformation of the structure can be compared with the experimental one by checking the displacement of node 17. As reported in Table 57, the numerical results show a very good match with the experimental ones.

Concerning the results of the pushover analysis with the triangular inverse force distribution, they are illustrated in Figure 6.66. In this case, the structure is still in the elastic phase of the seismic response during the load steps Petrovac 45, 75 and 100. Therefore, compared to the previous results, this behaviour doesn't find such good agreement with the experimental observations reported by [73]. Moreover, the capacity curve is characterized by less dissipative cycles with a "flag" shape due to the prevailing flexural behaviour of the masonry piers (Figure 6.68, Figure 6.69, Figure 6.70). In contrast with the previous results, the bottom piers of the P5 wall don't attain the DL3 limit by diagonal cracking (Figure 6.69) since the plastic deformations are also present in the masonry elements of the upper storey.

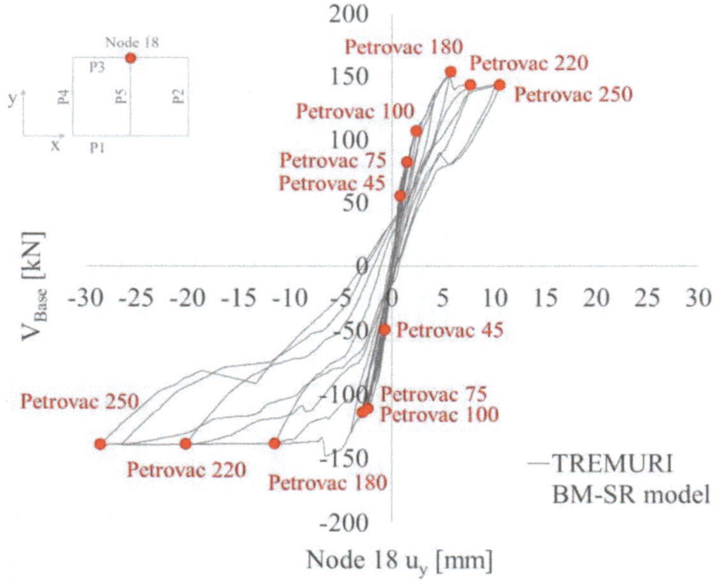

Figure 6.66: BM-SR model: cyclic pushover curve with the triangular inverse load pattern

In this regard, Figure 6.67 shows the percentages of the damaged masonry piers on the cyclic pushover curve. During the application of Petrovac 220, 55% of the piers have achieved the DL2. Even during the last seismic input, no elements attain the DL3 limit.

This behaviour is due to the triangular inverse load distribution, which generally increases actions at the upper storeys. In this case, the most damaged piers of the in-plane walls remain the lower ones, but the upper elements also show plastic deformations. Compared to the previous results, the damage is no more concentrated in the bottom components of the structure, which doesn't correspond to the observed behaviour.

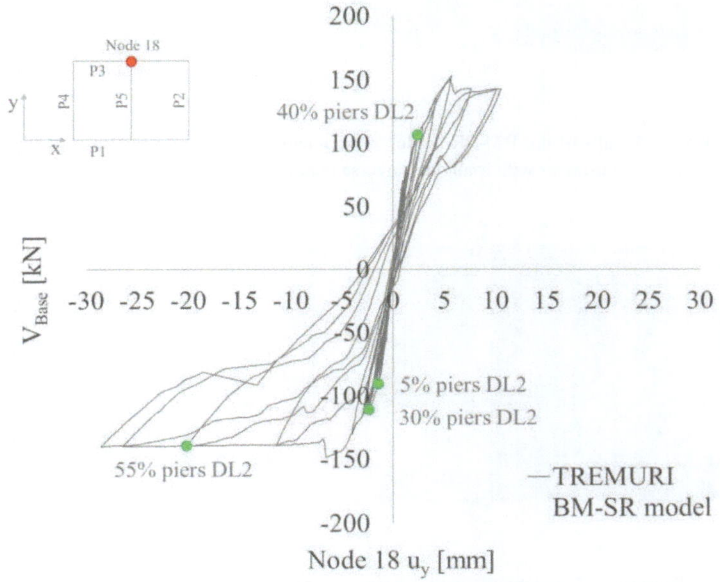

Figure 6.67: Cyclic pushover curve of the BM-SR model with the triangular inverse load pattern: DLs of the masonry piers

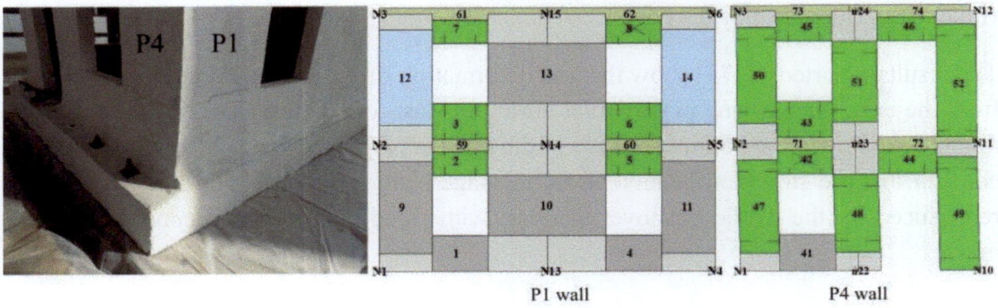

Figure 6.68: P4 - P1 corner of the BM-SR model: experimental [73] and numerical final DLs (cyclic pushover with triangular inverse load pattern)

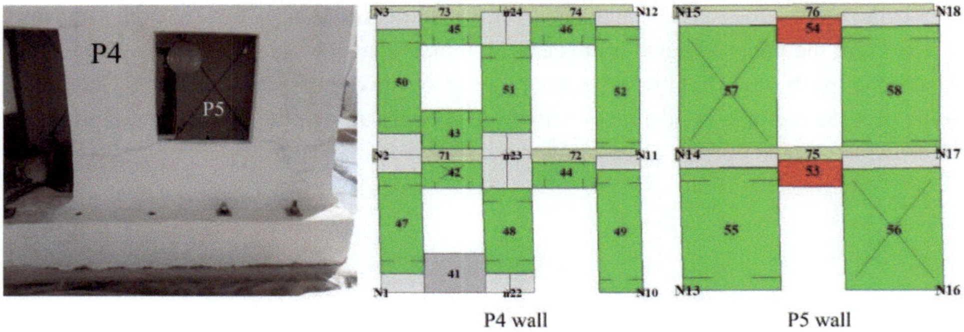

Figure 6.69: P4 and P5 walls of the BM-SR model: experimental [73] and numerical final DLs final damage state (cyclic pushover with triangular inverse load pattern)

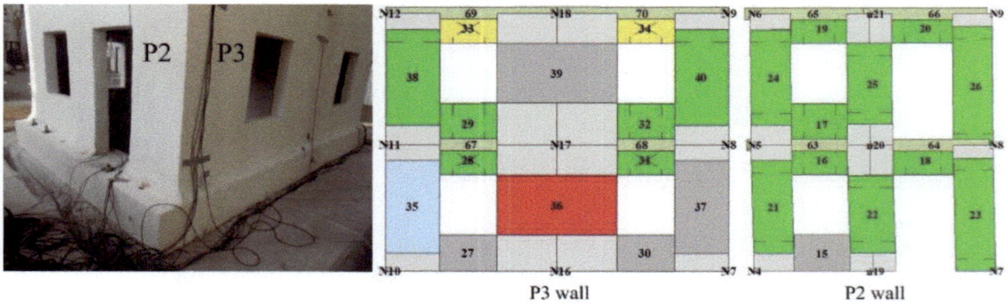

Figure 6.70: P2 - P3 corner of the BM-SR model: experimental [73] and numerical final DLs (cyclic pushover with triangular inverse load pattern)

Finally, the experimental horizontal displacement (LP03 - LP01) is also compared to the values of u_y calculated by TREMURI in node 17 (Table 58).

The results reported in [73] show that the deformation of the numerical model is different from the experimental one in particular during the last four earthquakes, in the retrofitted model is widely in the nonlinear phase of the seismic response. Therefore, it is possible to confirm that the structural behaviour and damage pattern of the BM-SR model is better reproduced by the cyclic pushover analysis with force distribution proportional to the masses.

Table 58: Horizontal displacement of the first storey of the BM-SR model in the y-direction (cyclic pushover analysis with triangular inverse load pattern)

Earthquake	span [%]	(LP03 - LP01) [mm]		TREMURI node 17 uy [mm]	
El Centro	50	0.423	-0.38	0.297	-0.368
Petrovac	36	0.438	-0.32	0.478	-0.735
Northridge	16	0.275	-0.271	0.198	-0.292
El Centro	65	0.579	-0.562	0.278	-0.527
Petrovac	40	0.629	-0.306	0.521	-0.337
El Centro	75	0.801	-0.613	0.347	-0.615
Petrovac	45	0.65	-0.463	0.427	-0.372
Northridge	20	0.449	-0.298	0.319	-0.275
El Centro	80	0.793	-0.679	0.401	-0.621
Northridge	25	0.556	-0.511	0.396	-0.415
Petrovac	50	0.751	-0.534	0.482	-0.451
Petrovac	70	1.22	-0.973	0.675	-1.205
Petrovac	75	1.113	-1.199	0.771	-0.855
Petrovac	100	1.981	-1.936	1.260	-1.568
Petrovac	120	2.378	-2.061	1.534	-1.638
El Centro	100	1.261	-0.845	0.607	-0.683
Petrovac	150	3.146	-3.575	2.204	-2.237
Petrovac	180	4.162	-11.044	3.016	-6.407
Petrovac	220	5.179	-19.742	3.890	-12.071
Petrovac	260	6.708	-26.033	5.229	-16.711
Petrovac	250	7.401	-27.983	5.510	-18.276

6.2.3 Nonlinear dynamic analyses of the BM-SR model

The last four acceleration-time histories of the experimental test program, i.e. Petrovac 180, Petrovac 220, Petrovac 260 and Petrovac 250 (Figure 6.71), are considered to perform the non-linear dynamic analyses in the y-direction of the BM-SR model. The material properties of the retrofitted structure and the parameters adopted for the cyclic nonlinear behaviour of the ML-BEAM elements (piers and spandrels) are the same as those assumed in § 6.2.1 and § 6.2.2 for the modal and pushover analyses. The coefficients α_1 and β_1 of the Rayleigh viscous damping are determined using the same procedure adopted for the BM model and described in § 6.1.3.

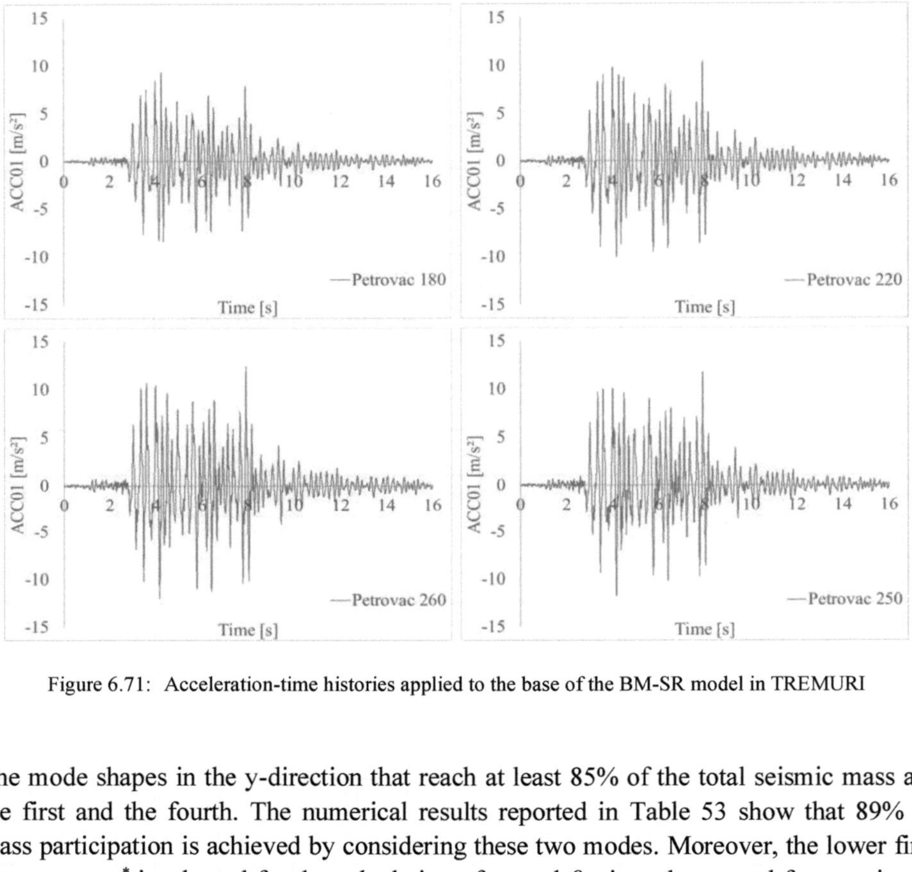

Figure 6.71: Acceleration-time histories applied to the base of the BM-SR model in TREMURI

The mode shapes in the y-direction that reach at least 85% of the total seismic mass are the first and the fourth. The numerical results reported in Table 53 show that 89% of mass participation is achieved by considering these two modes. Moreover, the lower first frequency ω_1^* is adopted for the calculation of α_1 and β_1 since the natural frequencies of the BM-SR model were also measured after testing. Table 59 summarizes the experimental first natural frequency recorded in the East-West direction before and after testing and the fourth one obtained from the modal analysis with TREMURI.

Table 59: Natural frequencies and periods of the BM-SR model

1st mode (before testing)	$\omega_1 = 79.33$ rad/s	$T_1 = 0.079$ s
1st mode (after testing)	$\omega_1^* = 37.07$ rad/s	$T_1^* = 0.17$ s
4th mode (modal analysis)	$\omega_4 = 205.62$ rad/s	$T_4 = 0.031$ s

Therefore, the coefficients α_1 and β_1 are calculated adopting the frequencies ω_1^* and ω_4 and assuming the elastic damping ratio ξ equal to 3%:

$$\begin{cases} \alpha_1 = \xi \dfrac{2\omega_1^* \omega_4}{\omega_1^* + \omega_4} = 0.03 \cdot \dfrac{(2 \cdot 37.07 \cdot 205.62)}{(37.07 + 205.62)} = 1.884 \\ \beta_1 = \xi \dfrac{2}{\omega_1^* + \omega_4} = 0.03 \cdot \dfrac{2}{(37.07 + 205.62)} = 0.000247 \end{cases} \tag{6.15}$$

As depicted in Figure 6.72, the elastic damping ratio ξ can be assumed nearly constant in the range between the 1st and 4th frequency. Therefore, the lower damping ratio ξ^* in the fundamental mode ω_1 is equal to 2.2%, and it is calculated through Eq. (6.16):

$$\xi^* = \frac{\alpha_1}{2\omega_1} + \frac{\beta_1}{2}\omega_1 = \frac{1.884}{2 \cdot 79.33} + \frac{0.000247}{2} \cdot 79.33 = 2.2\% \tag{6.16}$$

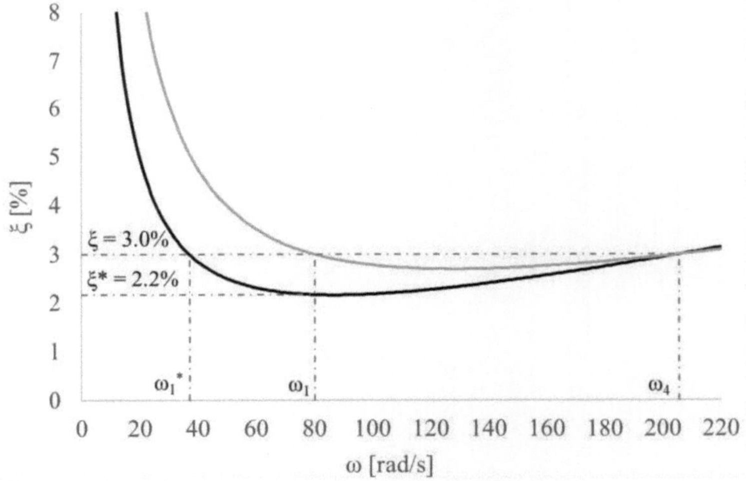

Figure 6.72: Rayleigh damping ratios ξ and ξ^* of the BM-SR model

The numerical results are compared to the experimental ones in terms of displacements and accelerations. In this regard, Figure 6.73 illustrates the results of the last performed nonlinear dynamic analysis, Petrovac 250. The other ones, i.e. Petrovac 180, 220 and 260, are reported in Chapter 8.

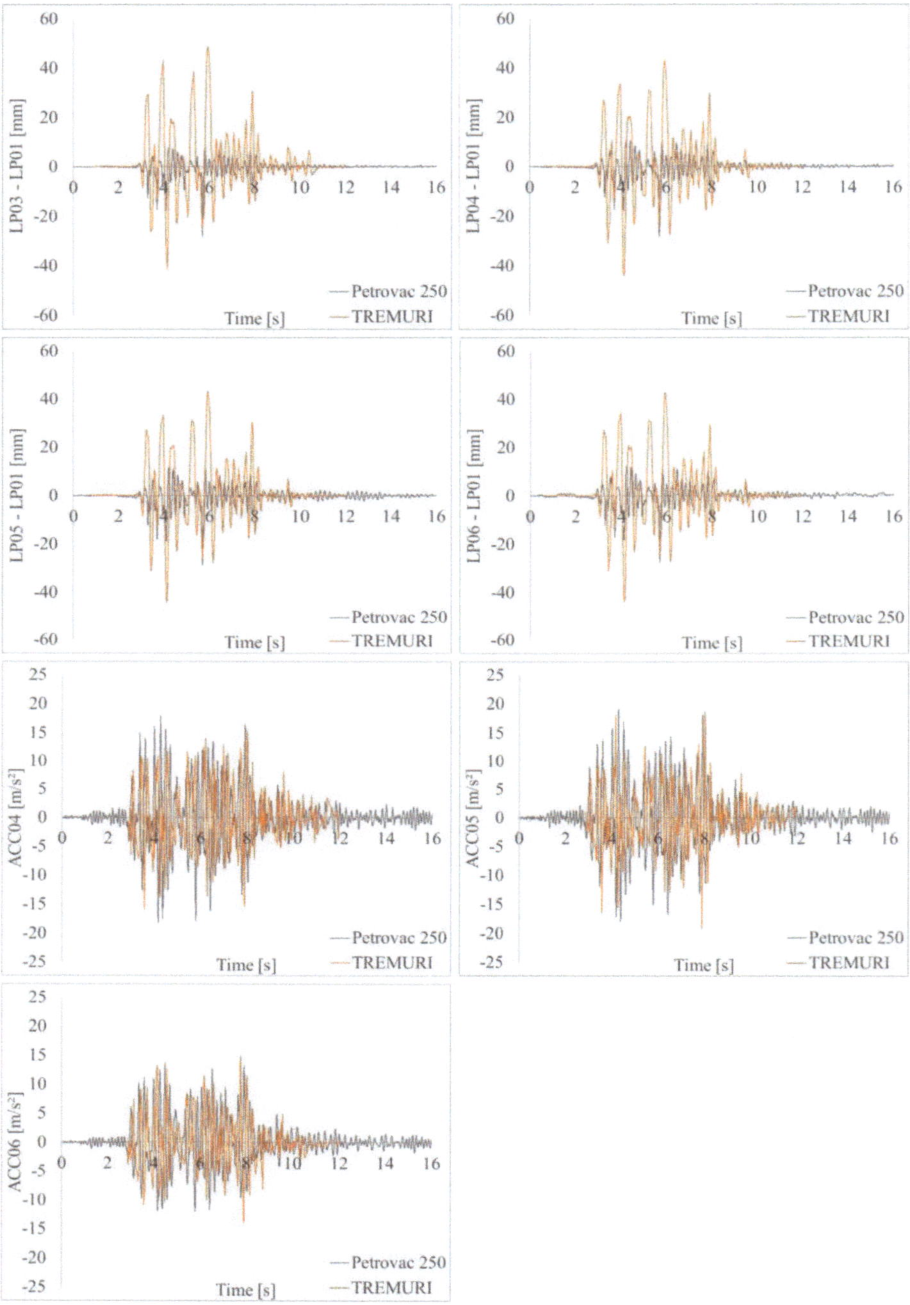

Figure 6.73: Results of the last nonlinear dynamic analysis Petrovac 250 performed on the BM-SR model with TREMURI

In detail, Figure 6.73 depicts the experimental horizontal relative displacements of the first and second storey, (LP03 – LP01), (LP04 – LP01), (LP05 – LP01), (LP06 – LP01), compared with the values calculated by TREMURI in the corresponding nodes of the model, i.e. 17, 18, 9 and 12. The numerical displacements indicate a more symmetrical structural response than observed during the tests. It is important to stress that the non-symmetrical experimental displacement values were probably due to the higher resistance of some masonry elements of the BM-SR model. As shown in Figure 6.75 and Figure 6.77, flexural cracks occurred in the P1 and P4 walls and not in the symmetrical ones, i.e. the P2 and P3. Therefore, it is reasonable to assume that some masonry panels of the P2 and P3 walls probably had a higher resistance and ductility. For this reason, the structural response of the BM-SR model lost symmetry when the weaker elements failed, and the experimental damage prevailed only in the already cracked panels. Regarding the absolute accelerations measured by the sensors ACC04, ACC05, ACC06, they are also compared in Figure 6.73 with those calculated by the program in the corresponding nodes of the model, i.e. 6, 12, 17[4]. In this case, a very good match between the results is evident.

Furthermore, Figure 6.74 depicts the base shear force-displacement diagram obtained from the dynamic analyses.

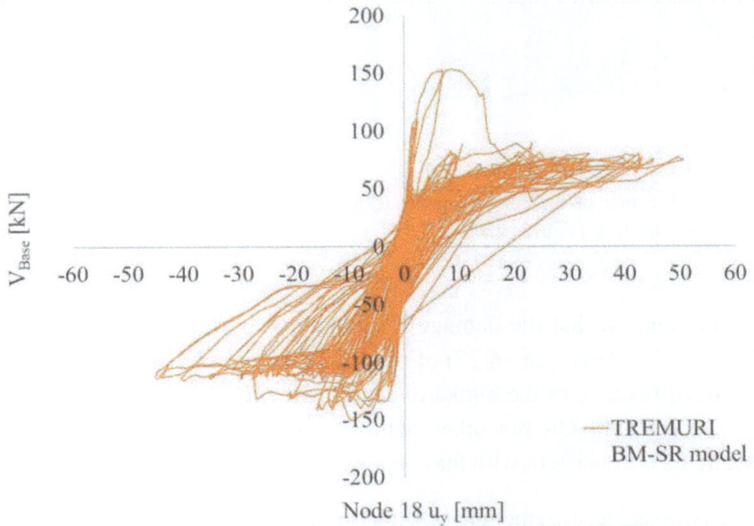

Figure 6.74: Base shear force - horizontal displacement curve resulting from the nonlinear dynamic analyses performed on the BM-SR model with TREMURI

[4] The complete arrangement of the adopted sensors is illustrated in [73].

The maximum and minimum shear force values are achieved during the seismic input Petrovac 180 and are equal to 153.8 kN and -150.1 kN, respectively. It is worth noting that the curve pattern, strength and displacement values attained in the negative quadrant are coherent with the cyclic pushover curve with the load distribution proportional to the masses (Figure 6.61). Moreover, the strength drop after the positive peak of base shear force (Figure 6.74) is due to attaining the ultimate tensile normal force of some masonry piers of the in-plane walls. Then, when these elements are compressed again during the analysis, their in-plane resistance equals the unstrengthened one. Therefore, the residual positive base shear force is very close to the value reached by the unstrengthened BM model (Figure 6.42).

Finally, the numerical results confirm that the bottom piers of the P2, P4 and P5 walls are more damaged. These elements achieve the DL3 limit for flexural, shear or hybrid failure, as experimentally observed and confirmed by the cyclic pushover analysis with load pattern proportional to the masses (§ 6.2.2.2).

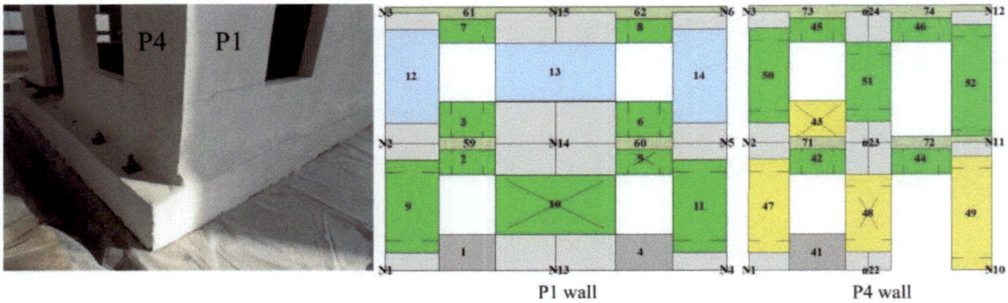

Figure 6.75: P4 - P1 corner of the BM-SR model: experimental [73] and numerical final DLs (nonlinear dynamic analysis Petrovac 250)

It is worth pointing out that the damage level attained by the piers 47, 49 (Figure 6.75, Figure 6.76), and 21, 23 (Figure 6.77) of the P2, P4 is higher than that obtained from the pushover analysis because of the higher displacement values calculated by the program in the positive y-direction. On the other hand, the damage pattern shown by the other masonry elements is consistent with that obtained from the pushover analysis in § 6.2.2.2.

Therefore, it is possible to conclude that the seismic response of the BM-SR model is well reproduced since the numerical results show a good correspondence with the experimental ones. The assumed material properties, the parameters defined for the cyclic nonlinear behaviour of the ML-BEAM elements (piers and spandrels) and the calculation method adopted for the EQ-GRID system are confirmed.

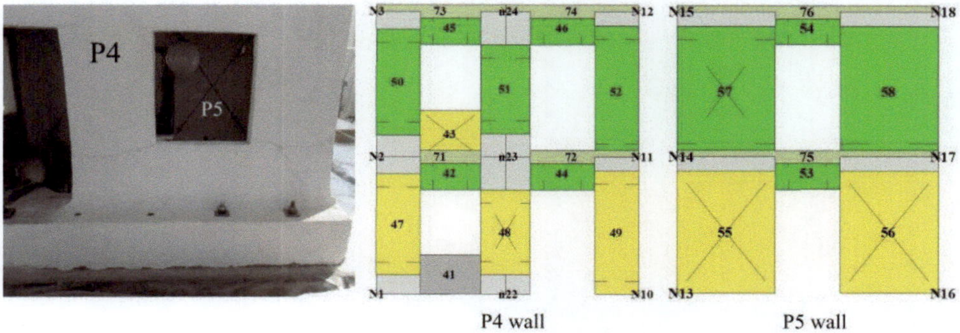

Figure 6.76: P4 and P5 walls of the BM-SR model: experimental [73] and numerical final DLs (nonlinear dynamic analysis Petrovac 250)

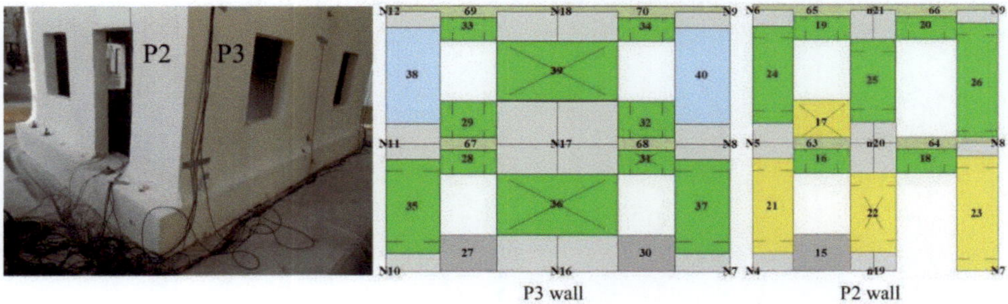

Figure 6.77: P2 - P3 corner of the BM-SR model: experimental [73] and numerical final DLs (nonlinear dynamic analysis Petrovac 250)

Chapter 7

Conclusions

This work concerns the experimental characterization and modelling of textile reinforced masonry structures with the Equivalent frame method. It is subdivided into six chapters. The first one gives an overview of the structural behaviour of unreinforced masonry buildings under seismic actions and the most common numerical modelling approaches. In particular, more attention is given to the Equivalent frame method implemented in the TREMURI program since it is adopted for the numerical analyses. This modelling approach is based on idealizing the masonry walls in a frame. Then, each load-bearing wall is subdivided into a set of deformable masonry panels, i.e. piers and spandrels, in which the deformation and nonlinear behaviour are concentrated.

Furthermore, the structural elements are connected by rigid nodes that correspond to the wall parts not generally subjected to damage. Consequently, a limited number of degrees of freedom is required. The analysis of complex three-dimensional models of masonry structures is obtained by assembling walls and floors, mainly referring to their in-plane strength and stiffness contributions. This approach is also suggested in some seismic codes because of the reasonable computational effort.

The second chapter concerns the seismic reinforcing and retrofitting of masonry buildings with composite materials. In detail, the limits of the fibre-reinforced polymers (FRP) in the application to the masonry structures are illustrated, and the innovative strengthening technique of the externally bonded textile mortar is presented. This technology originates from the ferrocement, in which the steel mesh is replaced with high-performance, open textile meshes applied with inorganic matrices. It is commonly named fabric-reinforced cementitious matrix (FRCM) or textile-reinforced mortar (TRM). In this work, only the name FRCM is used for the sake of simplicity. Concerning the open mesh, it generally consists of continuous fibre yarns of alkali-resistant (AR) glass, carbon, aramid, basalt, or polyparaphenylene benzobisoxazole (PBO). The threads can be preimpregnated, coated, or dry and are generally arranged in two or more directions. Furthermore, the matrix can be cement- or lime mortar-based. It has the function to protect the fibre yarns and ensure the stress transfer between masonry and textile layers.

The FRCM system studied in this work is named EQ-GRID, and it has been developed at the Karlsruhe Institute of Technology in Germany. The third chapter illustrates the results of a wide experimental campaign performed on this system. The textile mesh is a multi-axial hybrid grid made of alkali-resistant glass and polypropylene fibres. The matrix is a very fine-grained natural hydraulic lime mortar (NHL). Therefore, it can encapsulate the threads very well by penetrating the mesh openings. It is worth highlighting that this aspect is very important for the FRCM systems since the bond strongly influences the performance at the textile-matrix and matrix-support interface.

The performed experimental program included direct tensile tests on bare textile samples and composite specimens for each main direction of the grid, i.e. vertical, horizontal and diagonal. Two types of grips were adopted to test the composite samples: clamping and clevis-type grips. With the first ones, the free textile ends of the composite specimens were clamped between the machine wedges, and the observed failure was always the fibres' rupture. On the contrary, the tensile load was introduced in the specimens only by adhesion with the clevis-type grips. Therefore, the second test setup corresponds to the application cases in which no mechanical anchorage is adopted. The results show that the textile tensile strength's exploitation ratios obtained with both test setups are very similar for the horizontal and diagonal directions. On the contrary, the tensile response of the system in the vertical direction depends on the adopted test setup.

Furthermore, compression and bending tests were performed on the EQ-GRID matrix to identify its mechanical properties. Then, the bonding behaviour of the applied system was investigated through double- and single-lap shear-bond tests. Hollow and solid clay bricks were adopted as standard masonry support. The system was applied without any mechanical anchorage, and different bonding lengths were tested. No premature debonding from the masonry unit was observed. Only the textile tensile failure or the slippage of the grid within the matrix were observed.

Finally, twelve masonry panels were tested through cyclic lateral shear tests. Four specimens were in original conditions, and eight were strengthened with EQ-GRID. Two types of masonry units were adopted, i.e. the calcium-silicate and hollow clay bricks. Two different values of constant vertical load were applied, that is, 120 kN and 240 kN.

The structural performance of the strengthened specimens is compared to that of the URM specimens. The achieved damage level and the experimental shear-drift diagrams are shown. Moreover, equivalent multilinear curves are determined from the cyclic envelopes of the hysteretic curves. The obtained results confirm that the EQ-GRID system can improve the load-bearing and inelastic deformation capacity of a masonry panel.

Furthermore, the in-plane shear strength and failure domains of all the tested masonry panels are calculated in the fourth chapter of this work. The assumed mechanical properties for the URM specimens are consistent with Eurocode 6 ([81], [28]) and values typical for the tested masonry. The presence of the EQ-GRID system is considered through the proposed modification of the masonry initial shear strength and brick tensile strength as well as after the Italian technical standard CNR-DT 215/2018 "Guide for the Design and Construction of Externally Bonded Fibre Reinforced Inorganic Matrix Systems for Strengthening Existing Structures" [52]. The analytical results show that the experimental shear forces can be obtained with a good approximation. Moreover, a very good correspondence is observed between the values calculated with the proposed formulations and that suggested by CNR-DT 215/2018 [52].

The analysis of the panel's cross-section strengthened with EQ-GRID is also performed. In this regard, it is observed that the in-plane flexural domains of the reinforced specimens shift to tensile values of the normal force. Compared to the URM elements, the strengthened ones can carry tensile loads thanks to the EQ-GRID system. It is worth noting that this aspect is fundamental to individuate the correct type of failure and resistance, particularly in the domain parts where the shear and flexural resistance intersect.

Furthermore, the fifth chapter of this work shows the numerical modelling of all the tested masonry panels. The nonlinear beam element with piecewise behaviour and lumped inelasticity, implemented in the TREMURI program, is adopted. The parameters obtained from the calculated equivalent multilinear curves, e.g. secant stiffness, drift values at the ith DL achievement, are assumed. Therefore, the cyclic lateral shear tests are reproduced through sequential static analyses under displacement control. The results are expressed in the form of internal forces and moments. Thus, the numerical in-plane strength, the achieved DLs are compared to the experimental observations.

The obtained results confirm that the Equivalent frame method and the nonlinear beam element with multilinear constitutive law can well reproduce the cyclic behaviour of masonry panels strengthened with EQ-GRID. In this context, the implemented formulations after CNR-DT 215/2018 [52] allow individuating the correct behaviour, particularly when the system is applied on both sides. In this case, the consideration of the tensile load-bearing capacity in the flexural behaviour is fundamental.

Finally, the sixth chapter of this work presents the numerical modelling of two shaking table tests performed on a 1:2 scaled masonry building at the Institute of Earthquake Engineering and Engineering Seismology (IZIIS) in Skopje (2013). The original model was a two-storey solid clay masonry building, and it was tested under different scaled earthquakes until a considerable damage level was observed. Then, the model was repaired and retrofitted with the EQ-GRID system. Therefore, the strengthened building was tested to compare its structural performance with the original one.

Since the natural frequencies of the URM and reinforced model were experimentally measured, modal analyses are performed. In this way, the elastic material properties, such as the Young's and shear moduli, are well-calibrated. Then, monotonic and cyclic pushover analyses are carried out. To this aim, the nonlinear beam element with piecewise behaviour and lumped inelasticity is adopted. The assumed masonry material properties are consistent with the average values suggested by the explanatory Circular of the Italian Building Code [22] for existing solid clay masonry buildings. Therefore, the pushover analyses allow comparing the experimental observations and failure modes in detail with the numerical results. Moreover, the formulations after CNR-DT 215/2018 [52] are adopted since the flexural behaviour of the strengthened piers was dominant in the retrofitted structure.

Finally, time-history analyses are performed on both models with the TREMURI program. The measured displacements and accelerations are compared to the numerical ones, and the results obtained with the static analyses are also dynamically well reproduced.

Therefore, the Equivalent frame method and the adopted nonlinear beam element are validated as a technique suitable for modelling the textile reinforced masonry structures.

Chapter 8

Appendix

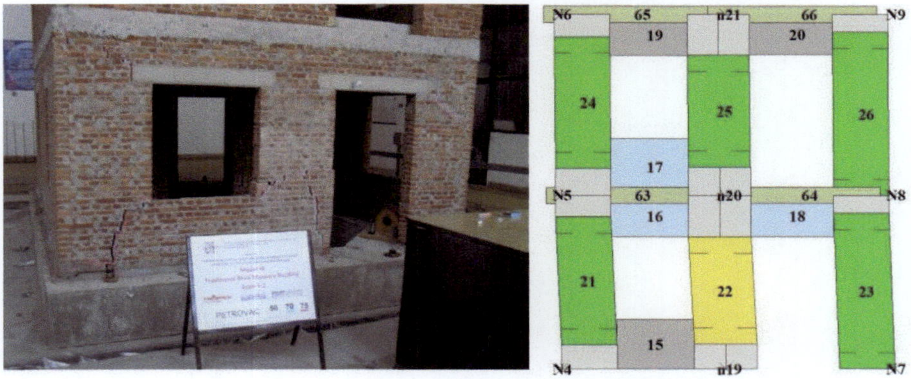

Figure 8.1: P2 wall of the BM model: experimental [73] and numerical final DLs (monotonic pushover analysis in the -y direction with load pattern proportional to the masses)

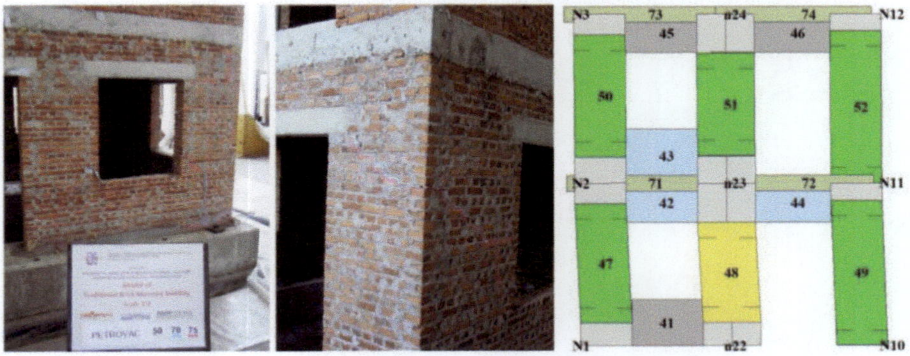

Figure 8.2: P4 wall of the BM model: experimental [73] and numerical final DLs (monotonic pushover analysis in the -y direction with load pattern proportional to the masses)

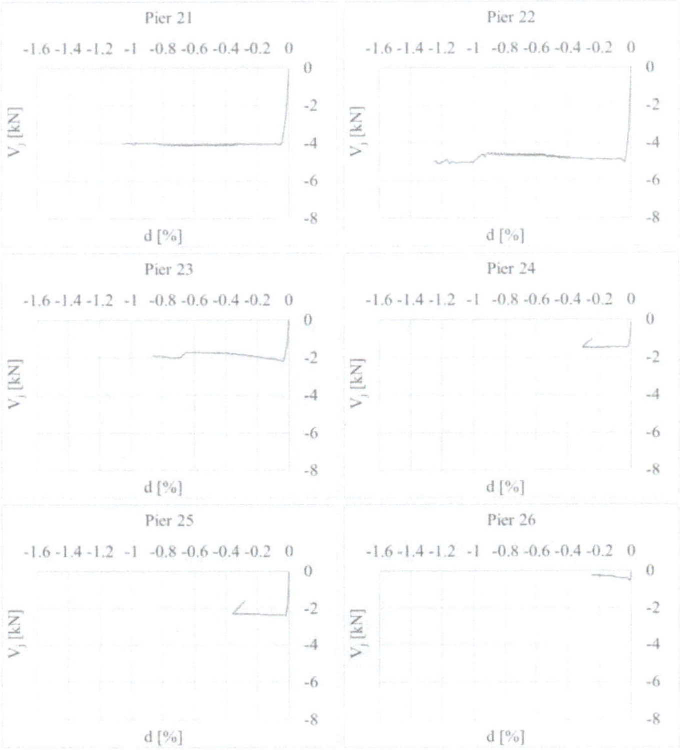

Figure 8.3: BM model: shear-drift diagrams of the masonry piers of the P2 wall (monotonic pushover in the -y direction with load pattern proportional to the masses)

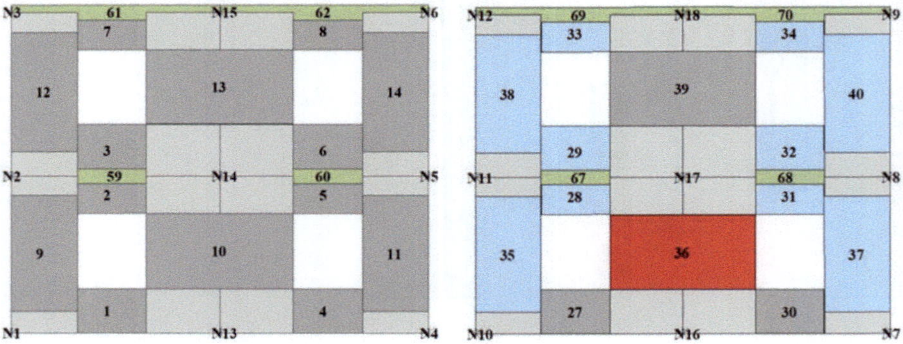

Figure 8.4: P1, P3 walls of the BM model: numerical final DLs (monotonic pushover in the -y direction with load pattern proportional to the masses)

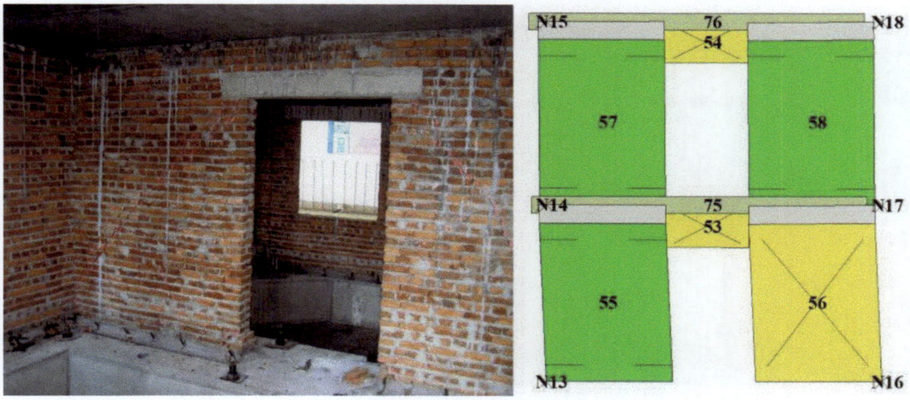

Figure 8.5: Wall P5 of the BM model: experimental [73] and numerical final DLs (monotonic pushover analysis in the -y direction with load pattern proportional to the masses)

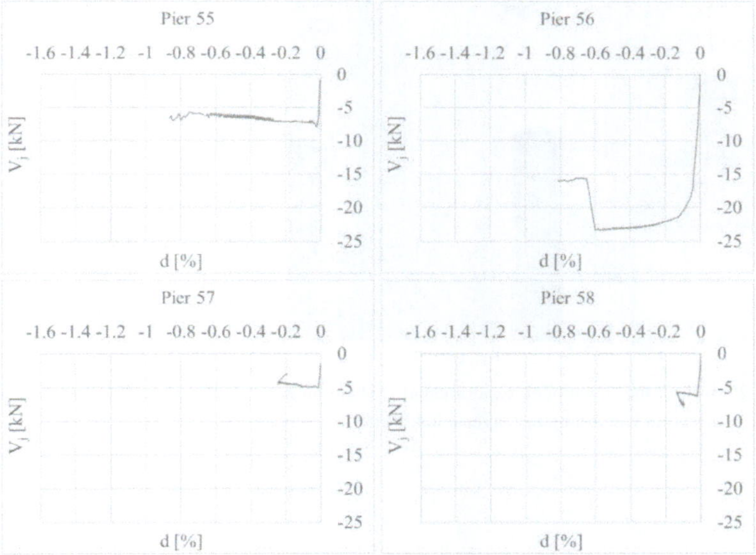

Figure 8.6: BM model: shear-drift diagrams of the masonry piers of the P5 wall (monotonic pushover in the -y direction with load pattern proportional to the masses)

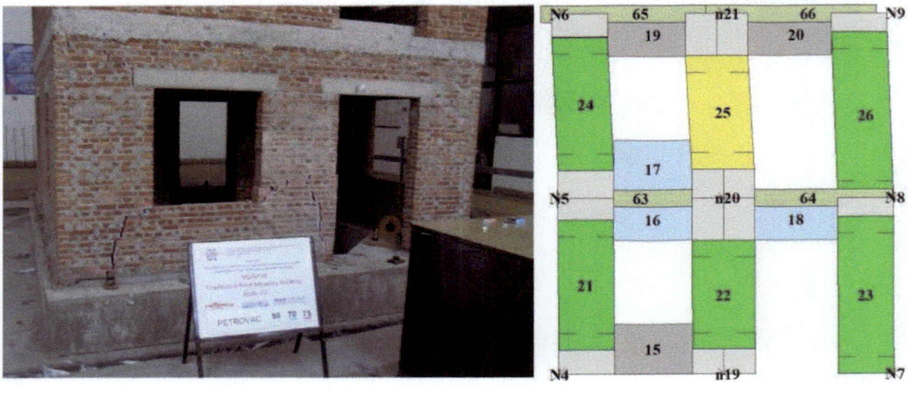

Figure 8.7: P2 wall of the BM model: experimental [73] and numerical final DLs (monotonic pushover in the -y direction with triangular inverse load pattern)

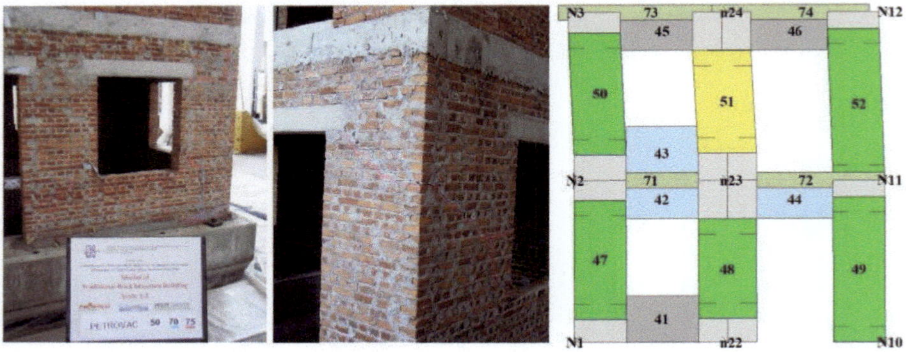

Figure 8.8: P4 wall of the BM model: experimental [73] and numerical final DLs (monotonic pushover in the -y direction with triangular inverse load pattern)

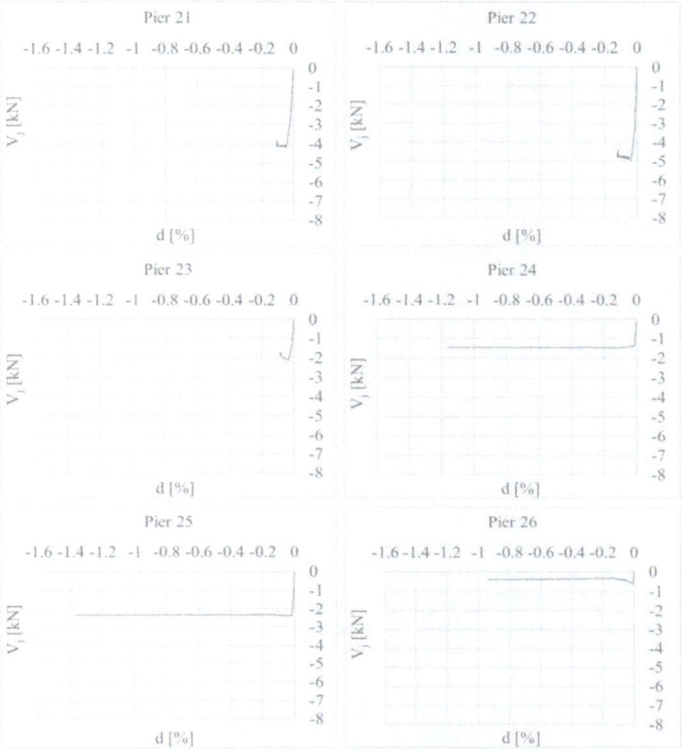

Figure 8.9: BM model: shear-drift diagrams of the masonry piers of the P2 wall (monotonic pushover in the -y direction with triangular inverse load pattern)

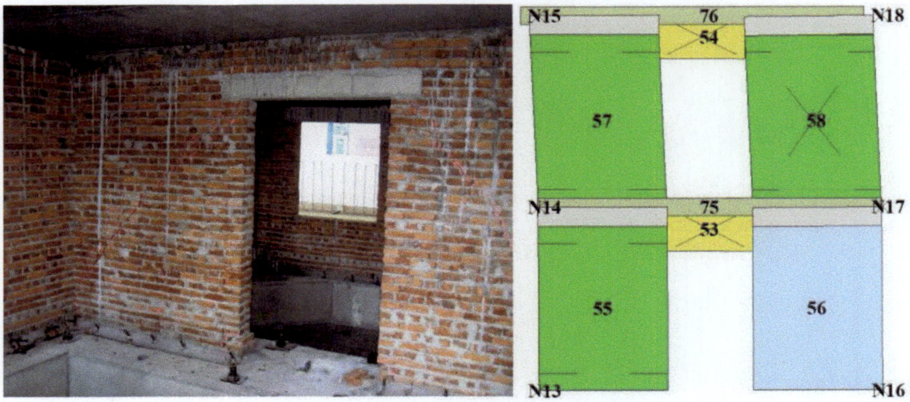

Figure 8.10: P5 wall of the BM model: experimental [73] and numerical final DLs (monotonic pushover in the -y direction with triangular inverse load pattern)

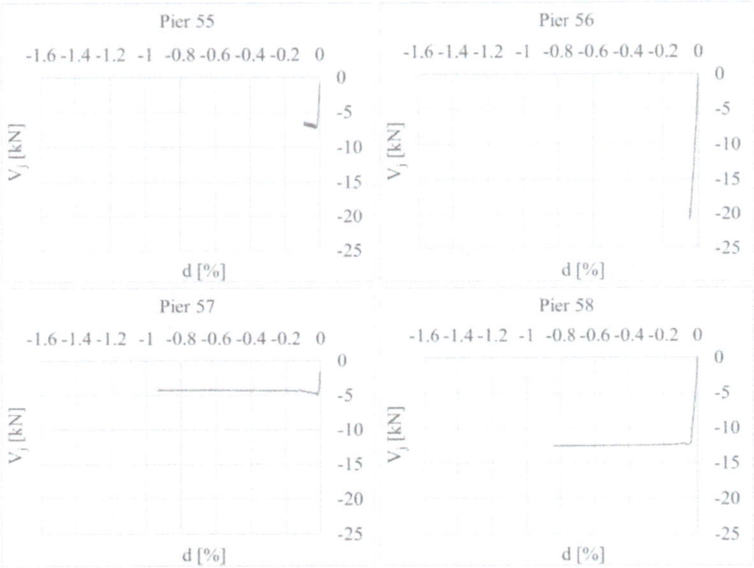

Figure 8.11: BM model: shear-drift diagrams of the masonry piers of the P5 wall (monotonic pushover in the -y direction with triangular inverse load pattern)

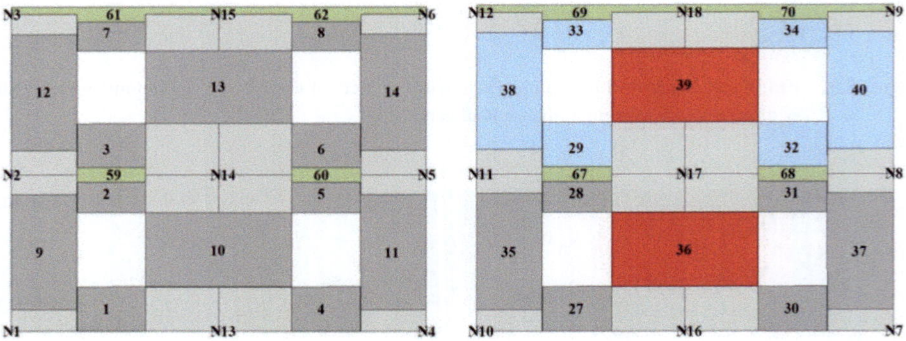

Figure 8.12: P1, P3 walls of the BM model: numerical final DLs (monotonic pushover in the -y direction with triangular inverse load pattern)

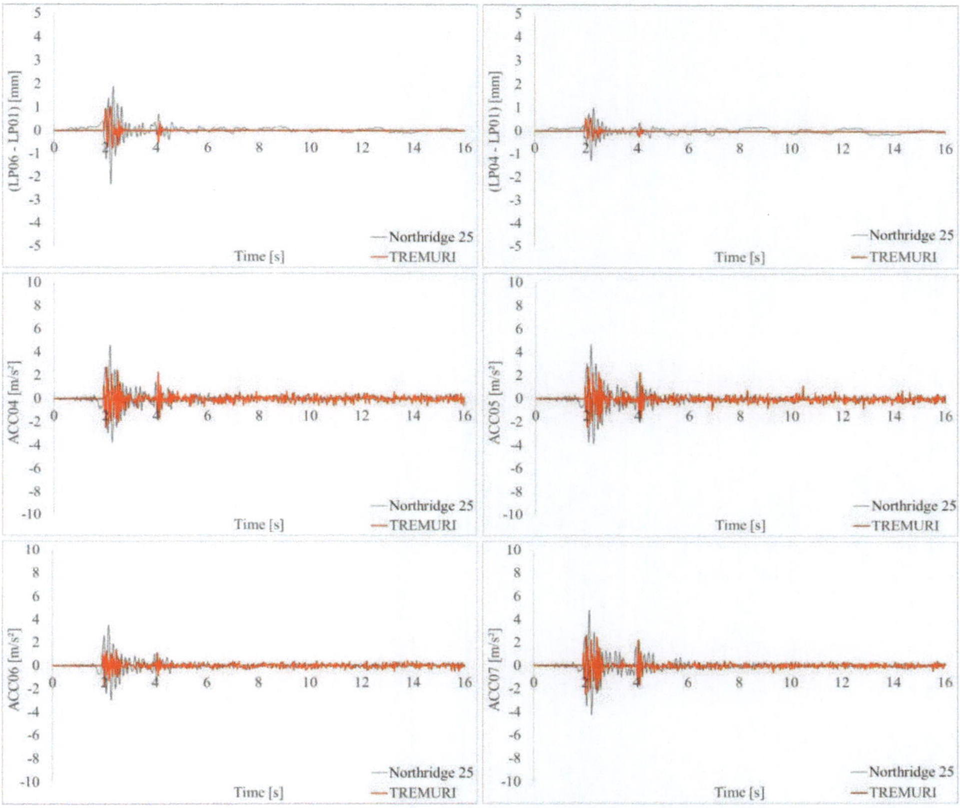

Figure 8.13: Results of the nonlinear dynamic analysis Northridge 25 performed on the BM model with TREMURI

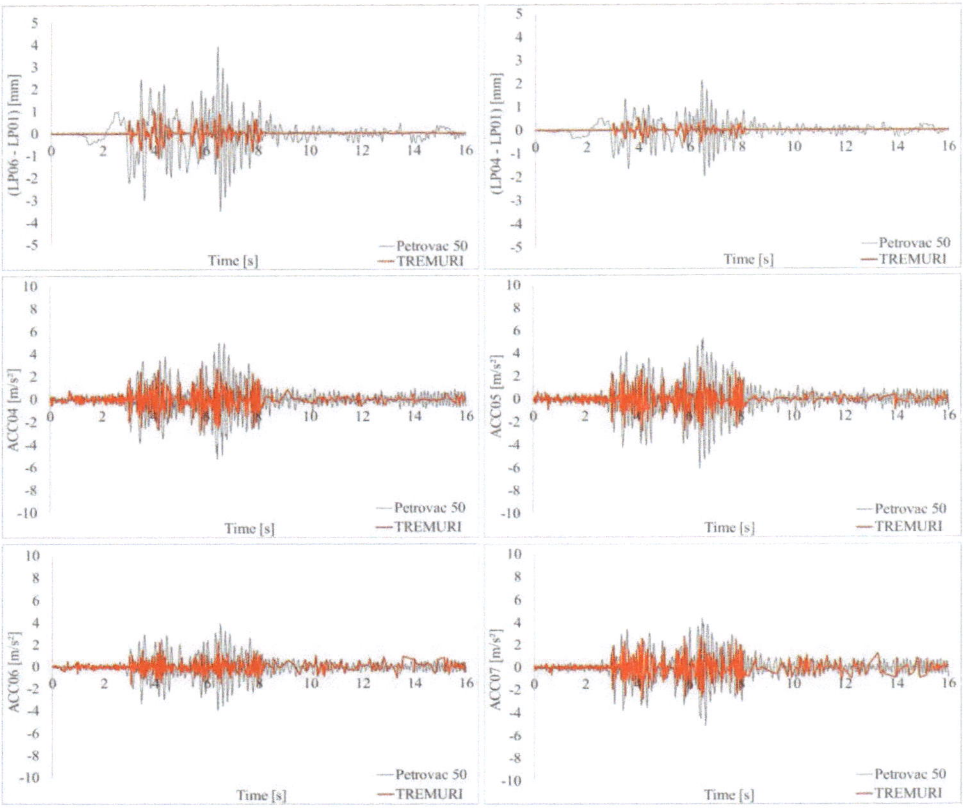

Figure 8.14: Results of the nonlinear dynamic analysis Petrovac 50 performed on the BM model with TREMURI

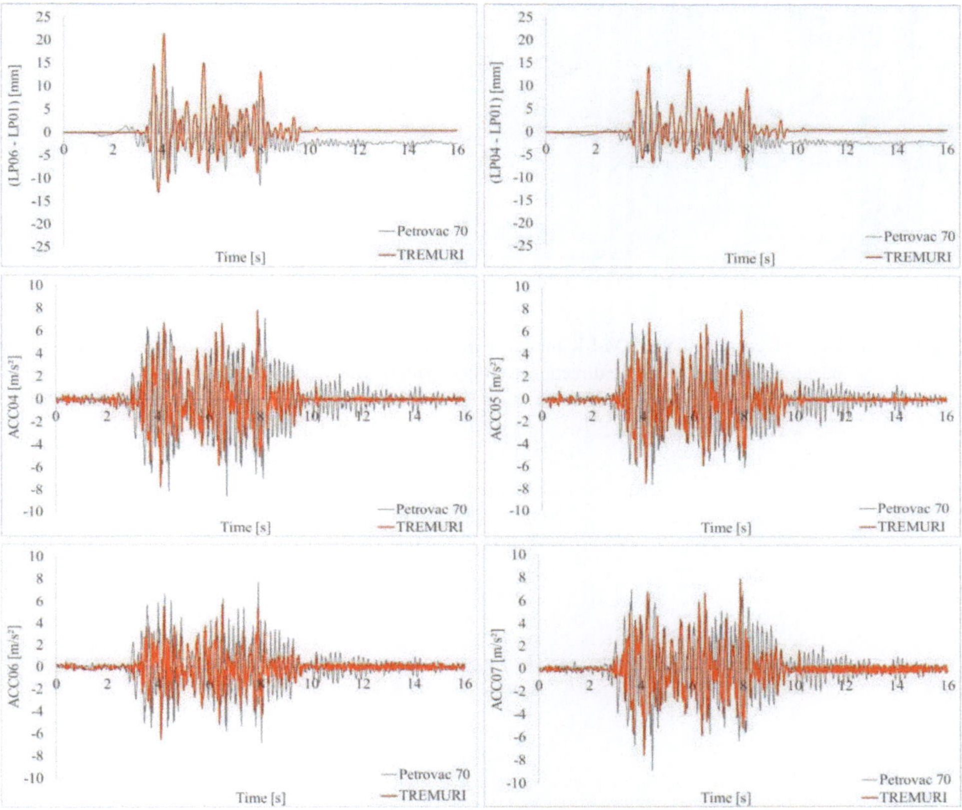

Figure 8.15: Results of the nonlinear dynamic analysis Petrovac 70 performed on the BM model with TREMURI

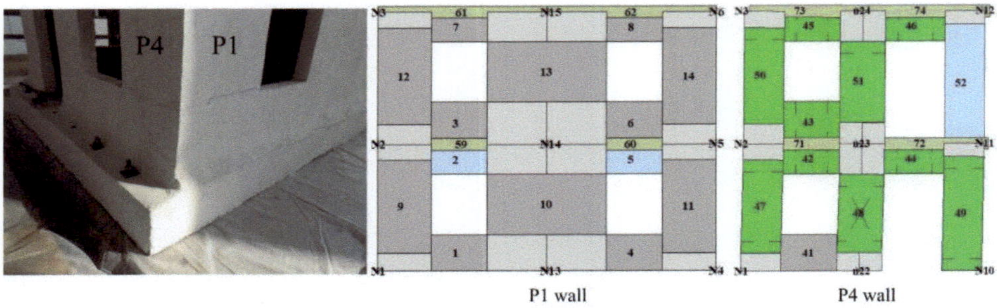

Figure 8.16: P4 - P1 corner of the BM-SR model: experimental [73] and numerical final DLs (monotonic pushover analysis in the +y direction with load pattern proportional to the masses)

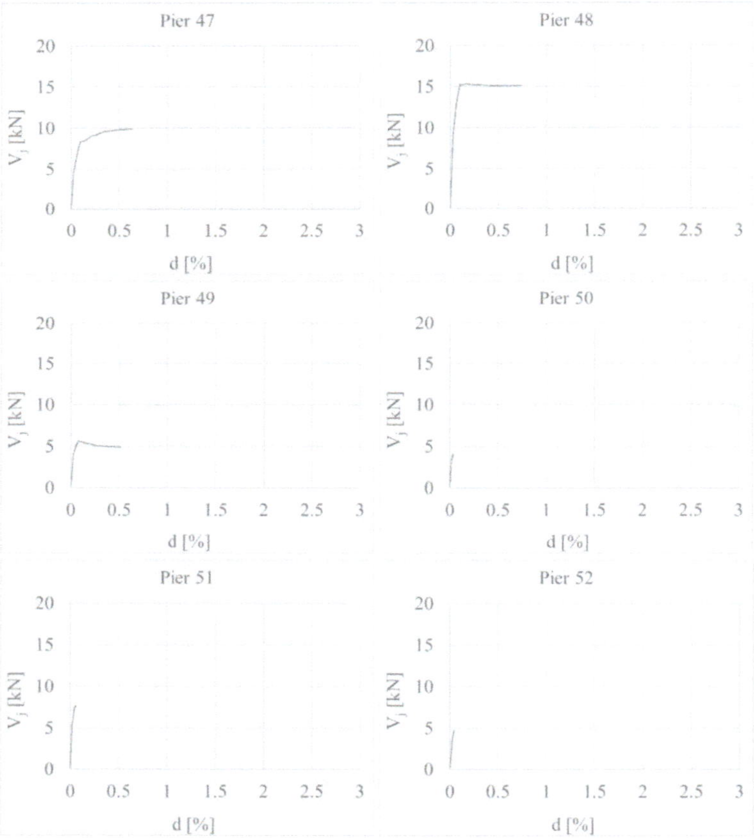

Figure 8.17: BM-SR model: shear-drift diagrams of the masonry piers of the P4 wall (monotonic pushover in the +y direction with load pattern proportional to the masses)

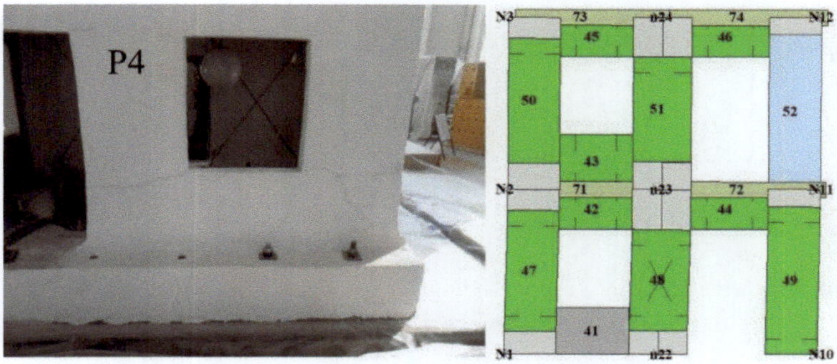

Figure 8.18: P4 wall of the BM-SR model: experimental [73] and numerical final DLs (monotonic pushover analysis in the +y direction with load pattern proportional to the masses)

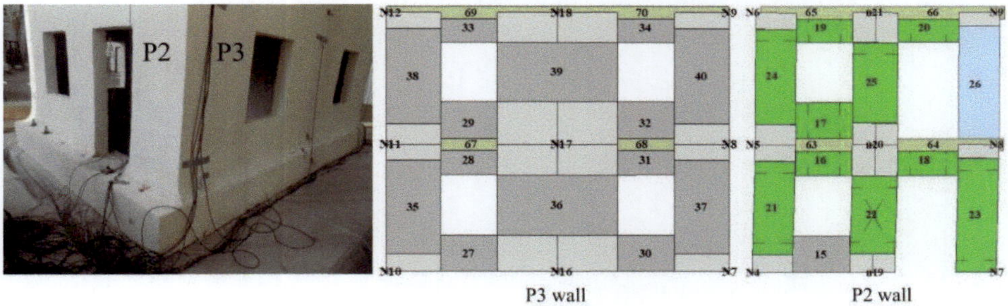

Figure 8.19: P2 - P3 corner of the BM-SR model: experimental [73] and numerical final DLs (monotonic pushover analysis in the +y direction with load pattern proportional to the masses)

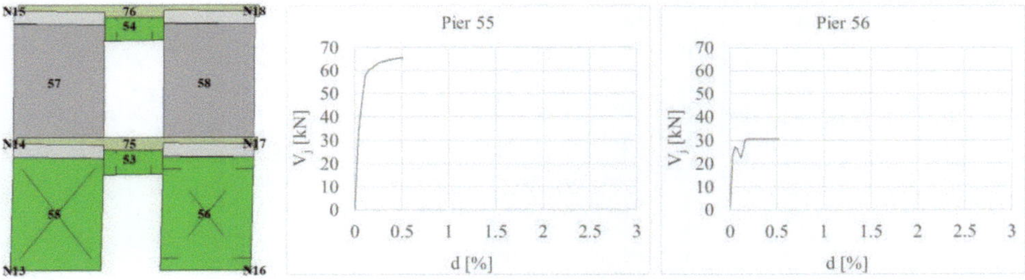

Figure 8.20: P5 wall of the BM-SR model: numerical final DLs and shear-drift diagrams of the masonry piers 55 and 56 (monotonic pushover analysis in the +y direction with load pattern proportional to the masses)

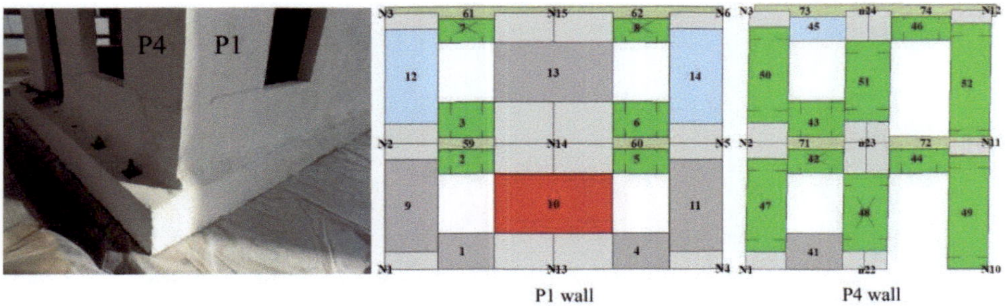

Figure 8.21: P4 - P1 corner of the BM-SR model: experimental [73] and numerical final DLs (monotonic pushover analysis in the +y direction with triangular inverse load pattern)

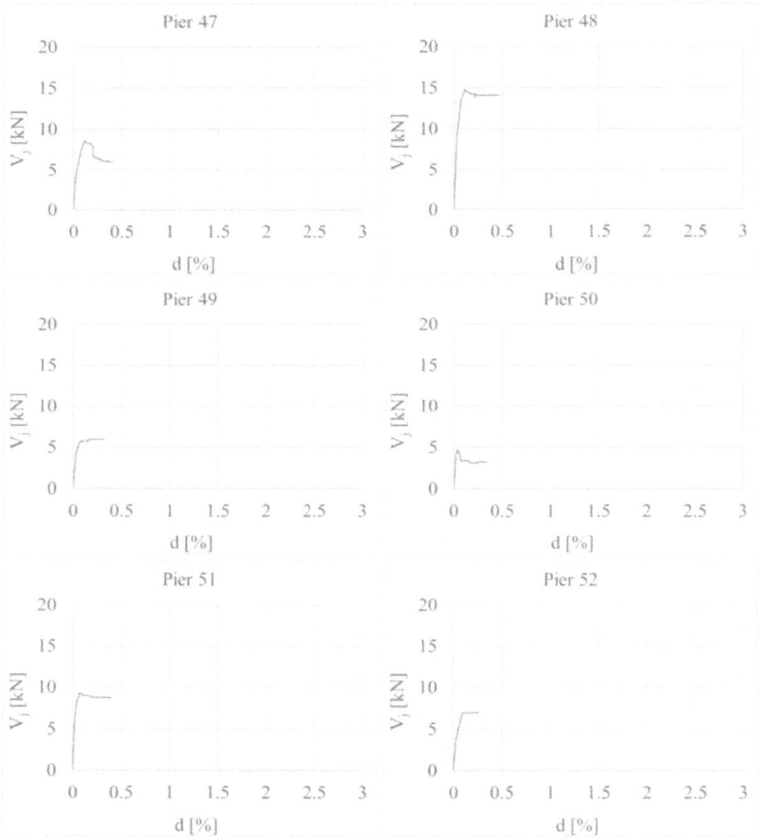

Figure 8.22: BM-SR model: shear-drift diagrams of the masonry piers of the P4 wall (monotonic pushover in the +y direction with triangular inverse load pattern)

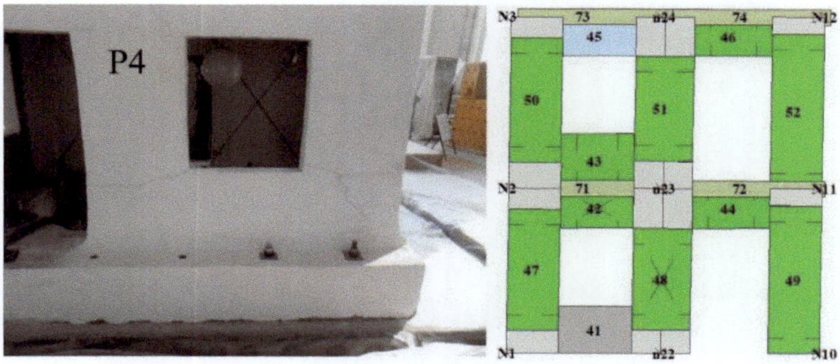

Figure 8.23: P4 wall of the BM-SR model: experimental [73] and numerical final DLs (monotonic pushover analysis in the +y direction with triangular inverse load pattern)

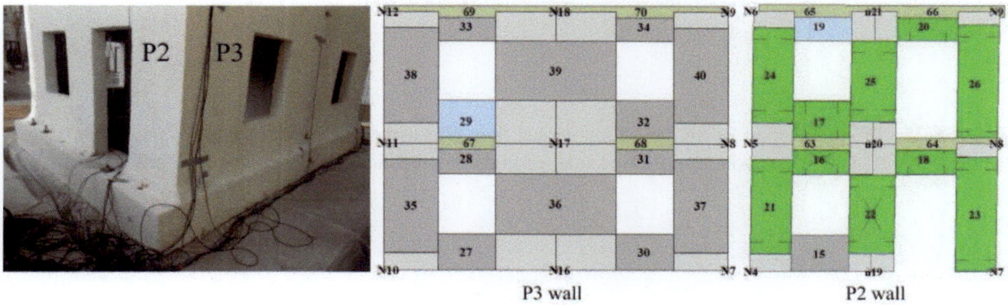

Figure 8.24: P2 - P3 corner of the BM-SR model: experimental [73] and numerical final DLs (monotonic pushover analysis in the +y direction with triangular inverse load pattern)

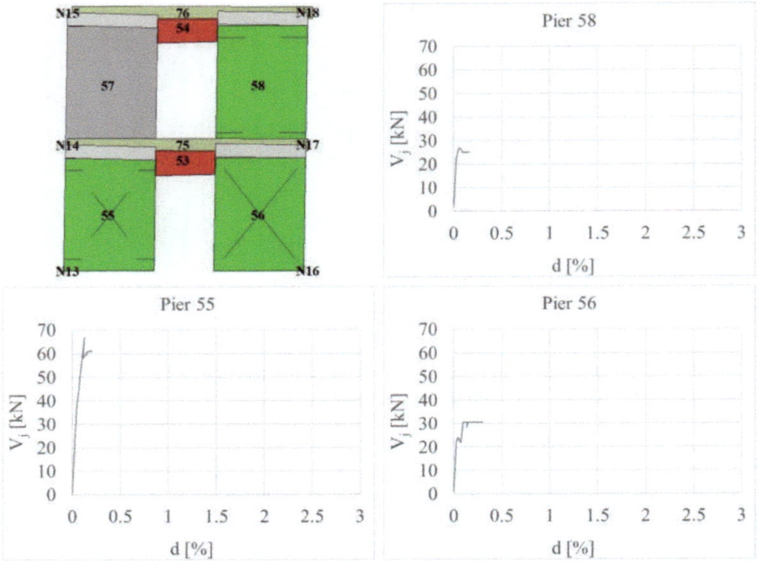

Figure 8.25: P5 wall of the BM-SR model: numerical final DLs and shear-drift diagrams of the masonry piers 55 and 56 (monotonic pushover analysis in the +y direction with triangular inverse load pattern)

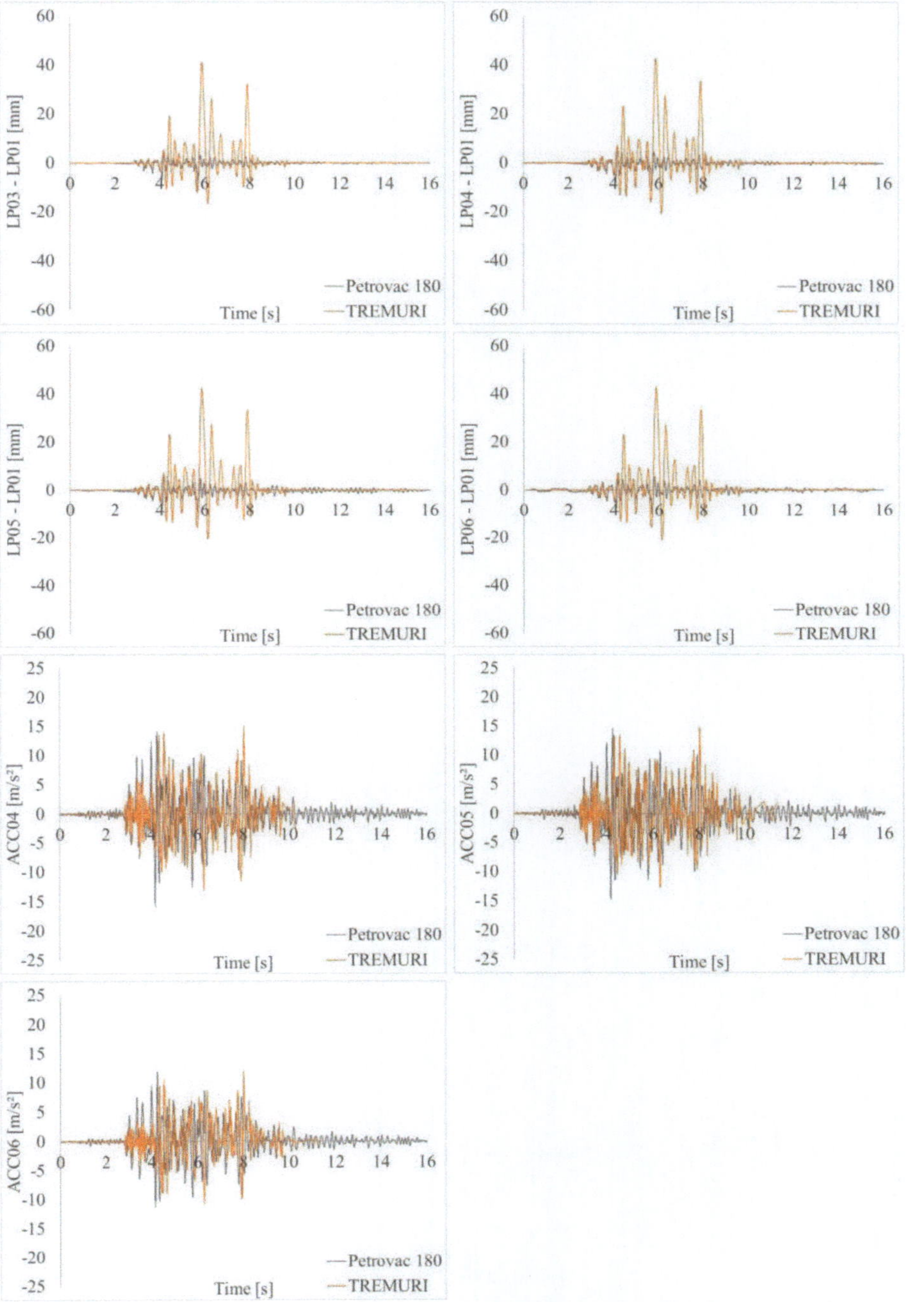

Figure 8.26: Results of the nonlinear dynamic analysis Petrovac 180 performed on the BM-SR model with TREMURI

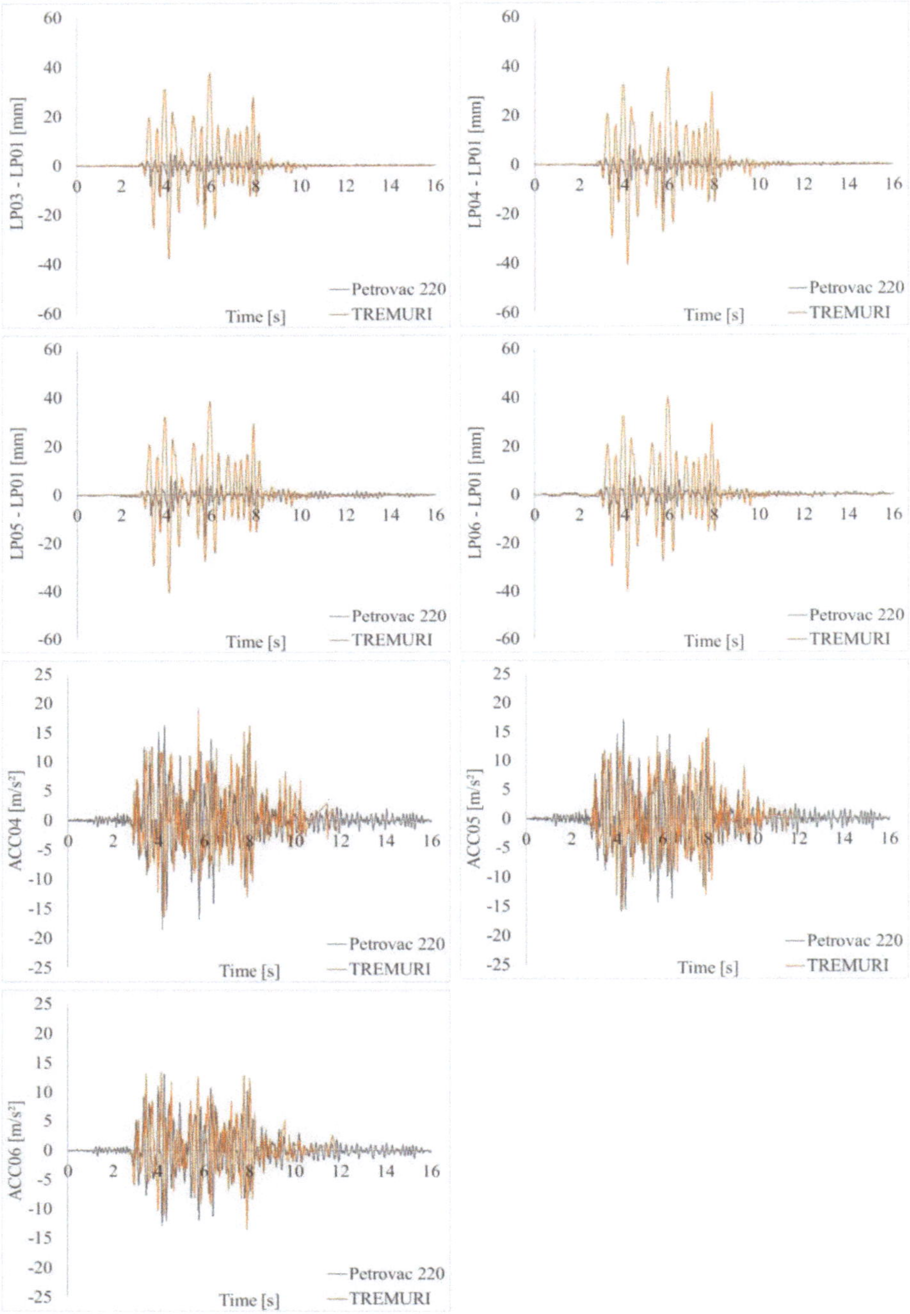

Figure 8.27: Results of the nonlinear dynamic analysis Petrovac 220 performed on the BM-SR model with TREMURI

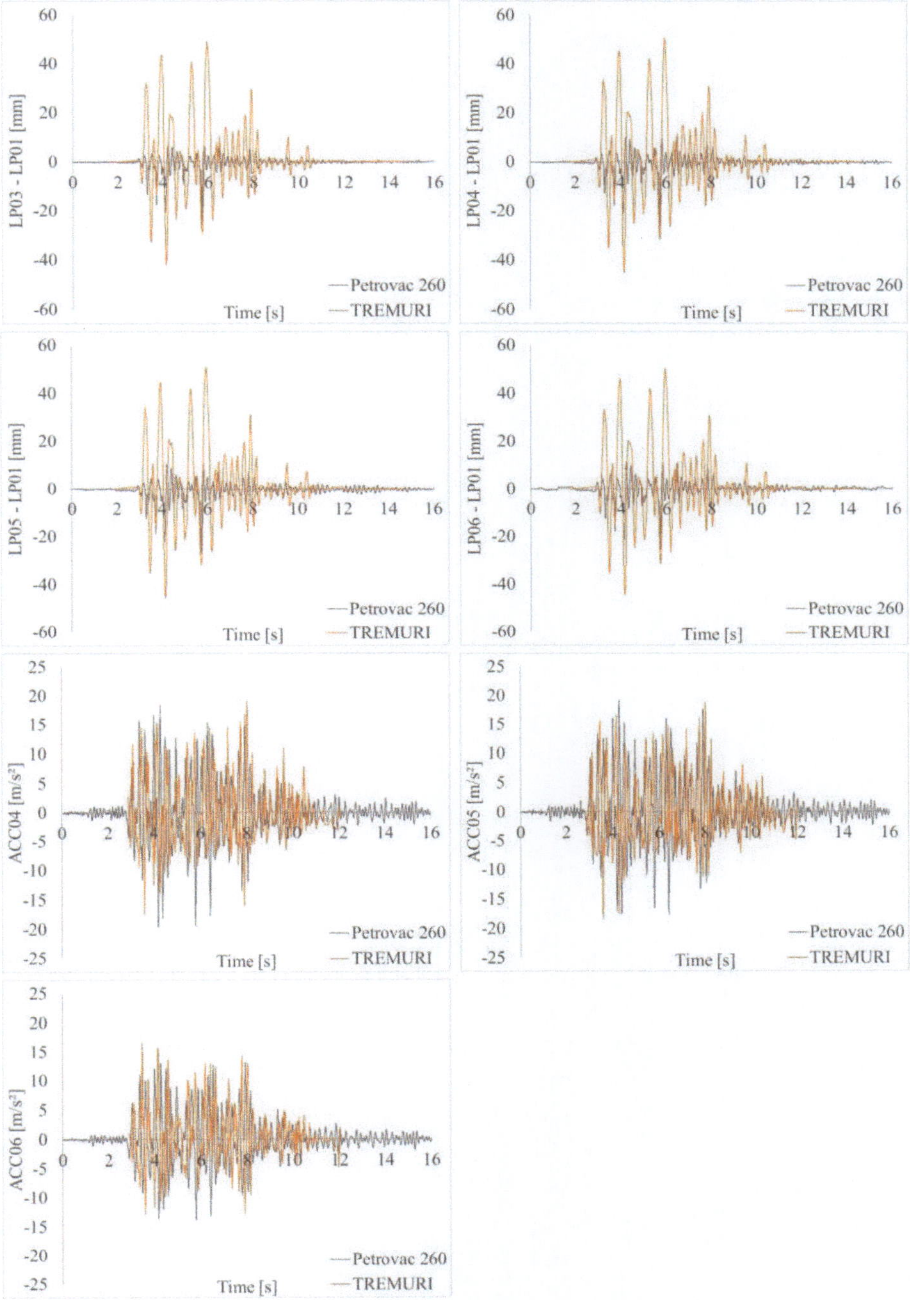

Figure 8.28: Results of the nonlinear dynamic analysis Petrovac 260 performed on the BM-SR model with TREMURI

List of figures

List of tables

Bibliography

[1] A. Giuffrè, *Sicurezza e conservazione dei centri storici in zona sismica. Il caso di Ortigia.* Bari, Italy: Laterza, 1993.

[2] P. G. Touliatos, "Seismic Behaviour of Traditionally-Built Constructions," in *Protection of the Architectural Heritage Against Earthquakes*, V. Petrini and M. Save, Eds. Springer, 1996, pp. 57–69.

[3] G. Macchi and G. Magenes, *Ingegneria delle strutture*, Third. Torino, Italy: UTET, 2002.

[4] A. Galasco, "ANALISI SISMICA DEGLI EDIFICI IN MURATURA," Università degli Studi di Genova, 2006.

[5] M. Como and A. Grimaldi, "A new method on the lateral strength evaluation of masonry walls and buildings," *Proc. 8th Eur. Conf. Earthq. Eng. LNEC, Lisbon*, 1986.

[6] D. Abruzzese, M. Como, and G. Lanni, "On the lateral strength of multistory masonry walls with openings and horizontal reinforcing connections," *10th World Conf. Earthq. Eng. Madrid, Ed. Balkema*, pp. 4525–4530, 1992.

[7] M. Como and A. Grimaldi, "A Unilateral Model for the Limit Analysis of Masonry Walls," in *Unilateral Problems in Structural Analysis*, Internatio., G. Del Piero and F. Maceri, Eds. Vienna: Springer, 1985.

[8] B. Calderoni, P. Marone, and M. Pagano, "Modelli per la verifica statica degli edifici in muratura in zona sismica," *Ing. Sismica*, vol. 3, pp. 19–27, 1987.

[9] B. Calderoni, P. Lenza, and M. Pagano, "Attuali prospettive per l'analisi sismica non lineare di edifici in muratura," *Atti del 4° Congr. Naz. ANIDIS, Milano*, 1989.

[10] F. Braga and M. Dolce, "Un metodo per l'analisi di edifici multipiano in muratura antisismici," *Proc. 6th I.B.Ma.C., Rome*, 1982.

[11] M. Tomazevic, "The computer program POR," *Rep. ZRMK*, 1978.

[12] M. Dolce, "Schematizzazione e modellazione per azioni nel piano delle pareti," *Corso sul consolidamento degli Edif. muratura Zo. sismica, Ordine degli Ing. Potenza*, 1989.

[13] M. Tomazevic and P. Weiss, "A rational, experimentally based method for the verification of earthquake resistance of masonry buildings," *Proc. 4th U.S. Natl. Conf. Earthq. Eng. Palm Springs*, vol. 2, pp. 349–359, 1990.

[14] S. Lagomarsino, A. Penna, A. Galasco, and S. Cattari, "TREMURI program: An equivalent frame model for the nonlinear seismic analysis of masonry buildings,"

Eng. Struct., vol. 56, pp. 1787–1799, 2013, doi: 10.1016/j.engstruct.2013.08.002.

[15] S. Cattari, "MODELLAZIONE A TELAIO EQUIVALENTE DI STRUTTURE ESISTENTI IN MURATURA E MISTE MURATURA-C.A.: FORMULAZIONE DI MODELLI SINTETICI," Università degli Studi di Genova, 2007.

[16] Ministero delle Infrastrutture e dei Trasporti, "Aggiornamento delle 'Norme tecniche per le costruzioni' (in Italian)." pp. 1–198, 2018.

[17] DIN EN 1998-3, *Eurocode 8: Auslegung von Bauwerken gegen Erdbeben – Teil 3: Beurteilung und Ertüchtigung von Gebäuden; Deutsche Fassung EN 1998-3:2005 + AC:2010*. 2010.

[18] S. Frumento, G. Magenes, P. Morandi, and G. M. Calvi, *Interpretation of experimental shear tests on clay brick masonry walls and evaluation of q-factors for seismic design*, no. May. 2009.

[19] M. Tomaževič, "Shear resistance of masonry walls and Eurocode 6: Shear versus tensile strength of masonry," *Mater. Struct. Constr.*, vol. 42, no. 7, pp. 889–907, 2009, doi: 10.1617/s11527-008-9430-6.

[20] V. Turnšek and F. Čačovič, "Some experimental results on the strength of brick masonry walls," *Proc. 2nd Int. brick-masonry Conf. Br. Ceram. Soc. Stoke-on-Trent*, pp. 149–156, 1971.

[21] D. Benedetti and M. Tomaževič, "Sulla verifica sismica di costruzioni in muratura," *Ing. Sismica I*, pp. 9–16, 1984.

[22] NTC 2018 explanatory Circular 21.01.2019 n. 7 C.S.LL.PP., "Istruzioni per l'applicazione dell'«Aggiornamento delle 'Norme tecniche per le costruzioni'» di cui al decreto ministeriale 17 gennaio 2018." Rome, Italy, 2019.

[23] W. Mann and H. Müller, "Failure of shear-stressed masonry – An enlarged theory, tests and application to shear-walls," *Proc. Int. Symp. Load-bearing Brickwork, London*, pp. 1–13, 1980.

[24] W. Jäger and G. Marzahn, *Mauerwerk Bemessung nach DIN 1053-100*. Ernst & Sohn, 2010.

[25] CNR national research council of Italy, *CNR-DT 212/2013: Guide for the Probabilistic Assessment of the Seismic Safety of Existing Buildings*. Rome, Italy, 2014.

[26] N. Augenti and F. Parisi, *Teoria e tecnica delle strutture in muratura*. Hoepli, 2019.

[27] P. T. Tassios, *Meccanica delle murature*. Liguori Editore, 1988.

[28] DIN EN 1996-1-1/NA, *Nationaler Anhang – National festgelegte Parameter – Eurocode 6: Bemessung und Konstruktion von Mauerwerksbauten – Teil 1-1: Allgemeine Regeln für bewehrtes und unbewehrtes Mauerwerk*. 2012.

[29] F. Parisi, N. Augenti, and A. Prota, "Implications of the spandrel type on the

lateral behavior of unreinforced masonry walls," *Earthq. Eng. Struct. Dyn.*, no. 43, pp. 1867–1887, 2014, doi: 10.1002/eqe.2441.

[30] K. Beyer and A. Dazio, "Quasi-static cyclic tests on masonry spandrels," *Earthq. Spectra*, vol. 28, no. 3, pp. 907–929, 2012, doi: 10.1193/1.4000063.

[31] N. Gattesco, L. Macorini, and A. Dudine, "Experimental Response of Brick-Masonry Spandrels under In-Plane Cyclic Loading," *J. Struct. Eng.*, vol. 142, no. 2, p. 4015146, 2016, doi: 10.1061/(asce)st.1943-541x.0001418.

[32] D. Benedetti, P. Carydis, and P. Pezzoli, "Shaking table tests on 24 simple masonry buildings," *Earthq. Eng. Struct. Dyn.*, vol. 27, pp. 67–90, 1998.

[33] J. Bothara, R. Dhakal, and J. Mander, "Seismic performance of an unreinforced masonry building: an experimental investigation," *Earthq. Eng. Struct. Dyn.*, vol. 39(1), pp. 45–68, 2010.

[34] S. Cattari and S. Lagomarsino, "A Strength Criterion for the Flexural Behaviour of Spandrels in Unreinforced Masonry Walls," *14 World Conf. Earthq. Eng.*, pp. 1–8, 2008, [Online]. Available: http://www.iitk.ac.in/nicee/wcee/article/14_05-04-0041.PDF.

[35] K. Beyer, A. Abo El Ezz, and A. Dazio, "Experimental investigation of the cyclic behaviour of unreinforced masonry spandrels," *9th US Natl. 10th Can. Conf. Earthq. Eng. 2010, Incl. Pap. from 4th Int. Tsunami Symp.*, vol. 7, no. 525, pp. 5118–5127, 2010.

[36] D. Benedetti and G. Magenes, "Correlazione tra tipo di danno ed energia dissipata negli edifici in muratura," *Ing. Sismica*, vol. 2, pp. 53–62, 2001.

[37] G. Magenes, D. Bolognini, and C. Braggio, "Metodi semplificati per l'analisi sismica non lineare di edifici in muratura," Rome, Italy: CNR-Gruppo Nazionale per la Difesa dei Terremoti, 2000.

[38] M. J. N. Priestley, G. M. Calvi, and M.]. Kowalsky, *Displacement-Based Seismic Design of Structures*. Pavia, Italy, 2007.

[39] C. Calderini and S. Lagomarsino, "A continuum model for in-plane anisotropic inelastic behaviour of masonry," *J. Struct. Eng. – ASCE*, vol. 134:2, pp. 209–220, 2008.

[40] M. Dolce, "Schematizzazione e modellazione degli edifici in muratura soggetti ad azioni sismiche," *L'Industria Delle Costr.*, vol. 25, no. 242, pp. 44–57, 1991.

[41] T. Yi, F. L. Moon, R. T. Leon, and L. F. Kahn, "Lateral load tests on a two-story unreinforced masonry building," *J. Struct. Eng.*, vol. 132, no. 5, pp. 643–652, 2006.

[42] S. Cattari and S. Lagomarsino, *Chapter 5 "Masonry structures" in: "Developments in the field of displacement based seismic assessment."* IUSS Press (PAVIA) and EUCENTRE, 2013.

[43] S. Cattari and K. Beyer, "Influence of spandrel modelling on the seismic

assessment of existing masonry buildings," *Proc. Tenth Pacific Conf. Earthq. Eng. Build. an Earthquake-Resilient Pacific*, vol. 88, no. January, 2015.

[44] K. Meskouris, K.-G. Hinzen, C. Butenweg, and M. Mistler, *Bauwerke und Erdbeben*. 2011.

[45] F. Graziotti, G. Magenes, A. Penna, and D. Fontana, "Experimental cyclic behaviour of stone masonry spandrels," *15th World Conf. Earthq. Eng.*, no. c, pp. 1–10, 2011.

[46] S. Cattari, D. Camilletti, S. Lagomarsino, S. Bracchi, M. Rota, and A. Penna, "Masonry Italian Code-Conforming Buildings. Part 2: Nonlinear Modelling and Time-History Analysis," *J. Earthq. Eng.*, vol. 22, no. S2, pp. 74–104, 2018, doi: 10.1080/13632469.2018.1541030.

[47] S. Cattari, A. Chioccariello, H. Degee, C. Doneaux, S. Lagomarsino, and C. Mordant, "Seismic Assessment of Masonry Buildings From Shaking Table Tests and Nonlinear Dynamic Simulations By the Proper Orthogonal Decomposition (POD)," *Second Eur. Conf. Earthq. Eng. Seismol.*, pp. 6–8, 2014.

[48] S. Cattari and S. Lagomarsino, "Analisi non lineari per la simulazione del danno di un fabbricato in San Felice sul Panaro (Emilia , 2012)," 2012.

[49] A. Nanni, "A New Tool for Concrete and Masonry Repair," *Concr. Int.*, no. april, pp. 43–49, 2012.

[50] S. De Santis, H. A. Hadad, F. De Caso Y Basalo, G. De Felice, and A. Nanni, "Acceptance Criteria for Tensile Characterization of Fabric-Reinforced Cementitious Matrix Systems for Concrete and Masonry Repair," *J. Compos. Constr.*, vol. 22, no. 6, pp. 1–14, 2018, doi: 10.1061/(ASCE)CC.1943-5614.0000886.

[51] W. Brameshuber, "Textile reinforced concrete. State-of-the-art report of RILEM Technical Committee 201-TRC," Bagneux, France, 2006.

[52] CNR national research council of Italy, "CNR-DT 215/2018: Istruzioni per la Progettazione, l'Esecuzione ed il Controllo di Interventi di Consolidamento Statico mediante l'utilizzo di Compositi Fibrorinforzati a Matrice Inorganica." 2020.

[53] E. S. ICC ES, *AC434: ACCEPTANCE CRITERIA FOR MASONRY AND CONCRETE STRENGTHENING USING FIBER-REINFORCED CEMENTITIOUS MATRIX (FRCM) COMPOSITE SYSTEMS*. 2013.

[54] American Concrete Institute (ACI) Committee 549, *ACI 549.4R-13: Guide to design and construction of externally bonded Fabric-Reinforced Cementitious Matrix (FRCM) systems for repair and strengthening concrete and masonry Structures*. Farmington Hills, Miami (U.S.A.), 2013.

[55] D. Arboleda, "Fabric reinforced cementitious matrix (FRCM) composites for infrastructure strengthening and rehabilitation: Characterization methods," University of Miami, 2014.

[56] G. P. Lignola *et al.*, "Performance assessment of basalt FRCM for retrofit applications on masonry," *Compos. Part B Eng.*, vol. 128, pp. 1–18, 2017, doi: 10.1016/j.compositesb.2017.05.003.

[57] F. G. Carozzi *et al.*, "Experimental investigation of tensile and bond properties of Carbon-FRCM composites for strengthening masonry elements," *Compos. Part B Eng.*, vol. 128, pp. 100–119, 2017, doi: 10.1016/j.compositesb.2017.06.018.

[58] M. Leone *et al.*, "Glass fabric reinforced cementitious matrix: Tensile properties and bond performance on masonry substrate," *Compos. Part B Eng.*, vol. 127, pp. 196–214, 2017, doi: 10.1016/j.compositesb.2017.06.028.

[59] C. Caggegi *et al.*, "Experimental analysis on tensile and bond properties of PBO and aramid fabric reinforced cementitious matrix for strengthening masonry structures," *Compos. Part B Eng.*, vol. 127, pp. 175–195, 2017, doi: 10.1016/j.compositesb.2017.05.048.

[60] G. de Felice *et al.*, "Recommendation of RILEM Technical Committee 250-CSM: Test method for Textile Reinforced Mortar to substrate bond characterization," *Mater. Struct. Constr.*, vol. 51, no. 4, pp. 1–9, 2018, doi: 10.1617/s11527-018-1216-x.

[61] L. Ascione, G. De Felice, and S. De Santis, "A qualification method for externally bonded Fibre Reinforced Cementitious Matrix (FRCM) strengthening systems," *Compos. Part B Eng.*, vol. 78, pp. 497–506, 2015, doi: 10.1016/j.compositesb.2015.03.079.

[62] S. De Santis, F. G. Carozzi, G. de Felice, and C. Poggi, "Test methods for Textile Reinforced Mortar systems," *Compos. Part B Eng.*, vol. 127, pp. 121–132, 2017, doi: 10.1016/j.compositesb.2017.03.016.

[63] A. Prota, G. Marcari, G. Fabbrocino, G. Manfredi, and C. Aldea, "Experimental in-plane behavior of tuff masonry strengthened with cementitious matrix-grid composite," *J. Compos. Constr.*, vol. 10, no. 3, pp. 223–233, 2006, doi: https://doi.org/10.1061/(ASCE)1090-0268(2006)10:3(223).

[64] C. G. Papanicolaou, T. C. Triantafillou, K. Karlos, and M. Papathanasiou, "Textile-reinforced mortar (TRM) versus FRP as strengthening material of URM walls: In-plane cyclic loading," *Mater. Struct. Constr.*, vol. 40, no. 10, pp. 1081–1097, 2007, doi: 10.1617/s11527-006-9207-8.

[65] C. Faella, E. Martinelli, E. Nigro, and S. Paciello, "Shear capacity of masonry walls externally strengthened by a cement-based composite material: An experimental campaign," *Constr. Build. Mater.*, vol. 24, no. 1, pp. 84–93, 2010, doi: 10.1016/j.conbuildmat.2009.08.019.

[66] S. Babaeidarabad and A. Nanni., "In-plane behavior of unreinforced masonry walls strengthened with fabric-reinforced cementitious matrix (FRCM)," *ACI Spec. Publ.*, vol. 299, pp. 69–80, 2015.

[67] G. Marcari, M. Basili, and F. Vestroni, "Experimental investigation of tuff masonry panels reinforced with surface bonded basalt textile-reinforced mortar,"

Compos. Part B Eng., vol. 108, pp. 131–142, 2017, doi: 10.1016/j.compositesb.2016.09.094.

[68] N. Ismail, T. El-Maaddawy, N. Khattak, and A. Najmal, "In-plane shear strength improvement of hollow concrete masonry panels using a fabric-reinforced cementitious matrix," *J. Compos. Constr.*, vol. 22, no. 2, 2018, doi: https://doi.org/10.1061/(ASCE)CC.1943-5614.0000835.

[69] S. Rizzo and L. Stempniewski, "EQ-grid: A Multiaxial Seismic Retrofitting System for Masonry Buildings," in *Earthquakes - Forecast, Prognosis and Earthquake Resistant Construction*, V. Svalova, Ed. IntechOpen, 2018.

[70] L. Garcia-Ramonda, L. Pelá, P. Roca, and G. Camata, "In-plane shear behaviour by diagonal compression testing of brick masonry walls strengthened with basalt and steel textile reinforced mortars," *Constr. Build. Mater.*, vol. 240, Apr. 2020, doi: 10.1016/j.conbuildmat.2019.117905.

[71] M. S. J. C. MSJC, "Building Code Requirements and Specification for Masonry Structures." 2011.

[72] L. Stempniewski and M. Urban, "Shaking Table Tests of a Full-Scale Two-Storey Pre-Damaged Natural Stone Building Retrofitted with the Multi-Axial Hybrid Textile System 'Eq-Grid,'" in *Seismic Evaluation and Rehabilitation of Structures*, A. Ilki and M. N. Fardis, Eds. 2014, pp. 155–169.

[73] M. Garevski and V. Shendova, "EXPERIMENTAL VERIFICATION OF INOVATIVE TECHNIQUE FOR SEISMIC RETROFITTING OF TRADITIONAL MASONRY BUILDING Final Report," 2013.

[74] N. Augenti, F. Parisi, A. Prota, and G. Manfredi, "In-Plane Lateral Response of a Full-Scale Masonry Subassemblage with and without an Inorganic Matrix-Grid Strengthening System," *J. Compos. Constr.*, vol. 15, no. 4, pp. 578–590, 2011, doi: 10.1061/(asce)cc.1943-5614.0000193.

[75] S. Babaeidarabad, F. De Caso, and A. Nanni, "URM Walls Strengthened with Fabric-Reinforced Cementitious Matrix Composite Subjected to Diagonal Compression," *J. Compos. Constr.*, vol. 18, no. 2, p. 4013045, 2014, doi: 10.1061/(asce)cc.1943-5614.0000441.

[76] M. Giaretton, D. Dizhur, E. Garbin, J. M. Ingham, and F. Da Porto, "In-Plane Strengthening of Clay Brick and Block Masonry Walls Using Textile-Reinforced Mortar," *J. Compos. Constr.*, vol. 22, no. 5, 2018, doi: 10.1061/(ASCE)CC.1943-5614.0000866.

[77] RÖFIX, "RÖFIX SismaCalce ®." pp. 1–3, 2020.

[78] RILEM TC 76-LUM, "Cyclic shear test for masonry panels designed to resist seismic forces 1991." 1991.

[79] J. C. Münich, "Hybride Multidirektionaltextilien zur Erdbebenverstärkung von Mauerwerk Experimente und numerische Untersuchungen mittels eines erweiterten Makromodells," 2010.

[80] G. Grünthal, *Cahiers du Centre Européen de Géodynamique et de Séismologie: Volume 15–European Macroseismic Scale 1998*, vol. 15. 1998.

[81] DIN EN 1996-1-1, *Eurocode 6: Bemessung und Konstruktion von Mauerwerksbauten – Teil 1-1: Allgemeine Regeln für bewehrtes und unbewehrtes Mauerwerk; Deutsche Fassung EN 1996-1-1:2005 + AC:2009*. 2010.

[82] D. Wolfram Jäger, *Mauerwerk Kalender 2019*. 2019.

[83] C. M. Urban, "Experimentelle Untersuchungen und Bemessungsansätze für faserverstärktes Mauerwerk unter Erdbebenbeanspruchungen," Karlsruher Institut für Technologie, 2014.

[84] P. Morandi, L. Albanesi, and G. Magenes, "In-plane test campaign on different load-bearing URM typologies with thin shell and web clay units," *Proceeding 16th Int. Brick Block Mason. Conf. Padua, Italy.*, 2016.

[85] P. Morandi, L. Albanesi, F. Graziotti, T. Li Piani, A. Penna, and G. Magenes, "Development of a dataset on the in-plane experimental response of URM piers with bricks and blocks," *Constr. Build. Mater.*, vol. 190, pp. 593–611, 2018, doi: 10.1016/j.conbuildmat.2018.09.070.

[86] G. Magenes, P. Morandi, and A. Penna, "Experimental in-plane cyclic response of masonry walls with clay units," *Proceeding 14th WCEE, Beijing, China*, vol. 95, p. 8, 2008.

[87] A. Galasco, S. Lagomarsino, and A. Penna, "On the use of pushover analysis for existing masonry buildings," *First Eur. Conf. Earthq. Eng. Seismol.*, no. September, pp. 3–8, 2006.

[88] S. Kallioras, F. Graziotti, and A. Penna, "Numerical assessment of the dynamic response of a URM terraced house exposed to induced seismicity," *Bull. Earthq. Eng.*, vol. 17, no. 3, pp. 1521–1552, 2019, doi: 10.1007/s10518-018-0495-5.

[89] M. J. N. Priestley and D. N. Grant, "Viscous damping in seismic design and analysis," *J. Earthq. Eng.*, vol. 9, no. SPEC. ISS. 2, pp. 229–255, 2005, doi: 10.1142/S1363246905002365.

Schriftenreihe des

Instituts für Massivbau und Baustofftechnologie (ISSN 0933-0461)

Heft 1 **Manfred Curbach**
Festigkeitssteigerung von Beton bei hohen Belastungsgeschwindigkeiten.

Heft 2 **Franz-Hermann Schlüter**
Dicke Stahlbetonplatten unter stoßartiger Belastung – Flugzeugabsturz.

Heft 3 **Marlies Schieferstein**
Der Zugflansch von Stahlbetonplattenbalken unter Längsschub und Querbiegung bei kritischer Druckbeanspruchung von Beton.

Heft 4 **Thomas Bier**
Karbonatisierung und Realkalisierung von Zementstein und Beton.

Heft 5 **Wolfgang Brameshuber**
Bruchmechanische Eigenschaften von jungem Beton.

Heft 6 **Bericht DFG-Forschungsschwerpunkt**
Durability of Non-Metallic Inanorganic Building Materials.

Heft 7 **Manfred Feyerabend**
Der harte Querstoß auf Stützen aus Stahl und Stahlbeton.

Heft 8 **Klaus F. Schönlin**
Permeabilität als Kennwert der Dauerhaftigkeit von Beton.

Heft 9 **Lothar Stempniewski**
Flüssigkeitsgefüllte Stahlbetonbehälter unter Erdbebeneinwirkung.

Heft 10 **Jörg Weidner**
Vergleich von Stoffgesetzen granularer Schüttgüter zur Silodruckermittlung.

Heft 11 **Pingli Yi**
Explosionseinwirkungen auf Stahlbetonplatten.

Heft 12 **Rainer Kunterding**
Beanspruchung der Oberfläche von Stahlbetonsilos durch Schüttgüter.

Heft 13 **Peter Haardt**
Zementgebundene und kunststoffvergütete Beschichtungen auf Beton.

Heft 14 **Günter Rombach**
Schüttguteinwirkungen auf Silozellen – Exzentrische Entleerung.

Heft 15 **Harald Garrecht**
Porenstrukturmodelle für den Feuchtehaushalt von Baustoffen mit und ohne Salzbefrachtung und rechnerische Anwendung auf Mauerwerk.

Heft 16 **Violandi Vratsanou**
Das nichtlineare Verhalten unbewehrter Mauerwerksscheiben unter Erdbebenbeanspruchung – Hilfsmittel zur Bestimmung der q-Faktoren.

Heft 17 **Carlos Rebelo**
Stochastische Modellierung menschenerzeugter Schwingungen.

Heft 18 **Seminar 29./30. März 1993**
Erdbebenauslegung von Massivbauten unter Berücksichtigung des Eurocode 8.

Heft 19 **Hubert Bachmann**
Die Massenträgheit in einem Pseudo-Stoffgesetz für Beton bei schneller Zugbeanspruchung.

Heft 20 **DBV/AiF-Forschungsbericht H. Emrich**
Zum Tragverhalten von Stahlbetonbauteilen unter Querkraft- und Längszugbeanspruchung.

Heft 21 **Robert Stolze**
Zum Tragverhalten von Stahlbetonplatten mit von den Bruchlinien abweichender Bewehrungsrichtung – Bruchlinien-Rotationskapazität.

Heft 22 **Jie Huang**
Extern vorgespannte Segmentbrücken unter kombinierter Beanspruchung aus Biegung, Querkraft und Torsion.

Heft 23 **Rolf Wörner**
Verstärkung von Stahlbetonbauteilen mit Spritzbeton.

Heft 24 **Ioannis Retzepis**
Schiefe Betonplatten im gerissenen Zustand.

Heft 25 **Frank Dahlhaus**
Stochastische Untersuchungen von Silobeanspruchungen.

Heft 26 **Cornelius Ruckenbrod**
Statische und dynamische Phänomene bei der Entleerung von Silozellen.

Heft 27 **Shishan Zheng**
Beton bei variierender Dehngeschwindigkeit, untersucht mit einer neuen modifizierten Split-Hopkinson-Bar-Technik.

Heft 28 **Yong-zhi Lin**
Tragverhalten von Stahlfaserbeton.

Heft 29 **DFG**
Korrosion nichtmetallischer anorganischer Werkstoffe im Bauwesen.

Heft 30 **Jürgen Ockert**
Ein Stoffgesetz für die Schockwellenausbreitung in Beton.

Heft 31 **Andreas Braun**
Schüttgutbeanspruchungen von Silozellen unter Erdbebeneinwirkung.

Heft 32 **Martin Günter**
Beanspruchung und Beanspruchbarkeit des Verbundes zwischen Polymerbeschichtungen und Beton.

Heft 33 **Gerhard Lohrmann**
Faserbeton unter hoher Dehngeschwindigkeit.

Heft 34 **Klaus Idda**
Verbundverhalten von Betonrippenstäben bei Querzug.

Heft 35 **Stephan Kranz**
Lokale Schwind- und Temperaturgradienten in bewehrten, oberflächennahen Zonen von Betonstrukturen.

Heft 36 **Gunther Herold**
Korrosion zementgebundener Werkstoffe in mineralsauren Wässern.

Heft 37 **Mostafa Mehrafza**
Entleerungsdrücke in Massefluss-Silos – Einflüsse der Geometrie und Randbedingungen.

Heft 38 **Tarek Nasr**
Druckentlastung bei Staubexplosionen in Siloanlagen. 2000

Heft 39 **Jan Akkermann**
Rotationsverhalten von Stahlbeton-Rahmenecken.

Heft 40 **Viktor Mechtcherine**
Bruchmechanische und fraktologische Untersuchungen zur Rißausbreitung in Beton.

Heft 41 **Ulrich Häußler-Combe**
Elementfreie Galerkin-Verfahren – Grundlagen und Einsatzmöglichkeiten zur Berechnung von Stahlbetontragwerken.

Heft 42 **Björn Schmidt-Hurtienne**
Ein dreiaxiales Schädigungsmodell für Beton unter Einschluß des Dehnrateneffekts bei Hochgeschwindigkeitsbelastung.

Heft 43 **Nazir Abdou**
Ein stochastisches nichtlineares Berechnungsverfahren für Stahlbeton mit finiten Elementen.

Heft 44 **Andreas Plokitza**
Ein Verfahren zur numerischen Simulation von Betonstrukturen beim Abbruch durch Sprengen.

Heft 45 **Timon Rabczuk**
Numerische Untersuchungen zum Fragmentierungsverhalten von Beton mit Hilfe der SPH-Methode.

Heft 46 **Norbert J. Krutzik**
Zu Anwendungsgrenzen von FE-Modellen bei der Simulation von Erschütterungen in Kernkraftbauwerken bei Stoßbelastungen.

Heft 47 **Thorsten Timm**
Beschuß von flüssigkeitsgefüllten Stahlbehältern.

Heft 48 **Slobodan Kasic**
Tragverhalten von Segmentbauteilen mit interner und externer Vorspannung ohne Verbund.

Heft 49 **Christoph Kessler-Kramer**
Zugtragverhalten von Beton unter Ermüdungsbeanspruchung.

Heft 50 **Nico Herrmann**
Experimentelle Verifizierung von Prognosen zur Sprengtechnik.

Heft 51 **Michael Baur**
Elastomerlager und nichtlineare Standorteffekte bei Erdbebeneinwirkung.

Heft 52 **Seminar 02. Juli 2004**
DIN 1045-1; Aus der Praxis für die Praxis.

Heft 53 **Abdelkhalek Saber Omar Mohamed**
Behaviour of Retrofitted Masonry Shear Walls Subjected to Cyclic Loading.

Heft 54 **Werner Hörenbaum**
Verwitterungsmechanismen und Dauerhaftigkeit von Sandsteinsichtmauenwerk.

Heft 55 **Seminar Februar 2006**
DIN 4149 – Aus der Praxis für die Praxis.

Heft 56 **Sam Foos**
Unbewehrte Betonfahrbahnplatten unter witterungsbedingten Beanspruchungen.

Heft 57 **Ramzi Maliha**
Untersuchungen zur Rissbildung in Fahrbahndecken aus Beton.

Heft 58 **Andreas Fäcke**
Numerische Simulation des Schädigungsverhaltens von Brückenpfeilern aus Stahlbeton unter Erdbebenlasten.

Heft 59 **Juliane Möller**
Rotationsverhalten von verbundlos vorgespannten Segmenttragwerken.

Heft 60 **Martin Larcher**
Numerische Simulation des Betonverhaltens unter Stoßwellen mit Hilfe des Elementfreien Galerkin-Verfahrens.

Heft 61 **Christoph Niklasch**
Numerische Untersuchungen zum Leckageverhalten von gerissenen Stahlbetonwänden.

Heft 62 **Halim Khbeis**
Experimentelle und numerische Untersuchungen von Topflagern.

Heft 63 **Sascha Schnepf**
Vereinfachte numerische Simulation des Tragverhaltens ebener mauerwerksausgefachter Stahlbetonrahmen unter zyklischer Belastung.

Heft 64 **Christian Wallner**
Erdbebengerechtes Verstärken von Mauerwerk durch Faserverbundwerkstoffe – experimentelle und numerische Untersuchungen.

Heft 65 **Niklas Puttendörfer**
Ein Beitrag zum Gleitverhalten und zur Sattelausbildung externer Spannglieder.

»»»»»»»»» **Bezug der Hefte 1 – 65 und 67**
Institut für Massivbau und Baustofftechnologie
Karlsruher Institut für Technologie (KIT)
Gotthard-Franz-Str. 3, 76131 Karlsruhe
www.betoninstitut.de

Fortführung als

Karlsruher Reihe Massivbau - Baustofftechnologie - Materialprüfung (ISSN 1869-912X)
Institut für Massivbau und Baustofftechnologie
Materialprüfungs- und Forschungsanstalt, MPA Karlsruhe

KIT Scientific Publishing

Heft 66	**Michael Haist** Zur Rheologie und den physikalischen Wechselwirkungen bei Zementsuspensionen. ISBN 978-3-86644-475-1
Heft 67	**Stephan Steiner** Beton unter Kontaktdetonation – neue experimentelle Methoden. (noch erschienen in der Schriftenreihe des Instituts für Massivbau und Baustofftechnologie, ISSN 0933-0461)
Heft 68	**Christian Münich** Hybride Multidirektionaltextilien zur Erdbebenverstärkung von Mauerwerk – Experimente und numerische Untersuchungen mittels eines erweiterten Makromodells. ISBN 978-3-86644-734-9
Heft 69	**Viktória Malárics** Ermittlung der Betonzugfestigkeit aus dem Spaltzugversuch an zylindrischen Betonproben. ISBN 978-3-86644-735-6
Heft 70	**Daniela Ruch** Bestimmung der Last-Zeit-Funktion beim Aufprall flüssigkeitsgefüllter Stoßkörper. ISBN 978-3-86644-736-3
Heft 71	**Marc Beitzel** Frischbetondruck unter Berücksichtigung der rheologischen Eigenschaften. ISBN 978-3-86644-783-7
Heft 72	**Michael Stegemann** Großversuche zum Leckageverhalten von gerissenen Stahlbetonwänden. ISBN 978-3-86644-860-5

Die Bände sind unter www.ksp.kit.edu als PDF frei verfügbar oder als Druckausgabe bestellbar.

Heft 73 **Isabel Anders**
Stoffgesetz zur Beschreibung des Kriech- und Relaxationsverhaltens junger normal- und hochfester Betone.
ISBN 978-3-7315-0043-8

Heft 74 **Jennifer C. Scheydt**
Mechanismen der Korrosion bei ultrahochfestem Beton.
ISBN 978-3-7315-0113-8

Heft 75 **Michael Auer**
Ein Verbundmodell für Stahlbeton unter Berücksichtigung der Betonschädigung.
ISBN 978-3-7315-0316-3

Heft 76 **Christian Moritz Urban**
Experimentelle Untersuchungen und Bemessungsansätze für faserverstärktes Mauerwerk unter Erdbebenbeanspruchungen.
ISBN 978-3-7315-0372-9

Heft 77 **Tobias Bacht**
Horizontaltragfähigkeit von Wänden aus Leichtbeton-Schalungssteinen – Experimente und numerische Modellierung.
ISBN 978-3-7315-0413-9

Heft 78 **Björn Haag**
Schadensidentifikation mit modalen Parametern: Anwendung auf extern vorgespannte Hohlkastenbrücken.
ISBN 978-3-7315-0458-0

Heft 79 **Engin Kotan**
Ein Prognosemodell für die Verwitterung von Sandstein.
ISBN 978-3-7315-0520-4

Heft 80 **Vladislav Kvitsel**
Zur Vorhersage des Schwindens und Kriechens von normal- und hochfestem Konstruktionsleichtbeton mit Blähtongesteinskörnung.
ISBN 978-3-7315-0521-1

Heft 81 **Michael Vogel**
Schädigungsmodell für die Hydroabrasionsbeanspruchung zur probabilistischen Lebensdauerprognose von Betonoberflächen im Wasserbau.
ISBN 978-3-7315-0522-8

Heft 82 **Georgios Maltidis**
Seismic soil structure interaction of navigation locks.
ISBN 978-3-7315-0718-5

Heft 83 **Steffen Siegel**
Zustandsbestimmung von externen Spanngliedern und Schrägseilen mit Frequenzanalysen.
ISBN 978-3-7315-0772-7

Heft 84 **Fernando Acosta Urrea**
Influence of elevated temperatures up to 100 °C on the mechanical properties of concrete.
ISBN 978-3-7315-0795-6

Heft 85 **Stephan Müller**
Ein Beitrag zum Energiedissipationsvermögen von Betonschrauben.
ISBN 978-3-7315-0943-1

Heft 86 **Bernhard Walendy**
Ertüchtigung von nicht tragendem Mauerwerk gegen seismische Einwirkungen mittels aufgeklebter Textilien.
ISBN 978-3-7315-0975-2

Heft 87 **Roman Sedlmair**
Theoretische und praktische Entwicklung einer aufgeklebten CFK Stahlbetonverstärkung unter Berücksichtigung der vollständigen Kompatibilität zum Betonstahl.
ISBN 978-3-7315-1048-2

Heft 88 **Eric Kirpal**
Verfahren zur Beurteilung der Schadensumfangs- und Zustandsentwicklung vorgespannter Brückenbauwerke.
ISBN 978-3-7315-1103-8

Heft 89 **Mareike Kohm**
Entwicklung eines Messsystems zur modalbasierten Schädigungsanalyse von Brückenüberbauten.
ISBN 978-3-7315-1186-1

Heft 90 **Stefania Rizzo**
Experimental characterization and modelling of textile reinforced masonry structures with the Equivalent frame method.
ISBN 978-3-7315-1214-1